Tobacco Smoke Exposure Biomarkers

Additional Contributions From

Xiaotao Zhang

Jianjun Xia

Wei Xiong

Huan Chen

Xiaojing Zhang

Yongfeng Tian

Chunlei Ren

Yaning Fu

Tobacco Smoke Exposure Biomarkers

Qingyuan Hu • Hongwei Hou

Chemical Industry Press

CRC Press
Taylor & Francis Group
Boca Raton London New York

CRC Press is an imprint of the
Taylor & Francis Group, an **informa** business

Published by Taylor & Francis Group under an exclusive license with Chemical Industry Press.

To the extent permissible under applicable laws, no responsibility is assumed by Taylor & Francis, nor by Chemical Industry Press for any injury and/or damage to persons or property as a result of any actual or alleged libelous statements, infringement of intellectual property or privacy rights. or product liability, whether resulting from negligence or otherwise, or from any use or operation of any ideas, instructions, procedures, products or methods contained in this book.

CRC Press
Taylor & Francis Group
6000 Broken Sound Parkway NW, Suite 300
Boca Raton, FL 33487-2742

First issued in paperback 2019

© 2015 by Chemical Industry Press
CRC Press is an imprint of Taylor & Francis Group, an Informa business

No claim to original U.S. Government works

ISBN-13: 978-1-4987-0559-2 (hbk)
ISBN-13: 978-0-367-37776-2 (pbk)

Library of Congress Cataloging-in-Publication Data

Hu, Qingyuan, author.
  Tobacco smoke exposure biomarkers / Qingyuan Hu and Hongwei Hou.
  p. ; cm.
  Includes bibliographical references and index.
  ISBN 978-1-4987-0559-2 (hardcover : alk. paper)
  I. Hou, Hongwei, author. II. Title.
  [DNLM: 1. Biological Markers. 2. Tobacco Smoke Pollution--analysis. 3. Epidemiologic Methods. 4. Inhalation Exposure--adverse effects. 5. Smoking--adverse effects. QV 137]

RA1226
615.9'02--dc23
                                                                            2014040859

Visit the Taylor & Francis Web site at
http://www.taylorandfrancis.com

and the CRC Press Web site at
http://www.crcpress.com

Contents

Preface

The causal relation between tobacco smoking and a variety of cancers is well documented and is attributable to the many carcinogens that smokers inhale. Federal regulation of cigarettes and other tobacco products has been recommended as an element of a broad-based approach to reduce the disease burden of tobacco use. One component of federal regulation is likely to be the power to mandate limits of emissions of tobacco smoke toxins. The Family Smoking Prevention and Tobacco Control Act (FSPTCA) enables the Food and Drug Administration to establish tobacco product standards. Similarly, Article 9 in the Framework Convention on Tobacco Control (FCTC) describes the regulation of content and emissions of tobacco products.

Developing simple and robust analytical methods to measure toxicants in cigarette smoke is of great importance for tobacco regulation. This has been an ongoing requirement by the World Health Organization (WHO) under the FCTC. National tobacco control organizations have been tasked to develop analytical methods to support this initiative. Biomarkers allow for a quantitative linking of biological exposure to tobacco or tobacco smoke to specific disease induction or progression prior to the advent of the clinically apparent disease. It provides a strong foundation by which to make scientific evaluations and regulatory decisions regarding tobacco products.

In this book, the toxicology, metabolic pathway, and biomarker of nicotine, tobacco-specific nitrosamines, polycyclic aromatic hydrocarbons, volatile organic compounds, aromatic amines, catechol and hydroquinone, hydrogen cyanide, carbon monoxide and NO$_x$, and heavy metals concerned by WHO are summarized, and their application in exposure assessment and/or environmental epidemiology of cigarette smoke exposure has been included. Ethics guidelines and guidelines for bioanalytical method validation, toxicity, and classification of targeted constitutes are provided in the appendices. The book will be very useful as a reference text for easy lookup of information related to biomarkers of individual tobacco smoke components. It may be useful for teaching graduate specialized courses and for professionals wishing to transition their focus area to tobacco research.

Qingyuan Hu
Hongwei Hou
Zhengzhou, Henan, People's Republic of China

1 Smoking and Health

1.1 GLOBAL TOBACCO EPIDEMIC

The World Health Organization (WHO) estimates that tobacco kills nearly 6 million people each year, and more than 5 million of those deaths are attributed to direct tobacco use and more than 600,000 to indirect exposure through secondhand smoke. This number could rise up to 8 million by 2030, according to the WHO. As of 2002, approximately 5.5 trillion cigarettes are smoked annually around the world. The global consumption of cigarettes in 2009, divided by the world's population at that time, results in an equivalent of 865 cigarettes smoked by each man, woman, and child on the planet in 1 year (Wipfli, 2012).

Thousands of years ago, Native Americans rolled their own tobacco in corn or palm leaves. The first factory-made cigarettes were produced 350 years later after Europeans began colonizing the Americas. Factory-made cigarettes grew in popularity in the early twentieth century. Consumption doubled in the first decade, between 1900 and 1910, from 50 to 100 billion cigarettes. By 1920, 300 billion cigarettes were being consumed, and in 1940 consumption reached 1 trillion cigarettes (Wipfli, 2012). The incidence of lung cancer also grew significantly in the twentieth century, up until the early 1970s, particularly in the United States and in the United Kingdom. The first paper on smoking and lung cancer was published by Hoffmann in 1931 under the title "Cancer and smoking habits" (Hoffman, 1931). In 1939, Mueller published *Tabakmissbrauch und Lungencarcinom*. Doll and Hill and Wynder and Graham were among the first to investigate the association of lung cancer with cigarette smoking after the Second World War. In 1950, Wynder and Graham published a paper titled "Tobacco smoking as a possible etiologic factor in bronchiogenic carcinoma," which subsequently became a landmark in cancer epidemiology. Also in 1950, Doll and Hill published a preliminary report on smoking and carcinoma of the lung in the United Kingdom. In 1954, they published a report on their long-term cohort study of 40,000 British doctors (a cohort that could be followed easily over time through the General Medical Council's register), the results of which led them to suggest that "tobacco has a significant adjuvant effect" (on the steady increase in mortality they observed). In the United States, the Surgeon General released his first report titled "Smoking and health" in 1964, in which he concluded that smoking causes cancer (Surgeon General's Advisory Committee on Smoking and Health, 1964). In the United States, the consumption per capita of factory-made cigarettes among the population at or above 18 years had more than tripled from 1920 to 1950. Upon declining in 1954, following the publication of the report by Doll and Hill, it rose again up to a record of 4345 cigarettes in 1963. From 1964 onward, it decreased gradually reaching 13% of the record incidence by 1981 (U.S. Department of Agriculture, 1981). This trend continued aided by aggressive antitobacco campaigns, increased taxes on cigarettes, legislation banning smoking

1

in public areas, restriction to youth access, provisions for tobacco dependence treatment, and prevention education (Bonnie, 2007). From 2000 to 2011, the sales volume of all combustible tobacco decreased from 450.7 billion cigarette equivalents to 326.6, a 27.5% decrease. The per capita sales of all combustible tobacco products declined from 2148 to 1374, a 36.0% decrease. However, while sales of cigarettes decreased 32.8% from 2000 to 2011, sales of cigars and loose tobacco increased 123.1% over the same period. As a result, the percentage of total combustible tobacco sales composed of cigars and loose tobacco increased from 3.4% in 2000 to 10.4% in 2011. The data suggest that certain smokers have switched from cigarettes to other combustible tobacco products, most notably since a 2009 increase in the federal tobacco excise tax that created tax disparities between product types (MMWR, 2012). The United States have made historic progress in combating the epidemic of tobacco-caused illness and death. Despite this progress, tobacco use remains the leading cause of preventable death in many countries all over the world. At present, it is estimated that tobacco is used by more than one-seventh of the global population, including 1200 million smokers, and hundreds of millions of smokeless tobacco (SLT) users (Khariwala et al., 2012).

Low- and middle-income economies, once mainly affected by basic food and sanitation challenges and infectious diseases, have started to face a double burden of disease and are increasingly suffering from noncommunicable diseases (NCDs), such as chronic respiratory illnesses, cancer, diabetes, and cardiovascular disorders. The health profile of low- and middle-income countries is therefore changing rapidly, precipitating increases in health-care costs and depriving families of income. Tobacco is one of the main risk factors driving this change. It is the single greatest preventable cause of NCDs, which are now the number one cause of premature death and disability worldwide, killing more people than HIV, tuberculosis, and malaria combined.

1.2   SMOKE-RELATED DISEASES

A variety of cancers are attributable to tobacco use, including those of the lung, head and neck, bladder, ureter, kidney and cervix, and myeloid leukemia. There are 69 identified carcinogens in tobacco smoke (Hoffmann and Hoffmann, 2001) and no less than 16 in processed tobacco leaf. Among these, tobacco-specific nitrosamines (TSNAs) (such as 4-(methylnitrosamino)-1-(3-pyridyl)-1-butanone [NNK] and N'-nitrosonornicotine [NNN]), polycyclic aromatic hydrocarbons (PAHs) (such as benzo[a]pyrene (B[a]P)), and aromatic amines (AAs) (such as 4-aminobiphenyl) are among the most potent carcinogens (Hecht, 2003). Lung cancer and head and neck cancer have similar genotoxic risk factors. While the vast majority of lung cancers are associated with cigarette smoking alone, smoking in combination with heavy drinking is the main etiological agents of head and neck cancer (Hunter et al., 2005).

The constituents of tobacco smoke most likely to promote carcinogenesis in different organs have been identified on the basis of the organ tropism observed in occupationally induced cancers and of the sites of tumor formation observed in animal studies. Thus, Hoffmann and Hecht (1990) postulated PAHs, acetaldehyde, and

formaldehyde (with polonium-210 as a possible minor factor) are the main carcinogens involved in the development of cancer of the lung and of the larynx. Bladder
cancer is mainly attributable to AAs. Cancers of the esophagus and pancreas are
attributable to TSNAs, and PAHs and nitrosamines are involved in the development
of cancer of the oral cavity (Phillips, 1996).

Extensive scientific research has been conducted for decades in the field
of tobacco carcinogenesis aimed at understanding the mechanisms by which
tobacco products induce carcinoma (Ding et al., 2008; LeMarchand et al., 2008).
Meanwhile, it has been demonstrated that tobacco carcinogens and their metabolites can bind covalently to DNA, resulting in the formation of DNA adducts. DNA
adducts, if unrepaired, can cause miscoding and permanent mutations, which can
activate oncogenes such as K-ras or inactivate tumor suppressor genes such as p53
(Figure 1.1; Ding et al., 2008; Hecht, 1999, 2003). These advances paved the way
for the use of tobacco carcinogen and toxicant metabolites and DNA adducts as
biomarkers of exposure to, and metabolism of, tobacco smoke constituents and/or
tobacco leaf components (Hecht et al., 2010).

1.3 SMOKE CONSTITUENTS OF CONCERN

Tobacco smoke is a very complex mixture of chemicals. By 1959, some 400 chemicals were known to be present in tobacco smoke. Meanwhile, more than 6000
tobacco smoke constituents and more than 2000 tobacco leaf components have
been identified. Moreover, 69 smoke constituents have been classified as carcinogens by the International Agency for Research on Cancer (IARC) (IARC Working
Group on the Evaluation of Carcinogenic Risks to Humans, 2007). In addition,
tobacco smoke contains many free radicals—those in the gaseous phase being
short lived and those in the particulate phase relatively long lived. In general, environmental tobacco smoke (ETS) contains higher levels of identified carcinogens
than mainstream smoke (MSS) (which is directly inhaled by smokers) (Narkowicz
et al., 2013).

In 2008, the working group tasked with the elaboration of the Guidelines for
Implementation of Articles 9 and 10 of the WHO's Framework Convention on
Tobacco Control (FCTC) identified nine cigarette smoke constituents for which
methods for testing and measuring in MSS (analytical chemistry) should be validated as a priority (WHO Framework Convention on Tobacco Control, 2008). Those
nine chemicals are included in the *initial list of priority toxicants* recommended by
the WHO Study Group on Tobacco Product Regulation (TobReg) for reporting and
regulation, also in 2008 (Series, 2008; Table 1.1). In 2009, Xie et al. (2009) developed a novel hazard index of mainstream cigarette smoke (MSS) for risk evaluation of cigarette products. The yields of 29 toxic chemicals (4 TSNAs, 3 PAHs,
8 carbonyls, 7 phenolic compounds, HCN, NO, NO_x, NH3, CO, Nic, and tar) in
MSS were determined. The toxicity of cigarette smoke was measured employing a
bacterial mutagenicity assay (Ames Salmonella mutagenicity assay) with cigarette
smoke condensate (CSC), the MTT cytotoxicity assay with CSC, the micronucleus
assay in vivo with CSC, and a mouse inhalation study. The results indicate that
contributions of seven toxic chemicals (CO, HCN, NNK, NH_3, B[a]P, phenol, and

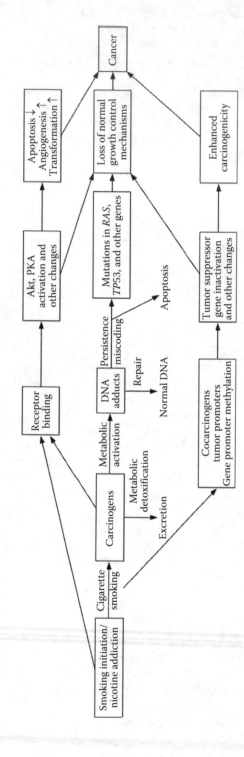

FIGURE 1.1 The development of carcinoma after carcinogenic damage induced by tobacco consumption. (From Khariwala, S.S. et al., *Head Neck*, 34, 441, 2012.)

TABLE 1.1
Priority Smoke Constituents

WHO TobReg	U.S. FDA HPHC List
NNN, NNK	NNN, NNK
Acetaldehyde, acrolein, formaldehyde	Acetaldehyde, acrolein, formaldehyde
Benzene, 1,3-butadiene, acrylonitrile	1,3-Butadiene, benzene, acrylonitrile
Carbon monoxide	Carbon monoxide
4-Aminobiphenyl	4-Aminobiphenyl
Cd	Cd
Catechol	
Crotonaldehyde	Crotonaldehyde
Hydrogen cyanide	
Hydroquinone	
2-Aminonaphthalene	2-Aminonaphthalene
Nitrogen oxides	
Others = 0	Others = 6 (73)
Total = 18	Total = 20 (93)

crotonaldehyde) to the toxicity of MSS were significant. In 2011, Health Canada published a list of *carcinogens in tobacco smoke* (Table 1.2), which was based on IARC, the U.S. Surgeon General's Report 2010, and research conducted to establish standards under the U.S. Family Smoking Prevention and Tobacco Control Act. In the United States, the Food and Drug Administration (FDA) established a list of 93 harmful and potentially harmful constituents (HPHCs) in tobacco

TABLE 1.2
Health Canada List of Carcinogens in Tobacco Smoke

Compounds	Number	Unit	Delivery (2R4F)	IARC Group
Inorganic compounds	4	µg/cigarette	11.02–251.56	
PAHs	1	ng/cigarette	6.96	1
TSNAs	4	ng/cigarette	16.6–146.01	1
AAs	4	ng/cigarette	1.73–15.06	1
Toxic metals	7	ng/cigarette	3.82–73.01	3
Carbonyls	8	µg/cigarette	16.18–560.48	1, 2B, 3
VOCs	5	µg/cigarette	8.28–297.68	1, 2B, 3
Semi-VOCs	3	µg/cigarette	0.23–7.02	2B, 3
Volatile phenols	7	µg/cigarette	0.91–37.9	2B, 3
Routine analytic components	Tar	mg/cigarette	8.91	Tar
	Nicotine	mg/cigarette	0.75	Nicotine
	CO	mg/cigarette	11.96	CO

Note: Group 1, carcinogenic to humans; Group 2B, possibly carcinogenic to humans; Group 3, not classifiable as to carcinogenicity to humans.

products sold in the United States and in tobacco smoke in 2012, as required by the Federal Food, Drug, and Cosmetic Act. All HPHCs included on the list cause or may cause serious health problems including cancer, lung disease, and addiction to tobacco products.

1.3.1 TSNAs

Tobacco and tobacco smoke contain three types of nitrosamines, which include volatile nitrosamines, TSNAs, and nitrosamines derived from residues of agricultural chemicals on tobacco (Monarca et al., 1996). TSNAs have been identified as some of the most potent carcinogens in tobacco (Hecht and Hoffmann, 1988). Extensive research has been conducted on the role of TSNAs in carcinogenesis (Hecht, 1998). So far, seven TSNAs have been identified in cigarettes, that is, NNK, NNN. *N*-nitrosoanabasine (NAB) and *N*-nitrosoanatabine (NAT) 4-(methylnitrosamino)-1-(3-pyridyl)-1-butanol (NNAL), *iso*-NNAL, and *iso*-NNAC (Hoffmann et al., 1994). The evidence that TSNAs play an important role in carcinogenesis is quite indicative (Wilbourn et al., 1986). The tobacco epidemic kills nearly six million people a year from lung cancer, heart disease, and other illnesses. By 2030, the death toll will exceed eight million a year, and 80% of those deaths will occur in the developing world (World Health Organization, 2011).

1.3.2 B[a]P

PAHs are a group of ubiquitous persistent organic pollutants released into the atmosphere during incomplete combustion and/or pyrolysis of fossil fuel, industrial or domestic coal, wood, cigarettes, and food items (Boström et al., 2002). B[a]P is among the PAHs that have been most extensively investigated in connection with lung cancer (IARC Working Group on the Evaluation of Carcinogenic Risks to Humans, 2007). It is classified as Group 1 (*carcinogenic to humans*) (IARC Working Group on the Evaluation of Carcinogenic Risks to Humans, 2007). Recent studies demonstrated that the concentration of PAH metabolites in urine or serum correlates significantly with the incidence of lung cancer in smokers (Yuan et al., 2011). But PAHs are not the only compounds suspected to promote lung cancer development in smokers.

1.3.3 Volatile Organic Compounds

Volatile organic compounds (VOCs) are emitted as gases from certain solids or liquids. VOCs include a variety of chemicals with short- and long-term adverse health effects, such as 1,3-butadiene, benzene, acrolein, and acrylonitrile. Typical levels in MSS of 1,3-butadiene (6.4–68.7 µg/cigarette), benzene (6.1–58.9 µg/cigarette), acrolein (54–155 µg/cigarette), and acrylonitrile (7.8–39.1 µg/cigarette) are 100–1000 times greater than those typically found for PAHs and NNK (Carmella et al., 2009; Hecht, 2012). Mice exposed to 1,3-butadiene by inhalation developed bronchiolar/alveolar adenomas and carcinoma of the lung in addition to other tumors (Huff et al., 1985). But lung tumors were not observed in rats exposed to 1,3-butadiene (IARC, 2008). Epidemiological studies suggested that workers exposed to 1,3-butadiene had an increased risk of leukemia and non-Hodgkin lymphoma, but not of lung cancer. 1,3-Butadiene is classified as Group 1 (*carcinogenic to humans*) (IARC, 2008).

Benzene has been shown to cause tumors at multiple sites including the lung when administered to mice by gavage and, in some cases, by inhalation (Farris et al., 1993; Yin et al., 1989). Occupational studies found increased risks of mortality from hematopoietic malignancies and sometimes lung cancer in workers who were exposed to benzene at the workplace (Hayes et al., 1996; Hecht et al., 2010). Benzene is considered to be a cause of various types of leukemia in humans. Benzene is classified as Group 1 (carcinogenic to humans) by the IARC since 1982.

Acrolein is toxic to the cilia of the lung and is a strong irritant (Hecht et al., 2010; IARC, 1995). Acrolein–DNA adducts are present in the human lung (Zhang et al., 2007). This observation has led to the hypothesis that they may react with the p53 gene at hot spots associated with lung cancer and thus play an important role in lung cancer etiology in smokers (Feng et al., 2006). Acrolein is a product of lipid peroxidation and may be involved in inflammation (Chung et al., 1999; Thompson and Burcham, 2008). It is classified as Group 3 (*not classifiable as to its carcinogenicity to humans*) (IARC, 1987).

Acrylonitrile, although not directly carcinogenic/mutagenic, is a potentially carcinogenic organic compound through its metabolites. In carcinogenesis bioassays of rats, acrylonitrile caused tumors in the brain, forestomach, and Zymbal's gland (Maltoni et al., 1988). A number of epidemiology studies have been conducted to assess the possible carcinogenic activity of acrylonitrile in humans (Blair et al., 1998; Cole et al., 2008). Acrylonitrile is classified as Group 2B carcinogen (*possibly carcinogenic to humans*) (IARC, 1999).

1.3.4 Aromatic Amines

AAs are widespread occupational and environmental pollutants (Richter and Branner, 2002). During the smoking process, carcinogenic primary AAs are generated, which are present in the MSS as well as sidestream smoke (SSS) (Hoffmann et al., 1969; Saha et al., 2009), such as 1-naphthylamine (1-NA), 2-naphthylamine (2-NA), 3-aminobiphenyl (3-ABP), and 4-aminobiphenyl (4-ABP). The main exposures of the general population to AAs are through cigarette smoke or through products that contain products synthesized from AAs. 4-Aminobiphenyl is a cigarette smoke AA classified as a class 1 IARC carcinogen and also present in the FDA HPHCs list (IARC, 2010). 4-Aminobiphenyl MSS yields ranged from 0.5 to 3.3 ng/cigarette in 48 commercial cigarettes smoked with three different regimes including ISO (Riedel et al., 2006). 2-NA has been classified as Group 1 (carcinogenic to humans) by IARC (IARC, 2004). The concentration of 2-NA in MSS was 1.53–13.8 ng/cigarette (Saha et al., 2009). Smoking, in particular from cigarettes, and occupational exposure to 2-NA and 4-ABP are well known risk factors for various diseases including bladder cancer (Letašiová et al., 2012).

1.3.5 CATECHOL AND HYDROQUINONE

Cigarette smoke contains more than 200 semivolatile phenols, while 1,4-dihydroxy-benzene (hydroquinone) and 1,2-dihydroxybenzene (catechol) occur in abundance (Leanderson and Tagesson, 1990). The pyrolysis of tobacco leaf pigments yields up to 100 μg of hydroquinone and 280 μg of catechol per cigarette (Hecht et al., 1981). Catechol and hydroquinone are separately classified in Groups 2B and 3 by the IARC, as possible human carcinogen, and their reactions with various biomolecules may lead to nonrepairable damage (Schweigert et al., 2001). It is reported that a mixture of catechol and hydroquinone can produce a synergistic genotoxic response in cultured human lymphocytes (Robertson et al., 1991).

1.3.6 HYDROGEN CYANIDE, CARBON MONOXIDE, AND NITROGEN OXIDES

Hydrogen cyanide (HCN) has been assigned as both reproductive and cardiovascular toxicants. An HCN concentration of 300 mg/m^3 in air will kill a human within 10–60 min. An HCN concentration of 3500 ppm (about 3200 mg/m^3) will kill a human in about 1 min. The precursors of HCN in smoke are predominantly tobacco nitrate and proteins. HCN is formed from the combustion of these species at the high temperatures generated during a puff as opposed to the lower temperatures during free smolder of a burning cigarette. In cigarette smoke, HCN is partitioned between gas and particulate phases of smoke at a ratio of ca 1:1 (Newsome et al., 1965; Stedman, 1968; Wynder and Hoffmann, 1967).

Carbon monoxide (CO) and nitrogen oxides (NOx) are present in the vapor phase of cigarette smoke. The most common symptoms of CO poisoning may resemble other types of poisonings and infections, including symptoms such as headache, nausea, vomiting, dizziness, fatigue, and a feeling of weakness. Neurological signs include confusion, disorientation, visual disturbance, syncope, and seizures (Blumenthal and Ivan, 2001). The oxides of nitrogen are risk factors for respiratory tract, which could induce cough, pneumonia, and other lesions in the lungs. NOx may also play a role in nitrosamine formation in cigarette smoke (Hoffmann and Hoffmann, 1997). Moreover, the affinity of hemoglobin for CO (or NO) is hundreds (or hundreds of thousand) times greater than the affinity for O$_2$, reducing the oxygen carrying capacity of the blood.

1.3.7 INORGANIC COMPOUNDS

Exposure to arsenic (As) is also a risk factor for lung diseases. Epidemiological studies conducted in Chile, Bangladesh, and the West Bengal region of India show that chronic exposure to As via drinking water correlates with the observed increased incidence of chronic cough, chronic bronchitis, shortness of breath, and obstructive or restrictive lung diseases (Mazumder, 2007; Mazumder et al., 2000; von Ehrenstein et al., 2005; Yoshida et al., 2004). However, very few studies have investigated As-induced lung lesions, except lung cancer, in animals. Recently, Singh et al. (2010) reported that oral administration of sodium arsenite over 180 days induced degenerative changes in the bronchiolar epithelium with emphysema in mice. It appears that the lung is one of the target organs for As exposure via drinking water.

Nickel is included in the FDA HPHC list as a carcinogen, suggesting that the latter applies to this element in its metallic form. In 1990 metallic nickel was considered to be possibly carcinogenic to humans (Group 2B) according to IARC Volume 49 (IARC, 1990). Interestingly, the TobReg Report 951 mentions nickel as a carcinogen (without providing further classification) with reference to IARC Monograph Volume 89. Seemingly in adherence to the same TobReg Report, the FDA refers to nickel as carcinogenic to humans. IARC provides evidence of repeated misclassification in the formal monographs, by listing nickel as Group 2B (possibly carcinogenic to humans) however with reference to the earlier, 1990 edition of Volume 49 of the monographs.

As shown also in the TobReg Report 951, the concentration of metallic nickel in cigarette smoke is typically below the level of detection. Most of the tobacco nickel is recovered in the ash. Not surprisingly, the nickel concentrations in blood plasma and urine have been found to be quite similar among smokers and nonsmokers (Torjussen et al., 2003).

1.4   BIOMARKERS

Estimating the exposure to chemicals in adult cigarette smokers and SLT consumers is a challenge in the health-risk assessment of tobacco products. The extent of individual consumer exposure to MSS and SLT constituents depends primarily on the chemical composition of MSS or SLT and consumption behavior (e.g., amount consumed, frequency, and duration of consumption). Exposure estimates are further complicated by the concentration range of individual chemical constituents in SLT and MSS, which vary by product type and large inter- and intraindividual variability in consumption patterns.

Most of what is known about harm caused by the use of tobacco products has resulted from epidemiology supported by in vitro studies, laboratory animal studies, and human experiments. However, the studies of diseases with long latency, such as cancer, heart disease, and chronic obstructive pulmonary disease, are problematic because of the long time they require to provide useful data. Biomarkers provide useful indications on long-term outcomes within a short time. This can be a basis upon for estimating specific effects long before direct evidence from epidemiological studies becomes available.

Biomarkers can be defined as measurements of any tobacco leaf component, tobacco smoke constituent, or effect of such chemical in a body fluid, exhaled air, or organ (Stratton et al., 2001). In order for biomarkers to be useful in predicting and quantifying exposure and dose, they must have adequate sensitivity, specificity, and limited complexity. They must also allow for a quantitative link of biological exposure to tobacco or tobacco smoke to specific disease induction or progression prior to the advent of clinically apparent disease. The validation and development of further biomarkers provides a strong foundation by which to make scientific evaluations and regulatory decisions regarding tobacco products.

1.4.1 CLASSIFICATION OF BIOMARKERS

Biomarkers are intuitively more informative and indicative of specific diseases when measured directly in the target tissue through biopsies (mucosa, lung, and bladder). However, biomarker assays are technically limited and biopsies can be difficult to obtain, especially in nondiseased smokers. Hence, biomarker assays have been developed for surrogate biological tissues and fluids (saliva, blood, and urine), hair, or exhaled breath.

1.4.1.1 Biomarkers of Exposure

Biomarkers of exposure are tobacco smoke constituents or their metabolites in a biological fluid or tissue, hair, or exhaled breath, which interact with biological macromolecules. They are not measurements of how tobacco smoke constituents interact with body functions or biological macromolecules to cause harm. Examples include cotinine in urine, carboxyhemoglobin in blood, PAHs in lung tissue, nicotine in hair, and CO in exhaled breath. Specific biomarkers of exposure are described in more detail in Chapter 2. This type of biomarker is generally technically feasible and capable of providing information about short-term and long-term exposures. Such indications allow researchers to anticipate exposure to a certain extent and prevent it from causing damage. By contrast, using biomarkers of potential harm scientists must determine by deduction if and what kind of exposure took place, and it may be too late to prevent further harm. On the other hand, biomarkers of exposure are relatively expensive and not always specific for tobacco products. Moreover, this type of marker does not necessarily reflect the biologically effective dose. Nevertheless, biomarkers of exposure to tobacco smoke are a useful tool for future studies on tobacco smoke constituents aimed at developing new or more stringent stricter laws or guidelines regarding specific chemical(s). Some specific suggestions are presented and discussed in the final chapters.

1.4.1.2 Biomarkers of Potential Harm

Biomarkers of potential harm are measurements of an effect resulting from exposure, such as an alteration in morphology, structure, or function of target or surrogate tissue. Examples in target tissue include changes in RNA or protein expression, somatic mutations, change in methylation or gene control, and mitochondrial mutations. Examples in surrogate tissue include leukocytosis, mutations in hypoxanthine phosphoribosyltransferase, chromosomal aberrations, and circulating lymphocytes. This type of biomarker allows for an assessment of the mechanistic pathway leading to disease. However, their use requires rather sophisticated technical equipment, and the target tissue is difficult to obtain. Moreover, the results of such measurements may be biased by harm that has already occurred. The main limitation is that no assays have been shown to have sufficient predictive power to allow for an adequate assessment of the risk of cancer.

1.4.1.3 Biomarkers of Susceptibility

A biomarker of susceptibility is an indicator of an inherent or acquired limitation of a host's ability to prevent harm from exposure to a specific toxicants.

Tobacco toxicants affect people to variable degrees, as there is large interindividual variation in cellular responses, for example, in metabolism and detoxification of toxicants and DNA repair. The study of genetic susceptibilities can improve the accuracy of estimates of disease associations (Khoury and Wagener, 1995). Biological markers of susceptibility are not discussed in detail in the next chapter but only insofar as they can serve as markers of exposure or effect.

1.5 BIOMARKERS OF EXPOSURE TO ENVIRONMENTAL TOBACCO SMOKE

ETS is a highly complex mixture of particles, droplets, and gases resulting from a burning cigarette and is composed mainly of exhaled MSS and SSS, the latter a sum of all emissions emerging from the cigarette between puffs. Approximately 15%–43% of the particulate matter of ETS and 13% of the vapor phase constituents can be traced back to exhaled MSS (Baker and Proctor, 1990). Exposure to ETS typically occurs at much lower levels than in smokers. This renders the biomarker assessment to this type of exposure difficult and subject to the limit of detection of available biomarker assays. The most consistently used biomarkers of this type reflect exposure to nicotine (via cotinine in serum, plasma, or urine). Urinary metabolites of TSNAs have also been found in persons exposed passively to tobacco smoke (Atawodi et al., 1998; Hecht et al., 1993; Parsons et al., 1998). Biomarkers of potential harm caused by ETS are available to a lesser extent.

1.6 BIOMARKERS APPLICATION IN TOBACCO REGULATION AND HEALTH POLICY

Exposure biomarkers are considered the gold standard for exposure assessment, as they provide direct evidence that both exposure and uptake into the body have occurred (Sexton, 2006). In the case of SLT and MSS exposure, exposure biomarkers play an important role in informing and/or validating the exposure assessment, generally considered the risk assessment phase with the greatest epistemic uncertainty. Specifically, using quantified, direct measurements of exposure and corporeal uptake has the potential to reduce uncertainty associated with the use of some surrogate measures of exposure such as those sometimes used in epidemiological or occupational studies (Santamaria et al., 2006). Data from biological monitoring of human populations are useful for estimating the exposure to chemicals of interest in risk assessment.

Reducing nicotine content in cigarettes and other combustible products to levels that are not reinforcing or addictive has the potential to substantially reduce tobacco-related morbidity and mortality (Hatsukami et al., 2013). The authority to reduce nicotine levels as a regulatory measure is provided in the U.S. Family Smoking Prevention and Tobacco Control Act and is consistent with the general regulatory powers envisioned under the relevant articles of the WHO's FCTC. Many experts have considered reducing nicotine in cigarettes to be a feasible national policy approach, but more research is necessary. One of the most important things is to develop and validate availability biomarkers of exposure and effect to assess exposure levels and the risk for nicotine addiction, cancer, and cardiovascular and pulmonary diseases, such as cotinine or urinary nicotine equivalents, acrolein (HPMA), and metabolites of butadiene (MHBMA).

REFERENCES

Atawodi, S. E., Lea, S., Nyberg, F., Mukeria, A., Constantinescu, V., Ahrens, W., Brueske-Hohlfeld, I., Fortes, C., Boffetta, P., Friesen, M. D. 4-Hydroxy-1-(3-pyridyl)-1-butanone-hemoglobin adducts as biomarkers of exposure to tobacco smoke: Validation of a method to be used in multicenter studies. *Cancer Epidemiology Biomarkers & Prevention*. 1998, 7, 817–821.

Baker, R. R., Proctor, C. J. The origins and properties of environmental tobacco smoke. *Environment International*. 1990, 16, 231–245.

Blair, A., Stewart, P. A., Zaebst, D. D., Pottern, L., Zey, J. N., Bloom, T. F., Miller, B., Ward, E., Lubin, J. Mortality of industrial workers exposed to acrylonitrile. *Scandinavian Journal of Work, Environment and Health*. 1998, 24, 25–41.

Blumenthal, I. Carbon monoxide poisoning. *Journal of the Royal Society of Medicine*. 2001, 94, 270–272.

Bonnie, R. J. *Ending the Tobacco Problem: A Blueprint for the Nation*. The National Academies Press, 2007, Washington, DC.

Boström, C.-E., Gerde, P., Hanberg, A., Jernström, B., Johansson, C., Kyrklund, T., Rannug, A., Törnqvist, M., Victorin, K., Westerholm, R. Cancer risk assessment, indicators, and guidelines for polycyclic aromatic hydrocarbons in the ambient air. *Environmental Health Perspectives*. 2002, 110, 451–488.

Carmella, S. G., Chen, M., Han, S., Briggs, A., Jensen, J., Hatsukami, D. K., Hecht, S. S. Effects of smoking cessation on eight urinary tobacco carcinogen and toxicant biomarkers. *Chemical Research in Toxicology*. 2009, 22, 734–741.

Chung, F., Zhang, L., Ocando, J., Nath, R. Role of 1, N2-propanodeoxyguanosine adducts as endogenous DNA lesions in rodents and humans. *IARC Scientific Publications*. 1999, (150), 45–54.

Cole, P., Mandel, J. S., Collins, J. J. Acrylonitrile and cancer: A review of the epidemiology. *Regulatory Toxicology and Pharmacology*. 2008, 52, 342–351.

Ding, L., Getz, G., Wheeler, D. A., Mardis, E. R., McLellan, M. D., Cibulskis, K., Sougnez, C., Greulich, H., Muzny, D. M., Morgan, M. B. Somatic mutations affect key pathways in lung adenocarcinoma. *Nature*. 2008, 455, 1069–1075.

Farris, G. M., Everitt, J. I., Irons, R. D., Popp, J. A. Carcinogenicity of inhaled benzene in CBA mice. *Toxicological Sciences*. 1993, 20, 503–507.

Feng, Z., Hu, W., Hu, Y., Tang, M.-S. Acrolein is a major cigarette-related lung cancer agent: Preferential binding at p53 mutational hotspots and inhibition of DNA repair. *Proceedings of the National Academy of Sciences*. 2006, 103, 15404–15409.

Hatsukami, D. K., Benowitz, N. L., Donny, E., Henningfield, J., Zeller, M. Nicotine reduction: Strategic research plan. *Nicotine & Tobacco Research*. 2013, 15, 1003–1013.

Hayes, R. B., Yin, S. N., Dosemeci, M., Li, G. L., Wacholder, S., Chow, W. H., Rothman, N., Wang, Y. Z., Dai, T. R., Chao, X.-J. Mortality among benzene-exposed workers in China. *Environmental Health Perspectives*. 1996, 104, 1349–1352.

Hecht, S. S. Biochemistry, biology, and carcinogenicity of tobacco-specific N-nitrosamines. *Chemical Research in Toxicology*. 1998, 11, 559–603.

Hecht, S. S., Carmella, S., Mori, H., et al. A study of tobacco carcinogenesis. XX. Role of catechol as a major cocarcinogen in the weakly acidic fraction of smoke condensate. *Journal of the National Cancer Institute*. 1981, 66, 163–169.

Hecht, S. S., Carmella, S. G., Murphy, S. E., Akerkar, S., Brunnemann, K. D., Hoffmann, D. A tobacco-specific lung carcinogen in the urine of men exposed to cigarette smoke. *New England Journal of Medicine*. 1993, 329, 1543–1546.

Hecht, S. S., Hoffmann, D. Tobacco-specific nitrosamines, an important group of carcinogens in tobacco and tobacco smoke. *Carcinogenesis*. 1988, 9, 875–884.

Hecht, S. S. Research opportunities related to establishing standards for tobacco products under the family smoking prevention and tobacco control act. *Nicotine & Tobacco Research*. 2012, 14, 18–28.

Hecht, S. S. Tobacco smoke carcinogens and lung cancer. *Journal of the National Cancer Institute*. 1999, 91, 1194–1210.

Hecht, S. S. Tobacco carcinogens, their biomarkers and tobacco-induced cancer. *Nature Reviews Cancer*. 2003, 3, 733–744.

Hecht, S. S., Yuan, J.-M., Hatsukami, D. Applying tobacco carcinogen and toxicant biomarkers in product regulation and cancer prevention. *Chemical Research in Toxicology*. 2010, 23, 1001–1008.

Hoffmann, D., Brunnemann, K. D., Prokopczyk, B., Djordjevic, M. V. Tobacco-specific N-nitrosamines and ARECA-derived N-nitrosamines: Chemistry, biochemistry, carcinogenicity, and relevance to humans. *Journal of Toxicology and Environmental Health, Part A Current Issues*. 1994, 41, 1–52.

Hoffmann, D., Hecht, S. Advances in tobacco carcinogenesis. In: *Chemical Carcinogenesis and Mutagenesis I*, C.S. Cooper and P.L. Grover (eds.), Springer-Verlag, Berlin, 1990, pp. 63–102.

Hoffmann, D., Hoffmann, I. The changing cigarette, 1950–1995. *Journal of Toxicology and Environmental Health*. 1997, 50, 307–364.

Hoffmann, D., Hoffmann, I. The changing cigarette: Chemical studies and bioassays. *Smoking and Tobacco Control Monograph*. 2001, 13, 159–192.

Hoffmann, D., Masuda, Y., Wynder, E. L. α-Naphthylamine and β-naphthylamine in cigarette smoke. *Nature*. 1969, 221, 254.

Hoffman, F. L. Cancer and smoking habits. *Annals of Surgery*. 1931, 93, 50.

Huff, J., Melnick, R., Solleveld, H., Haseman, J., Powers, M., Miller, R. Multiple organ carcinogenicity of 1, 3-butadiene in B6C3F1 mice after 60 weeks of inhalation exposure. *Science*. 1985, 227, 548–549.

Hunter, K. D., Parkinson, E. K., Harrison, P. R. Profiling early head and neck cancer. *Nature Reviews Cancer*. 2005, 5, 127–135.

IARC. *IARC Monographs on the Evaluation of Carcinogenic Risk of Chemicals to Humans*. 1987, IARC Scientific Publications, Lyon, France, pp. 1–42.

IARC. *Chromium, Nickel and Welding*, Vol. 49, *IARC Monographs on the Evaluation of Carcinogenic Risks to Humans*. 1990, IARC Scientific Publications, Lyon, France, pp. 257–445.

IARC. *Acrolein*, Vol. 63, *IARC Monographs on the Evaluation of Carcinogenic Risk of Chemicals to Humans*. 1995, IARC Scientific Publications, Lyon, France, pp. 337–372.

IARC. *Re-Evaluation of Some Organic Chemicals, Hydrazine and Hydrogen Peroxide*, Vol. 71, *IARC Monographs on the Evaluation of Carcinogenic Risks to Humans*. 1999, IARC, Lyon, France, pp. 43–108.

IARC. *1,3-Butadiene, ethylene oxide and vinyl halides (vinyl fluoride, vinyl chloride and vinyl bromide)*, Vol. 97, *IARC Monographs on the Evaluation of Carcinogenic Risks to Humans*. 2008, IARC Scientific Publications, Lyon, France, pp. 45–309.

IARC Working Group on the Evaluation of Carcinogenic Risks to Humans. *IARC Monographs on the Evaluation of Carcinogenic Risks to Humans*. 2007, IARC, Lyon, France.

IARC. Some aromatic amines, organic dyes, and related exposures. *IARC Monographs on the Evaluation of Carcinogenic Risks to Humans*. http://monographs.iarc.fr/ENG/Monographs/vol99/mono99.pdf. 2010.

Khariwala, S. S., Hatsukami, D., Hecht, S. S. Tobacco carcinogen metabolites and DNA adducts as biomarkers in head and neck cancer: Potential screening tools and prognostic indicators. *Head & Neck*. 2012, 34, 441–447.

Khoury, M. J., Wagener, D. K. Epidemiological evaluation of the use of genetics to improve the predictive value of disease risk factors. *American Journal of Human Genetics*. 1995, 56, 835–844.

Leanderson, P., Tagesson, C. Cigarette smoke-induced DNA-damage: Role of hydroquinone and catechol in the formation of the oxidative DNA-adduct, 8-hydroxydeoxyguanosine. *Chemico-Biological Ineractions*. 1990, 75, 71–81.

Le Marchand, L., Derby, K. S., Murphy, S. E., Hecht, S. S., Hatsukami, D., Carmella, S. G., Tiirikainen, M., Wang, H. Smokers with the CHRNA lung cancer-associated variants are exposed to higher levels of nicotine equivalents and a carcinogenic tobacco-specific nitrosamine. *Cancer Research*. 2008, 68, 9137–9140.

Letašiová, S., Medve'ová, A., Šovčíková, A., et al. Bladder cancer, a review of the environmental risk factors. *Environmental Health*. 2012, 11, S11.

Maltoni, C., Ciliberti, A., Cotti, G., Perino, G. Long-term carcinogenicity bioassays on acrylonitrile administered by inhalation and by ingestion to Sprague-Dawley rats. *Annals of the New York Academy of Sciences*. 1988, 534, 179–202.

Mazumder, D. G. Arsenic and non-malignant lung disease. *Journal of Environmental Science and Health Part A*. 2007, 42, 1859–1867.

Mazumder, D. N. G., Haque, R., Ghosh, N., Binay, K. D., Santra, A., Chakraborti, D., Smith, A. H. Arsenic in drinking water and the prevalence of respiratory effects in West Bengal, India. *International Journal of Epidemiology*. 2000, 29, 1047–1052.

MMWR, Consumption of Cigarettes and Combustible Tobacco — United States, 2000–2011, Morbidity and Mortality Weekly Report, Centers for Disease Control and Prevention, August 3, 2012, Atlanta, GA, pp. 565–569.

Monarca, S., Scassellati-Sforzolini, G., Donato, F., Angeli, G., Spiegelhalder, B., Fatigoni, C., Pasquini, R. Biological monitoring of workers exposed to N-nitrosodiethanolamine in the metal industry. *Environmental Health Perspectives*. 1996, 104, 78–82.

Narkowicz, S., Polkowska, Ż., Kiełbratowska, B., Namieśnik, J. Environmental tobacco smoke: Exposure, health effects, and analysis. *Critical Reviews in Environmental Science and Technology*. 2013, 43, 121–161.

Newsome, J., Norman, V., Keith, C. Vapor phase analysis of tobacco smoke. *Tobacco Science*. 1965, 9, 102–110.

Parsons, W. D., Carmella, S. G., Akerkar, S., Bonilla, L. E., Hecht, S. S. A metabolite of the tobacco-specific lung carcinogen 4-(methylnitrosamino)-1-(3-pyridyl)-1-butanone in the urine of hospital workers exposed to environmental tobacco smoke. *Cancer Epidemiology Biomarkers & Prevention*. 1998, 7, 257–260.

Phillips, D. H. DNA adducts in human tissues: Biomarkers of exposure to carcinogens in tobacco smoke. *Environmental Health Perspectives*. 1996, 104, 453–458.

Riedel, K., Scherer, G., Engl, J., et al. Determination of three carcinogenic aromatic amines in urine of smokers and nonsmokers. *Journal of Analytical Toxicology*. 2006, 30, 187–195.

Richter, E., Branner, B. Biomonitoring of exposure to aromatic amines: Haemoglobin adducts in humans. *Journal of Chromatography B*. 2002, 778, 49–62.

Robertson, M. L., Eastmond, D. A., Smith, M. T. Two benzene metabolites, catechol and hydroquinone, produce a synergistic induction of micronuclei and toxicity in cultured human lymphocytes. *Mutation Research*. 1991, 249, 201–209.

Saha, S., Mistri, R., Ray, B. Rapid and sensitive method for simultaneous determination of six carcinogenic aromatic amines in mainstream cigarette smoke by liquid chromatography/ electrospray ionization tandem mass spectrometry. *Journal of Chromatography A*. 2009, 1216, 3059–3063.

Santamaria, A., Ferriby, L., Harris, M., Paustenbach, D., DeCaprio, A. Use of biomarkers in health risk assessment. In: *Toxicologic Biomarkers*, A. P. DeCaprio (ed.), Informa Healthcare, London, 2006, p. 85–87.

Schweigert, N., Hunziker, R. W., Escher, B. I., et al. Acute toxicity of (chloro-)catechols and (chloro-)catechol-copper combinations in *Escherichia coli* corresponds to their membrane toxicity in vitro. *Environmental Toxicology and Chemistry*. 2001a, 20, 239–447.

Schweigert, N., Zehnder, A. J., Eggen, R. I. Chemical properties of catechols and their molecular modes of toxic action in cells, from microorganisms to mammals. *Environmental Microbiology*. 2001b, 3, 81–91.

Sexton, K. Biomarkers of toxicant exposure. In: DeCaprio, A. P. (Ed.), *Toxicologic Biomarkers*. 2006, Marcel Dekker, New York, pp. 39–63.

Singh, N., Kumar, D., Lal, K., Raisuddin, S., Sahu, A. P. Adverse health effects due to arsenic exposure: Modification by dietary supplementation of jaggery in mice. *Toxicology and Applied Pharmacology*. 2010, 242, 247–255.

Stedman, R. L. Chemical composition of tobacco and tobacco smoke. *Chemical Reviews*. 1968, 68, 153–207.

Stratton, K., Shetty, P., Wallace, R., Bondurant, S. *Clearing the Smoke: Assessing the Science Base for Tobacco Harm Reduction*. 2001, The National Academies Press, Washington, DC.

Thompson, C. A., Burcham, P. C. Genome-wide transcriptional responses to acrolein. *Chemical Research in Toxicology*. 2008, 21, 2245–2256.

Torjussen, W., Zachariasen, H., Andersen, I. Cigarette smoking and nickel exposure. *Journal of Environmental Monitoring*. 2003, 5, 198–201.

Surgeon General's Advisory Committee on Smoking and Health. Smoking and health: Report of the Advisory Committee to the Surgeon General of the Public Health Service. 1964, US Public Health Service.

U.S. Department of Agriculture. Economic Research Service, Tobacco situation and outlook report. TS-175. 1981.

von Ehrenstein, O. S., Mazumder, D. G., Yuan, Y., Samanta, S., Balmes, J., Sil, A., Ghosh, N., Hira-Smith, M., Haque, R., Purushothamam, R. Decrements in lung function related to arsenic in drinking water in West Bengal, India. *American Journal of Epidemiology*. 2005, 162, 533–541.

WHO Framework Convention on Tobacco Control. Elaboration of guidelines for implementation of Articles 9 and 10 of the WHO Framework Convention on Tobacco Control. http://apps.who.int/gb/fctc/PDF/cop3/FCTC_COP3_6-en.pdf. 2008.

WHO Technical Report Series no. 951: The scientific basis of tobacco product regulation, The study group's fourth meeting, Stanford, CA, July 25–27, 2007. WHO Press, Geneva, Switzerland, 2008, http://www.who.int/tobacco/global_interaction/tobreg/publications/9789241209519.pdf.

Wilbourn, J., Haroun, L., Heseltine, E., Kaldor, J., Partensky, C., Vainio, H. Response of experimental animals to human carcinogens: An analysis based upon the IARC Monographs programme. *Carcinogenesis*. 1986, 7, 1853–1863.

Wipfli, H. The tobacco atlas. *American Journal of Epidemiology*. 2012, 176, 1193.

World Health Organization. WHO report on the global tobacco epidemic, 2011: Warning about the dangers of tobacco. 2011, Geneva, Switzerland, WHO.

Wynder, E. L., Graham, E. A. Tobacco smoking as a possible etiologic factor in bronchiogenic carcinoma: A study of 684 proved cases. *Journal of the American Medical Association*. 1950, 143, 329–336.

Xie, J., Liu, H., Zhu, M., Zhong, K., Dai, Y., Du, W., Xie Fu, W., Miao, M. Development of a novel hazard index of mainstream cigarette smoke and its application on risk evaluation of cigarette products. *Tobacco Science & Technology*. 2009, 2, 5–15.

Yin, S. N., Li, G. L., Tain, F. D., et al. A retrospective cohort study of leukemia and other cancers in benzene workers. *Environmental Health Perspectives*. 1989, 82, 207–213.

Yoshida, T., Yamauchi, H., Fan Sun, G. Chronic health effects in people exposed to arsenic via the drinking water: Dose–response relationships in review. *Toxicology and Applied Pharmacology*. 2004, 198, 243–252.

Yuan, J.M., Gao, Y.T., Murphy, S. E., Carmella, S. G., Wang, R., Zhong, Y., Moy, K. A., Davis, A. B., Tao, L., Chen, M. Urinary levels of cigarette smoke constituent metabolites are prospectively associated with lung cancer development in smokers. *Cancer Research*. 2011, 71, 6749–6757.

Zhang, S., Villalta, P. W., Wang, M., Hecht, S. S. Detection and quantitation of acrolein-derived 1, N2-propanodeoxyguanosine adducts in human lung by liquid chromatography-electrospray ionization-tandem mass spectrometry. *Chemical Research in Toxicology*. 2007, 20, 565–571.

2 Biomonitoring and Biomarkers

2.1 INTRODUCTION

Today, alternative evaluation of environmental tobacco smoke (ETS) exposure by biological measures is commonly used. These methods involve the measurement of concentrations of smoke components in body fluids of an exposed individual, called biomarkers (Benowitz, 1996; Tutka et al., 2002), which defined the risk of exposure, the effects of exposure, or individual susceptibility to exposure (Scherer, 2006). Human biomonitoring is a means of acquiring data on the exposure of people to environmental toxicants and the effects of such exposure by means of the analysis of cells, tissues, and biological fluids (Polkowska et al., 2004). Biomarkers reveal the presence of the pollutants by the occurrence of typical symptoms or measurable responses. They are even suitable when assessing long-term exposure to cigarette smoke over days or months (Jaakkola and Jaakkola, 1997). They can be classified as a measure of (a) chemical exposure, that is, a direct or indirect measure of a tobacco-derived constituent or metabolite, which ideally can provide a quantitative estimate of tobacco exposure; (b) toxicity, including biologically effective dose, that is, "the amount that a tobacco constituent or metabolite binds to or alters a macromolecule either in target or surrogate tissue"; (c) injury or potential harm, that is, "a measurement of an effect due to exposure; these include early biological effects, alterations in morphology, structure or function, and clinical symptoms consistent with harm"; and (d) direct measures of health outcome (Hatsukami et al., 2006).

The application of biomarkers in epidemiological studies (*molecular epidemiology*) seems promising, as these represent steps in an exposure–disease continuum. However, to become a predictor of disease, biomarkers must be validated. Validation criteria include the specificity and sensitivity of the biomarker assays, as well as the degree to which dose–response relationships exist, inter- and intraindividual responses vary, and knowledge of the kinetics and the confounding factors involved. In addition, the characteristics of the analytical equipment and of the sampling procedure must be adequate. The latter includes constraints and noninvasiveness of sampling, sample stability. The predictive power of biomarkers must be properly validated to allow for an adequate assessment of exposure, susceptibility, and risk of disease. Biomonitoring of the exposure to complex mixtures, such as tobacco smoke, is particularly challenging given that more than 5000 constituents have been detected in tobacco smoke (Layten Davis and Nielsen, 1999; Talhout et al., 2011) and that many of these are also present in other complex mixtures that people are commonly exposed to, such as polluted ambient air and diesel exhausts, which may therefore be confounding factors.

Biomarkers are helpful because of their potential to assess the actual and internal dose of cigarettes exposure. Other advantages of biomarkers are as follows:

- Confirms absorption of cigarette smoke into the human body. Measure integrated exposure—not dependent upon models or assumptions
- Very low-level exposures detectable for analytical methods have become exquisitely sensitive over the past several decades
- Help to test and validate cigarette smoke exposure models when the results of modeling predictions are compared to internal doses actually measured in exposed individuals
- Help to follow cigarette smoke exposure trends when individuals or representative samples of groups are followed with serial biomarker testing over time
- Help to evaluate smoker health interventions, that is, withdraw smoke and smoke status testing

The application of the biomarker method for biomonitoring also has some limitations:

- It does not define sources or pathways of exposure—because it is a snapshot and an integrated measure, it tells us nothing about where the chemical came from or how it got into the body.
- It is susceptible to inferior or unscrupulous analytical laboratories—because by definition environmental chemicals and pollutants tend to be ubiquitous and many chemicals of interest are used in everyday products including laboratory equipment, it is possible that samples will be contaminated during collection and processing. Proper procedures are imperative to ensure interpretable results.
- Unqualified commercial laboratories may not only have poor techniques; they may also test for the wrong things in the wrong kinds of samples.

A number of biomarkers of exposure to ETS have been proposed. Among them, nicotine and cotinine, a major metabolite of nicotine, have been used most widely as surrogate measures of consumed nicotine dose (Benowitz, 1999; Sepkovic and Haley, 1985). However, nicotine assessed in biological fluids has a short half-life of 2–3 h, and thus does not seem to be suitable as a marker for cigarette smoking. At present, cotinine appears to be the best available biomarker of ETS exposure (Benowitz, 1996; Chang et al., 2000; Etzel, 1990). Cotinine levels in the body, derived primarily from tobacco smoke, can be measured with extremely high sensitivity. They also reflect exposure to a variety of cigarette types over days. Importantly, cotinine levels highly correlate with nicotine intake (Benowitz and Jacob, 1994).

2.2 PREPARATION OF SAMPLES OF BIOLOGICAL MATERIALS

In order to monitor hazard chemicals of exposure, a complete procedure used for analyzing samples of biological materials with a complex matrix composition is important. Those procedures included sampling, transport, storage, sample preparation, and instrumental analysis (Figure 2.1).

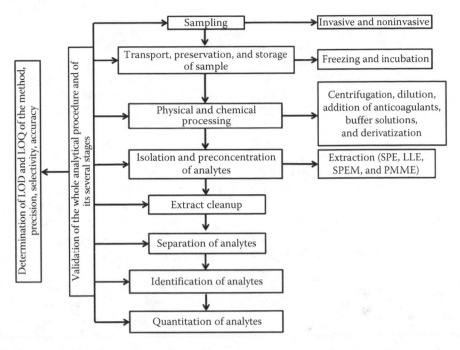

FIGURE 2.1 General scheme of the procedures used for analyzing samples of biological materials with a complex matrix composition.

2.2.1 SAMPLING OF BIOLOGICAL MATERIAL

The term *biological material* includes both physiological fluids and solid tissues. Toxic substances emitted in cigarette smoke that are absorbed by the human body can circulate in it together with physiological fluids, accumulate in tissues, or be excreted as polar metabolites or in unchanged form. Biological fluids are materials with a complex matrix, which therefore require special attention at the sample preparation stage (Polkowska et al., 2004). Real-time monitoring is carried out by analyzing blood, urine, and saliva; long-term monitoring involves the analysis of hair, nails, and tissue (Iyengar and Rapp, 2001). Among all of the biomarkers, the measurements of urine, saliva, hair, and blood are the most sensitive and specific methods, which are commonly used to monitor tobacco smoke constituents. But whole blood, plasma, and serum require invasive procedures, as opposed to urine and saliva. The proper selection of biological sample types is a difficult task; important criteria for the adequate sample selection (Narkowicz et al., 2012) must include

- Metabolic changes of particular elements or compounds
- Quantity of a sample that can be collected
- Analytical quality of the sample, which is affected by the methods of storage and preservation

2.2.2 Sample Storage

The analytical quality of samples depends on the conditions of storage and preservation. Due to the requirement of experimental design, we have to collect a large number of samples in a certain period of time, but we cannot do sampling and analyzing
at the same time by the limit of analysis speed, so we must store part of the samples
appropriately.

Cryopreservation is the most commonly used method, which not only terminates
the activity of the enzyme but also stores the samples. In some cases, if the collection
of samples cannot freeze in time, they must be placed in the flake ice at first, then
to freeze storage. If glass containers are used to store and freeze samples, we would
need to pay attention to vessels rupturing when temperatures dropped rapidly, which
would cause loss of sample or pollution. Plastic containers often contain plasticizer
with a high boiling point, which cannot only be released into the sample because of
pollution and also absorbed target objects in the sample, which may cause analysis
error. Therefore, we should make silicon alkylation treatment to glass containers if
necessary.

Besides that, some biological samples must be stored without oxygen or in a
dark place; otherwise sample components may be affected by air oxidation or photo
oxidation, such as catechol in cigarette smoke, which can easily produce quinone
impurities through air oxidation. If collected urine samples are stored under –15 °C
directly without antioxidants, they can remain stable for only 4 weeks, but with the
addition of ascorbic acid, they can last 10 weeks.

2.2.3 Techniques of Sample Preparation prior to Analysis

Sample preparation may seriously affect the quality of the measurement also. Sample
matrix, endogenous substances, and metabolites may interfere with the assay so
that we may get a low sensitivity and poor results. For that reason, biological samples were generally needed to purify before analysis and determination. For most
samples, preprocessing was necessary; only a handful of samples may be analyzed
directly.

Biological matrix containing biomarkers requires removing the bulk of proteins,
inorganic salts, and other sample components that can subsequently interfere with
analysis (Tretyakova et al., 2012). Some examples of sample enrichment methods
used in biomarker analyses include liquid–liquid extraction, solid-phase extraction,
solid-phase microextraction, and supercritical fluid extraction.

2.2.3.1 Liquid–Liquid Extraction

Liquid–liquid extraction is one of the classic extraction methods; it is based on the
distribution of target object in the two kinds of undissolved solvents. In the process of
extraction, sometimes adding some strong ion inorganic salts (such as sodium chloride), using the salting-out effect, promotes the component access into the organic
phase. Sometimes, we select different organic solvents to improve the selectivity.
Through liquid–liquid extraction, endogenous substances can be separated from the
sample matrix, which reduces the impurity interference for measurement; besides
that, liquid–liquid extraction is simple, rapid, economical, and practical, but it is easy

to bring about emulsification when extracting, has low recovery, and needs a large amount of sample and solvent that restrict its application in the biological sample preparation.

2.2.3.2 Solid-Phase Extraction

Solid-phase extraction is a kind of sample pretreatment technique that was developed in the 1970s; it is a separation and purification method, which is based on liquid chromatography (LC) separation mechanism—elective adsorption and elution. With its characteristics of high efficiency, selectivity, and automation, it has been widely used in all kinds of separation and purification of biological samples (Broich et al., 1971). Solid-phase extraction generally has four steps: activation, sampling, wash, and elution. Water Oasis HLB, a kind of column, has a wide pH range and can extract the hydrophilic, hydrophobic, acidic, alkaline, or neutral components, especially suitable for plasma, urine, and other biological samples (Huitema et al., 2001; Yoshihara et al., 2001).

2.2.3.3 Solid-Phase Microextraction

Solid-phase microextraction (SPME) is a relatively new type of sample pretreatment technology, which was founded in 1990 by Pawliszyn from Canada University of Waterloo. It was developed from solid-phase extraction, which is based on an equilibrium principle of gas–solid and liquid–solid adsorption, separation, and enrichment that was accomplished by the adsorption affinity between target object and active solid surface. In 1993, Supelco Corporation from the United States introduced a commercial SPME device, which caused a great response in analytical chemistry field, but now it has been already widely used in the environmental and biological analysis. This is a simple, efficient, high-sensitivity, and high-precision technology, integrating preparation and separation altogether; simplifies the traditional sampling, extraction, concentration, and sample operation, especially with less solvent consumption; and avoids environmental pollution.

2.2.3.4 Supercritical Fluid Extraction

Supercritical fluid extraction is an environment-friendly separation and enrichment technology, with high extraction efficiency and selectivity, time saving, less pollution, and easy-to-change operating condition characteristics. The principle of supercritical fluid extraction technology is to control the supercritical fluid above the critical temperature and pressure conditions to make it extract the target from the ingredient; when returned to atmospheric pressure and room temperature, the dissolved components immediately separate from the supercritical fluid.

2.3 METHODOLOGY OF DETERMINATION OF BIOMARKERS

Modern analytical techniques detect the contaminants and specific biomarkers of ETS at very low concentration (ng/mL). There are numerous scientific papers published about the determination of nicotine and cotinine in human fluids, including the Jaffe method (Bonsnes and Taussky, 1945; Jaffé, 1886), enzymatic method (Jacobs et al., 1991), flow injection analysis (Jacobs et al., 1991), high-performance

LC (HPLC) (Hewavitharana and Bruce, 2003; Patel and George, 1981; Tsikas et al., 2004), capillary electrophoretic (Ruiz-Jiménez et al., 2007; Shi et al., 1995; Zinellu et al., 2006), gas chromatography (GC)–mass spectrometry (MS) (Welch et al., 1986), or LC–MS (Felitsyn et al., 2004; Park et al., 2007; Takahashi et al., 2006). Each of these techniques has its advantages and disadvantages, and care must be taken in comparing results obtained in one laboratory with those obtained in another laboratory (Biber et al., 1987).

2.3.1 Colorimetric Method

The colorimetric method is the simplest and cheapest method for urinary cotinine determination. Since it quantifies the totality of metabolites for which the pyridine ring remains intact, this method has low specificity. Both cotinine and *trans*-3'-hydroxycotinine are measured in this method; the results, expressed as cotinine equivalents, are therefore higher than those obtained by more specific methods such as GC or LC (Byrd et al., 1992; Jacob et al., 1993; Kolonen and Puhakainen, 1991; Neurath et al., 1987). However, the measured levels are well correlated with daily cigarette consumption. Moreover, the high detection threshold makes this method inappropriate for monitoring of ETS exposure (Haufroid and Lison, 1998; Kolonen and Puhakainen, 1991).

2.3.2 Test Strip Assay

Test strip assay provides a simple and inexpensive method for the detection of cotinine in urine as a screening tool to verify current smoking status in a nonlaboratory environment. The test strip is immersed in urine and the color reaction visually determined. However, similar to colorimetric assays, a number of other endogenous compounds and drug metabolites in urine contain an unsubstituted pyridine ring (e.g., nicotinic acid, isoniazid, and nicotinamide) that may give rise to a false-positive test result (Thomas Karnes et al., 2001). While test strip assays are often considered to be an attractive low-cost screening method for determining current smoking status, several limitations have to be taken into consideration in order to minimize the number of false-positive and false-negative results. But the high detection threshold makes these methods inappropriate for the determination of ETS exposure too.

2.3.3 Immunoassay

Immunoassay techniques used for the determination of nicotine and cotinine concentrations in biological media include radio immunoassay (RIA) (Knight et al., 1985; Langone et al., 1973), enzyme-linked immunosorbent assay (Bernert et al., 1997; Bjercke et al., 1986; Langone et al., 1988), and fluorescence immunoassay (Hansel et al., 1986; Pickert et al., 1993). Immunoassays have been used to determine nicotine concentrations in amniotic fluid, blood plasma and serum, cervix mucus, hair, meconium, saliva, and urine. The International Agency for Research on Cancer has often recommended an RIA developed for nicotine and cotinine for exposure assessment of these metabolites in blood, saliva, and urine.

HPLC (Barlow and Thompson, 1987; Hariharan and VanNoord, 1991; Horstmann, 1985; Kolonen and Puhakainen, 1991) and GC (Feyerabend and Russell, 1990) are more specific, especially when the identity of the peaks can be verified by MS (Daenens et al., 1985; Jacob III et al., 1992; McAdams and Cordeiro, 1993), but this requires the use of costly material and skilled personnel for routine analysis. Another advantage of chromatography methods is the simultaneous quantification of nicotine and cotinine in a single analysis (Watts et al., 1990). The reagent costs are also generally low and the limits of detection are low, which makes these methods very suitable for passive smoking studies.

2.3.4 GAS CHROMATOGRAPHY/GAS CHROMATOGRAPHY–MASS SPECTROMETRY

GC methods are well established for the quantitation of nicotine, cotinine, and *trans*-3′-hydroxycotinine in biological media (Scherer, 1999). Initially, either flame ionization detectors or more selective nitrogen–phosphorus detectors (NPDs) were the detectors of choice (Curvall et al., 1982; Feyerabend and Russell, 1980). Most laboratories now use GC interfaced to a MS, which provides a more sensitive and specific method for the measurement of nicotine and multiple nicotine-derived metabolites (Dhar, 2004). A number of simple benchtop capillary GC–MS methods using selected ion monitoring have been developed for the determination of nicotine (Davoli et al., 1998; Petrakis et al., 1978), cotinine (Daenens et al., 1985; Skarping et al., 1988), and *trans*-3′-hydroxycotinine (Jacob III et al., 1992) in various biological matrices. A GC–MS method for the simultaneous determination of nicotine, cotinine, and thiocyanate concentrations in urine within a single run has also been developed (Toraño and van Kan, 2003). This method has also been applied to the determination of nicotine and cotinine concentrations in saliva.

Sorensen et al. (2007) developed GC–MS-based assays for low concentrations of nicotine and cotinine in small samples of hair and plasma from infants. The limits of quantification (LOQs) were 0.05 ng cotinine and 0.1 ng nicotine in 2 mg hair and 0.2 ng cotinine and 0.1 ng nicotine in 1 mL plasma. On the basis of 10 repeated measurements of a hair sample (mean concentration of 0.74 and 0.08 ng cotinine per mg), the intraday coefficient of variation (CV) was calculated to 9.5% for nicotine and to 8.4% for cotinine. The interday CV was 10.6% for nicotine and 12.1% for cotinine (mean concentration of 3.2 and 0.10 ng cotinine per mg). Ten repeated measurements of a plasma sample (mean concentration of 5.5 and 5.3 ng cotinine per mL) resulted in an intraday CV on 4.1% for nicotine and on 4.0% for cotinine. The interday CV was 7.2% for nicotine and 4.7% for cotinine (mean concentration of 5.3 and 5.3 ng cotinine per mL).

2.3.5 LIQUID CHROMATOGRAPHY/LIQUID
    CHROMATOGRAPHY–MASS SPECTROMETRY

LC is more versatile than GC and can be applied to a greater variety of nicotine-derived metabolites since only metabolite dissolution in the mobile phase rather than volatilization is required. Nicotine has been determined by HPLC with electrochemical detection (Chien et al., 1988; Mahoney and Al-Delaimy, 2001;

Mousa et al., 1985), while cotinine analysis is performed mainly by HPLC with UV detection prior to derivatization (Greaves et al., 2001; Machacek and Jiang, 1986; Shen et al., 1997) or as the barbituric acid derivative (Kolonen and Puhakainen, 1991). Simultaneous determination of nicotine and cotinine is possible using HPLC with DAD (Baranowski et al., 1998) and UV detection prior to derivatization (Hariharan et al., 1988; Page-Sharp et al., 2003) or after precolumn DETBA (1,3-diethyl-2-thiobarbituric acid) derivatization (Parviainen and Barlow, 1988).

The real advantage of using HPLC compared to GC separation is in its application to the simultaneous analysis of nicotine plus multiple nicotine metabolites. HPLC–DAD analysis of DETBA derivatives of nicotine plus 12 metabolites is reported in rodent and human urine (Rustemeier et al., 1993). HPLC–MS has been used for the determination of nicotine, cotinine, and *trans*-3′-hydroxycotinine concentrations in blood serum and seminal plasma (Pacifici et al., 1993). Determination of urinary nicotine, cotinine, *trans*-3′-hydroxycotinine, their glucuronide conjugates as aglycones, nicotine-*N*-1′-oxide, and cotinine-*N*-1-oxide has been reported using β-glucuronidase treatment of urine followed by HPLC separation of metabolites, reversed-gradient addition to the column eluent, and thermospray LC–MS (Byrd et al., 1992). A major advance in analytical technology over the last few years has been the analysis of nicotine metabolites by LC interfaced to a tandem mass spectrometer (LC–MS–MS) using improved atmospheric pressure ionization sources. Cotinine and *trans*-3′-hydroxycotinine concentrations in saliva can also be determined by automated solid-phase extraction and reversed-phase LC–MS–MS using an electrospray ionization interface (Bentley et al., 1999; Byrd et al., 2005). Following solid-phase extraction of saliva, cotinine has also been determined by LC with an atmospheric pressure ionization interface to a tandem mass spectrometer (LC–API–MS–MS) (Bernert et al., 2000). LC–API–MS–MS methods have been widely applied to the analysis of cotinine concentrations in blood serum (Bernert et al., 1997; Kellogg et al., 2004) and nicotine and cotinine concentrations in blood plasma (Taylor et al., 2004; Xu et al., 1996). Simultaneous analysis of nicotine, cotinine, *trans*-3′-hydroxycotinine, anabasine, and nornicotine concentrations in human serum and urine is reported by solid-phase extraction and LC–API–MS–MS (Moyer et al., 2002). Several methods report the analysis of nicotine plus multiple metabolites in urine using LC–API–MS–MS (Heavner et al., 2005; Xu et al., 2004). The LOQs for nicotine and its metabolites in these methods range from low μg/L to 10–20 μg/L urine, which is sufficient for the determination of nicotine metabolites in clinical studies.

2.4 BIOMARKERS IN TISSUES AND BODY FLUIDS

2.4.1 URINE

Urine specimens are commonly employed in biological monitoring because urine collection is noninvasive and poses minimal infectious disease risk to participants and researchers. Continuous and complete 24 h urine collection yields more accurate results, because spot urine sampling may not provide a valid overview of the entire toxicant exposure profile. Urine sample integrity and completeness is essential to exposure assessment research, and absence of compliance with the collection protocol is a fundamental concern to the researcher. Urinary biomarkers for some of the principal carcinogens are available.

Carcinogens and their metabolites and related compounds that have been quantified in the urine of smokers or nonsmokers exposed to environment tobacco smoke include cotinine, NNAL (metabolites of NNK), 1-hydroxypyrene(1-OHP, metabolites of PAH or benzo[a]pyrene), N-acetyl-S-(3,4-dihydroxybutyl)-L-cysteine (DHBMA, metabolites of 1,3-butadiene), S-phenylmercapturic acid (SPMA, metabolites of benzene), 3-hydroxypropyl)-L-cysteine (3-HPMA, metabolites of acrolein), N-acetyl-S-2-cyanoethyl-cysteine (CEMA, metabolites of acrylonitrile), aromatic amines (AAs), catechol, hydroquinonehydrogen cyanide (HCN), carbon monoxide (CO), nitrogen oxides, and heavy metals. Nitrosamines and their metabolites have also been quantified in the urine of smokeless tobacco users. The utility of these assays to provide information about carcinogen dose, delineation of exposed versus nonexposed individuals, and carcinogen metabolism in humans is discussed. NNAL and NNAL-Gluc are exceptionally useful biomarkers because they are derived from a carcinogen-4-(methylnitrosamino)-1-(3-pyridyl)-1-butanone (NNK) that is specific to tobacco products (Hatsukami et al., 2003). The NNAL assay has been applied widely in assays of urine from smokers and nonsmokers. It has also been used to investigate ethnic differences in NNK metabolism and the effects of diet and chemopreventive agents on NNK metabolism in humans. The NNAL assay has been particularly useful for studies in nonsmokers exposed to ETS. In several studies, levels of NNAL and its glucuronides in the urine of people exposed to ETS were significantly elevated compared with control subjects. The NNAL assay has high sensitivity and specificity, which are particularly important for studies on ETS exposure. Other useful assays that have been widely applied involve quantitation of 1-HOP and mercapturic acids such as SPMA, CEMA, and HPMA. Urinary carcinogen metabolite biomarkers will be the critical components of future studies on tobacco and human cancer, particularly with respect to strategies for harm reduction, the role of metabolic polymorphisms in cancer, and further evaluation of human carcinogen exposure from ETS (Hecht, 2002).

2.4.2 BLOOD

Blood specimens include plasma and serum, which are not commonly employed in biological monitoring that compares with urine because blood collection is invasive and poses potential infectious disease risk to participants and researchers. Plasma cotinine concentrations correlate better to various measures of biological effects of cigarette smoking than do self-reported cigarettes per day (Benowitz and Sharp, 1989; Perezstable et al., 1995). At present, cotinine, measured in blood, saliva, or urine, appears to be the most specific and the most sensitive biomarker. Chemicals in tobacco smoke such as carbon monoxide or cyanide (the latter metabolized in the body to thiocyanate) can be measured in blood. However, the levels of these chemicals are nonspecific, that is, there are significant sources of carbon monoxide and cyanide, including the body's own metabolism, other than ETS. Thus, these markers are both nonspecific and insensitive markers of ETS exposure. Other markers that have been proposed to quantitate tobacco exposure include adducts of 4-aminobiphenyl to hemoglobin in red blood cells (Bartsch et al., 1990; Hammond et al., 1993; Maclure et al., 1989), adducts of benzo[a]pyrene and other potential carcinogens to DNA in white blood cells (Binkova et al., 1995; Jahnke et al., 1990; Savela and Hemminki, 1991; Van Maanen et al., 1994), and adducts of polycyclic aromatic hydrocarbons to plasma albumin (Crawford et al., 1994).

2.4.3 SALIVA

The thiocyanate concentration in saliva is a biochemical measure, frequently used as an objective indicator of tobacco consumption. Myosmine is a minor tobacco alkaloid with widespread occurrence in the human diet. It is genotoxic in human cells and is readily nitrosated and peroxided yielding reactive intermediates with carcinogenic potential. For biomonitoring of short-term and long-term exposure, analytical methods were established for determination of myosmine together with nicotine and cotinine in saliva by GC–MS (Schütte-Borkovec et al., 2009).

2.4.4 TISSUES

Tissue samples are difficult to collect in experiment when compared with urine and saliva. Like blood, tissue collection is invasive too. Tobacco-related biomarkers like 1,3-butadiene adducts, TSNA DNA adducts, and benzene adducts are always detected in human tissues. Zhang et al. (2008) conducted a proteomic analysis to characterize the differential protein expression in lung tissue of rats exposed to cigarette smoke. The results presented in his study demonstrate the identification of proteomic pattern as an early indicator of lung damage induced by cigarette smoke. The differentially expressed proteins may be applied as exposure biomarkers in future experimental as well as epidemiologic investigations upon confirmation by a greater sample size and more validate study design for the proteomic research.

2.5 CONCLUSIONS

The use of biomarkers improves the assessment of exposure to mainstream tobacco smoke as it provides a relative contribution of individual carcinogens and estimates the total burden of a particular exposure where there are numerous sources (e.g., B[a] P from air, tobacco, diet, and occupation) (Vineis and Porta, 1996). Biomarkers can also establish differences in individual susceptibilities and help determine whether there are differences in response depending on dose. They are also useful in assessing exposure to ETS, as they enable the characterization of low-dose exposures. Technical limitations to the use of biomarkers remain, however, particularly in assessing the potential harm caused by tobacco. Target tissue can be difficult to obtain, especially in nondiseased smokers. The use of surrogate tissues and fluids, such as exhaled breath, saliva, blood, and urine, which are technically simpler to collect, often reduces the ability to prove a predictive value for the potential harm. Moreover, several nonspecific biomarkers of potential harm are related to smoking, such as leukocyte count (Parry et al., 1997; Phillips et al., 1992; Sunyer et al., 1996), which decreases with cessation (Green and Harari, 1995; Sunyer et al., 1996) and increases again with resumed smoking (Sunyer et al., 1996). Levels in former smokers can be significantly higher than in those who have never smoked (Parry et al., 1997). Whether these findings are independent predictors of disease risk remains to be proven, and the differences found may be covariants and/or due to disease unrelated to smoking (Wald et al., 1989).

REFERENCES

Baranowski, J. et al. Determination of nicotine, cotinine and caffeine in meconium using high-performance liquid chromatography. *Journal of Chromatography B: Biomedical Sciences and Applications.* 1998, 707(1), 317–321.

Barlow, R. D., Thompson, P. A. Simultaneous determination of nicotine, cotinine and five additional nicotine metabolites in the urine of smokers using pre-column derivatisation and high-performance liquid chromatography. *Journal of Chromatography B: Biomedical Sciences and Applications.* 1987, 419, 375–380.

Bartsch, H. et al. Carcinogen hemoglobin adducts, urinary mutagenicity, and metabolic phenotype in active and passive cigarette smokers. *Journal of the National Cancer Institute.* 1990, 82(23), 1826–1831.

Benowitz, N. L. Cotinine as a biomarker of environmental tobacco smoke exposure. *Epidemiologic Reviews.* 1996, 18(2), 188–204.

Benowitz, N. L. Biomarkers of environmental tobacco smoke exposure. *Environmental Health Perspectives.* 1999, 107(Suppl. 2), 349–355.

Benowitz, N. L., Jacob, P. Metabolism of nicotine to cotinine studied by a dual stable isotope method. *Clinical Pharmacology & Therapeutics.* 1994, 56(5), 483–493.

Benowitz, N. L., Sharp, D. S. Inverse relation between serum cotinine concentration and blood pressure in cigarette smokers. *Circulation.* 1989, 80(5), 1309–1312.

Bentley, M. C. et al. Validation of an assay for the determination of cotinine and 3-hydroxycotinine in human saliva using automated solid-phase extraction and liquid chromatography with tandem mass spectrometric detection. *Journal of Chromatography B: Biomedical Sciences and Applications.* 1999, 723(1), 185–194.

Bernert, J. T. et al. Development and validation of sensitive method for determination of serum cotinine in smokers and nonsmokers by liquid chromatography/atmospheric pressure ionization tandem mass spectrometry. *Clinical Chemistry.* 1997, 43(12), 2281–2291.

Bernert, J. T. et al. Comparison of serum and salivary cotinine measurements by a sensitive high-performance liquid chromatography-tandem mass spectrometry method as an indicator of exposure to tobacco smoke among smokers and nonsmokers. *Journal of Analytical Toxicology.* 2000, 24(5), 333–339.

Biber, A. et al. Determination of nicotine and cotinine in human serum and urine: An interlaboratory study. *Toxicology Letters.* 1987, 35(1), 45–52.

Binkova, B. et al. DNA adducts and personal air monitoring of carcinogenic polycyclic aromatic hydrocarbons in an environmentally exposed population. *Carcinogenesis.* 1995, 16(5), 1037–1046.

Bjercke, R. J. et al. Stereospecific monoclonal antibodies to nicotine and cotinine and their use in enzyme-linked immunosorbent assays. *Journal of Immunological Methods.* 1986, 90(2), 203–213.

Bonsnes, R. W., Taussky, H. H. On the colorimetric determination of creatinine by the Jaffe reaction. *Journal of Biological Chemistry.* 1945, 158(3), 581–591.

Broich, J. et al. Liquid—Solid extraction of lyophilized biological material for forensic analysis: I. Application to urine samples for detection of drugs of abuse. *Journal of Chromatography A.* 1971, 63, 309–312.

Byrd, G. et al. A rapid LC-MS-MS method for the determination of nicotine and cotinine in serum and saliva samples from smokers: Validation and comparison with a radioimmunoassay method. *Journal of Chromatographic Science.* 2005, 43(3), 133–140.

Byrd, G. D. et al. Evidence for urinary excretion of glucuronide conjugates of nicotine, cotinine, and *trans*-3'-hydroxycotinine in smokers. *Drug Metabolism and Disposition.* 1992, 20(2), 192–197.

Chang, M. Y. et al. Salivary cotinine levels in children presenting with wheezing to an emergency department. *Pediatric Pulmonology.* 2000, 29(4), 257–263.

Chien, C. Y. et al. High-performance liquid chromatography with electrochemical detection for the determination of nicotine in plasma. *Journal of Pharmaceutical Sciences.* 1988, 77(3), 277–279.

Crawford, F. G. et al. Biomarkers of environmental tobacco smoke in preschool children and their mothers. *Journal of the National Cancer Institute.* 1994, 86(18), 1398–1402.

Curvall, M. et al. Simultaneous determination of nicotine and cotinine in plasma using capillary column gas chromatography with nitrogen-sensitive detection. *Journal of Chromatography B: Biomedical Sciences and Applications* 1982, 232(2), 283–293.

Daenens, P. et al. Determination of cotinine in biological fluids by capillary gas chromatography–mass spectrometry–selected-ion monitoring. *Journal of Chromatography B: Biomedical Sciences and Applications.* 1985, 342, 79–87.

Davoli, E. et al. Rapid solid-phase extraction method for automated gas chromatographic–mass spectrometric determination of nicotine in plasma. *Journal of Chromatography B: Biomedical Sciences and Applications.* 1998, 707(1), 312–316.

Dhar, P. Measuring tobacco smoke exposure: Quantifying nicotine/cotinine concentration in biological samples by colorimetry, chromatography and immunoassay methods. *Journal of Pharmaceutical and Biomedical Analysis.* 2004, 35(1), 155–168.

Etzel, R. A. A review of the use of saliva cotinine as a marker of tobacco smoke exposure. *Preventive Medicine.* 1990, 19(2), 190–197.

Felitsyn, N. M. et al. Liquid chromatography–tandem mass spectrometry method for the simultaneous determination of δ-ALA, tyrosine and creatinine in biological fluids. *Clinica Chimica Acta* 2004, 350(1), 219–230.

Feyerabend, C., Russell, M. Assay of nicotine in biological materials: Sources of contamination and their elimination. *Journal of Pharmacy and Pharmacology.* 1980, 32(1), 178–181.

Feyerabend, C., Russell, M. A rapid gas-liquid chromatographic method for the determination of cotinine and nicotine in biological fluids. *Journal of Pharmacy and Pharmacology.* 1990, 42(6), 450–452.

Greaves, R. et al. A simple high-pressure liquid chromatography cotinine assay: Validation of smoking status in pregnant women. *Annals of Clinical Biochemistry.* 2001, 38(4), 333–338.

Green, M. S., Harari, G. A prospective study of the effects of changes in smoking habits on blood count, serum lipids and lipoproteins, body weight and blood pressure in occupationally active men. The Israeli CORDIS Study. *Journal of Clinical Epidemiology.* 1995, 48(9), 1159–1166.

Hammond, S. K. et al. Relationship between environmental tobacco smoke exposure and carcinogen-hemoglobin adduct levels in nonsmokers. *Journal of the National Cancer Institute.* 1993, 85(6), 474–478.

Hansel, M. et al. Single-reagent polarisation fluoroimmunoassay for cotinine (a nicotine metabolite) in urine. *Annals of Clinical Biochemistry.* 1986, 23, 596–602.

Hariharan, M. et al. A high-performance liquid-chromatographic method for routine simultaneous determination of nicotine and cotinine in plasma. *Clinical Chemistry.* 1988, 34(4), 724–729.

Hariharan, M., VanNoord, T. Liquid-chromatographic determination of nicotine and cotinine in urine from passive smokers: Comparison with gas chromatography with a nitrogen-specific detector. *Clinical Chemistry.* 1991, 37(7), 1276–1280.

Hatsukami, D. K., Hecht, S. S., Hennrikus, D. J., Joseph, A. M., Pentel, P. R. Biomarkers of tobacco exposure or harm: Application to clinical and epidemiological studies. *Nicotine & Tobacco Research.* 2003, 5, 387–396.

Hatsukami, D. K. et al. Biomarkers to assess the utility of potential reduced exposure tobacco products. *Nicotine & Tobacco Research.* 2006, 8(2), 169–191.

Haufroid, V., Lison, D. Urinary cotinine as a tobacco-smoke exposure index: A minireview. *International Archives of Occupational and Environmental Health.* 1998, 71(3), 162–168.

Heavner, D. L. et al. Validation and application of a method for the determination of nicotine and five major metabolites in smokers' urine by solid-phase extraction and liquid chromatographyase extraction and liquid. *Biomedical Chromatography.* 2005, 19(4), 312–328.

Hecht, S. S. Human urinary carcinogen metabolites: Biomarkers for investigating tobacco and cancer. *Carcinogenesis.* 2002, 23(6), 907–922.

Hewavitharana, A. K., Bruce, H. L. Simultaneous liquid chromatographic determination of creatinine and pseudouridine in bovine urine and the effect of sample pH on the analysis. *Journal of Agricultural and Food Chemistry.* 2003, 51(17), 4861–4865.

Horstmann, M. Simple high-performance liquid chromatographic method for rapid determination of nicotine and cotinine in urine. *Journal of Chromatography B: Biomedical Sciences and Applications.* 1985, 344, 391–396.

Huitema, A. D. et al. Sensitive gas chromatographic determination of the cyclophosphamide metabolite 2-dechloroethylcyclophosphamide in human plasma. *Journal of Chromatography B: Biomedical Sciences and Applications.* 2001, 757(2), 349–357.

Iyengar, G., Rapp, A. Human placenta as a 'dual' biomarker for monitoring fetal and maternal environment with special reference to potentially toxic trace elements. Part 3: Toxic trace elements in placenta and placenta as a biomarker for these elements. *Science of the Total Environment.* 2001, 280(1), 221–238.

Jaakkola, M., Jaakkola, J. Assessment of exposure to environmental tobacco smoke. *European Respiratory Journal.* 1997, 10(10), 2384–2397.

Jacob, P. et al. Determination of the nicotine metabolite *trans*-3'-hydroxycotinine in urine of smokers using gas chromatography with nitrogen-selective detection or selected ion monitoring. *Journal of Chromatography B: Biomedical Sciences and Applications.* 1992, 583, 145–154.

Jacob, P. et al. Gas chromatographic–mass spectrometric method for determination of anabasine, anatabine and other tobacco alkaloids in urine of smokers and smokeless tobacco users. *Journal of Chromatography B: Biomedical Sciences and Applications* 1993, 619(1), 49–61.

Jacobs, R. M. et al. Effects of interferents on the kinetic Jaffé reaction and an enzymatic colorimetric test for serum creatinine concentration determination in cats, cows, dogs and horses. *Canadian Journal of Veterinary Research.* 1991, 55(2), 150.

Jaffé, M. Ueber den Niederschlag, welchen Pikrinsäure in normalem harn erzeugt und über eine neue reaction des kreatinins. *Zeitschrift für physiologische Chemie.* 1886, 10(5), 391–400.

Jahnke, G. D. et al. Multiple DNA adducts in lymphocytes of smokers and nonsmokers determined by 32P-postlabeling analysis. *Carcinogenesis.* 1990, 11(2), 205–211.

Kellogg, M. D. et al. Rapid and simple tandem mass spectrometry method for determination of serum cotinine concentration. *Clinical Chemistry.* 2004, 50(11), 2157–2159.

Knight, G. et al. Improved 125I radioimmunoassay for cotinine by selective removal of bridge antibodies. *Clinical Chemistry.* 1985, 31(1), 118–121.

Kolonen, S. A., Puhakainen, E. V. Assessment of the automated colorimetric and the high-performance liquid chromatographic methods for nicotine intake by urine samples of smokers' smoking low-and medium-yield cigarettes. *Clinica Chimica Acta.* 1991, 196(2), 159–166.

Langone, J. J. et al. Nicotine and its metabolites. Radioimmunoassays for nicotine and cotinine. *Biochemistry.* 1973, 12(24), 5025–5030.

Langone, J. J. et al. Monoclonal antibody ELISA for cotinine in saliva and urine of active and passive smokers. *Journal of Immunological Methods.* 1988, 114(1), 73–78.

Layten Davis, D., Nielsen, M. T. *Tobacco: Production, Chemistry and Technology.* Blackwell Science Ltd., 1999, Hoboken, NJ.

Machacek, D. A., Jiang, N. Quantification of cotinine in plasma and saliva by liquid chromatography. *Clinical Chemistry.* 1986, 32(6), 979–982.

Maclure, M. et al. Elevated blood levels of carcinogens in passive smokers. *American Journal of Public Health.* 1989, 79(10), 1381–1384.

Mahoney, G. N., Al-Delaimy, W. Measurement of nicotine in hair by reversed-phase high-performance liquid chromatography with electrochemical detection. *Journal of Chromatography B: Biomedical Sciences and Applications.* 2001, 753(2), 179–187.

McAdams, S., Cordeiro, M. Simple selected ion monitoring capillary gas chromatographic–mass spectrometric method for the determination of cotinine in serum, urine and oral samples. *Journal of Chromatography B: Biomedical Sciences and Applications* 1993, 615(1), 148–153.

Mousa, S. et al. High-performance liquid chromatography with electrochemical detection for the determination of nicotine and N-methylnicotinium ion. *Journal of Chromatography A*, 1985, 347, 405–410.

Moyer, T. P. et al., Simultaneous analysis of nicotine, nicotine metabolites, and tobacco alkaloids in serum or urine by tandem mass spectrometry, with clinically relevant metabolic profiles. *Clinical Chemistry.* 2002, 48(9), 1460–1471.

Narkowicz, S. et al. Analysis of markers of exposure to constituents of Environmental Tobacco Smoke (ETS). *Critical Reviews in Analytical Chemistry.* 2012, 42(1), 16–37.

Neurath, G. B. et al. *Trans*-3′-hydroxycotinine as a main metabolite in urine of smokers. *International Archives of Occupational and Environmental Health.* 1987, 59(2), 199–201.

Pacifici, R. et al. Nicotine, cotinine, and *trans*-3-hydroxycotinine levels in seminal plasma of smokers: Effects on sperm parameters. *Therapeutic Drug Monitoring.* 1993, 15(5), 358–363.

Page-Sharp, M. et al. Measurement of nicotine and cotinine in human milk by high-performance liquid chromatography with ultraviolet absorbance detection. *Journal of Chromatography B.* 2003, 796(1), 173–180.

Park, E.-K. et al. Creatinine measurements in 24 h urine by liquid chromatography—Tandem mass spectrometry. *Journal of Agricultural and Food Chemistry.* 2007, 56(2), 333–336.

Parry, H. et al. Smoking, alcohol consumption, and leukocyte counts. *American Journal of Clinical Pathology.* 1997, 107(1), 64–67.

Parviainen, M. T., Barlow, R. D. Assessment of exposure to environmental tobacco smoke using a high-performance liquid chromatographic method for the simultaneous determination of nicotine and two of its metabolites in urine. *Journal of Chromatography B: Biomedical Sciences and Applications.* 1988, 431, 216–221.

Patel, C., George, R. Liquid chromatographic determination of creatinine in serum and urine. *Analytical Chemistry.* 1981, 53(4), 734–735.

Perezstable, E. J. et al. Is serum cotinine a better measure of cigarette-smoking than self-report? *Preventive Medicine.* 1995, 24(2), 171–179.

Petrakis, N. L. et al. Nicotine in breast fluid of nonlactating women. *Science.* 1978, 199(4326), 303–305.

Phillips, A. N. et al. The leukocyte count and risk of lung cancer. *Cancer*. 1992, 69(3), 680–684.

Pickert, A. et al. Comparison of a mechanized version of the 'König' reaction and a fluorescence polarization immunoassay for the determination of nicotine metabolites in urine. *Clinica Chimica Acta*. 1993, 217(2), 143–152.

Polkowska, Ż. et al. Biological fluids as a source of information on the exposure of man to environmental chemical agents. *Critical Reviews in Analytical Chemistry*. 2004, 34(2), 105–119.

Ruiz-Jiménez, J. et al. On-line automatic SPE-CE coupling for the determination of biological markers in urine. *Electrophoresis*. 2007, 28(5), 789–798.

Rustemeier, K. et al. High-performance liquid chromatographic determination of nicotine and its urinary metabolites via their 1, 3-diethyl-2-thiobarbituric acid derivatives. *Journal of Chromatography B: Biomedical Sciences and Applications*. 1993, 613(1), 95–103.

Savela, K., Hemminki, K. DNA adducts in lymphocytes and granulocytes of smokers and nonsmokers detected by the 32P-postlabelling assay. *Carcinogenesis*. 1991, 12(3), 503–508.

Scherer, G. Smoking behaviour and compensation: A review of the literature. *Psychopharmacology*. 1999, 145(1), 1–20.

Scherer, G. Carboxyhemoglobin and thiocyanate as biomarkers of exposure to carbon monoxide and hydrogen cyanide in tobacco smoke. *Experimental and Toxicologic Pathology*. 2006, 58(2–3), 101–124.

Schütte-Borkovec, K. et al. Analysis of myosmine, cotinine and nicotine in human toenail, plasma and saliva. *Biomarkers*. 2009, 14(5), 278–284.

Sepkovic, D., Haley, N. Biomedical applications of cotinine quantitation in smoking related research. *American Journal of Public Health*. 1985, 75(6), 663–665.

Shen, H.-M. et al. Detection of oxidative DNA damage in human sperm and the association with cigarette smoking. *Reproductive Toxicology*. 1997, 11(5), 675–680.

Shi, H. et al. A simple and fast method to determine and quantify urinary creatinine. *Analytica Chimica Acta*. 1995, 312(1), 79–83.

Skarping, G. et al. Determination of cotinine in urine using glass capillary gas chromatography and selective detection, with special reference to the biological monitoring of passive smoking. *Journal of Chromatography A*. 1988, 454, 293–301.

Sorensen, M. et al. Biomarkers of exposure to environmental tobacco smoke in infants. *Biomarkers*. 2007, 12(1), 38–46.

Sunyer, J. et al. Longitudinal relation between smoking and white blood cells. *American Journal of Epidemiology*. 1996, 144(8), 734–741.

Takahashi, N. et al. Tandem mass spectrometry measurements of creatinine in mouse plasma and urine for determining glomerular filtration rate. *Kidney International* 2006, 71(3), 266–271.

Talhout, R. et al. Hazardous compounds in tobacco smoke. *International Journal of Environmental Research and Public Health*. 2011, 8(2), 613–628.

Taylor, P. J. et al. The measurement of nicotine in human plasma by high-performance liquid chromatography-electrospray-tandem mass spectrometry. *Therapeutic Drug Monitoring*. 2004, 26(5), 563–568.

Thomas Karnes, H. et al. Assessment of nicotine uptake from cigarette smoke: Comparison of a colorimetric test strip (NicCheck I™) and gas chromatography/mass selective detector. *Biomarkers*. 2001, 6(6), 388–399.

Toraño, J. S., van Kan, H. J. Simultaneous determination of the tobacco smoke uptake parameters nicotine, cotinine and thiocyanate in urine, saliva and hair, using gas chromatography-mass spectrometry for characterisation of smoking status of recently exposed subjects. *Analyst*. 2003, 128(7), 838–843.

Tretyakova, N. et al. Quantitation of DNA adducts by stable isotope dilution mass spectrometry. *Chemical Research in Toxicology*. 2012, 25(10), 2007–2035.

Tsikas, D. et al. Simplified HPLC method for urinary and circulating creatinine. *Clinical Chemistry.* 2004, 50(1), 201–203.

Tutka, P. et al. Exposure to environmental tobacco smoke and children health. *International Journal of Occupational Medicine and Environmental Health.* 2002, 15(4), 325–335.

Van Maanen, J. et al. DNA adduct and mutation analysis in white blood cells of smokers and nonsmokers. *Environmental and Molecular Mutagenesis.* 1994, 24(1), 46–50.

Vineis, P., Porta, M. Causal thinking, biomarkers, and mechanisms of carcinogenesis. *Journal of Clinical Epidemiology.* 1996, 49(9), 951–956.

Wald, N. et al. Serum cholesterol and subsequent risk of cancer: Results from the BUPA study. *British Journal of Cancer.* 1989, 59(6), 936.

Watts, R. R. et al. Cotinine analytical workshop report: Consideration of analytical methods for determining cotinine in human body fluids as a measure of passive exposure to tobacco smoke. *Environmental Health Perspectives.* 1990, 84, 173.

Welch, M. J. et al. Determination of serum creatinine by isotope dilution mass spectrometry as a candidate definitive method. *Analytical Chemistry.* 1986, 58(8), 1681–1685.

Xu, A. S. et al. Determination of nicotine and cotinine in human plasma by liquid chromatography-tandem mass spectrometry with atmospheric-pressure chemical ionization interface. *Journal of Chromatography B: Biomedical Sciences and Applications.* 1996, 682(2), 249–257.

Xu, X. et al. Simultaneous and sensitive measurement of anabasine, nicotine, and nicotine metabolites in human urine by liquid chromatography–tandem mass spectrometry. *Clinical Chemistry.* 2004, 50(12), 2323–2330.

Yoshihara, K. et al. Determination of urinary pyrraline by solid-phase extraction and high performance liquid chromatography. *Biological and Pharmaceutical Bulletin* 2001, 24(8), 863–866.

Zhang, S. et al. Proteomic alteration in lung tissue of rats exposed to cigarette smoke. *Toxicology Letters.* 2008, 178(3), 191–196.

Zinellu, A. et al. Assay for the simultaneous determination of guanidinoacetic acid, creatinine and creatine in plasma and urine by capillary electrophoresis UV-detection. *Journal of Separation Science.* 2006, 29(5), 704–708.

3 Biomarkers of Pyridine Alkaloids
Nicotine and Cotinine

3.1 INTRODUCTION

Tobacco alkaloids are the active principal components in all tobacco products. Nicotine is the major alkaloid present in tobacco (Benowitz, 1997). Nicotine accounts for 98% of the total alkaloids present in tobacco, and owing to its addictive properties, it has widespread usage by humans all over the world (Kataoka et al., 2009). In addition, nicotine is a pharmacological substance used to treat human diseases, such as reproductive disorders and cardiovascular disease (Benowitz, 1997). There is good evidence that most smokers are addicted to nicotine, and the severity of dependence on tobacco is related to the amount of nicotine intake. Nicotine is absorbed rapidly in humans through the skin and mucosal lining of the mouth and nose or by inhalation in the lungs and exerts a number of physiological effects in both active and passive smokers, defined as cigarette smokers and nonsmokers exposed to environmental tobacco smoke (ETS). About 70%–80% of the nicotine absorbed into body is metabolized to cotinine (Bramer and Kallungal, 2003). Nicotine and its metabolite cotinine can be detected in all kinds of biological fluids, including blood, saliva, and urine (Tricker, 2006). Therefore, these compounds have been widely used as biological markers to determine tobacco smoking status and estimate exposure to ETS.

3.2 TOXICOLOGICAL EVALUATION

Nicotine is an unpleasant, bitter, colorless oily liquid, is highly volatile, and can be easily oxidized in air. It can be rapidly dissolved in water and alcohol and easily adsorbed by bronchial mucosa through the nose and mouth. Nicotine adsorbed to the skin surface also may be absorbed into the body. Nicotine is not only harmful to humans but also harmful for the zootic. Therefore, nicotine had been used as the main component of agricultural pesticides. Nicotine accounts for about 95% of the total alkaloid content and 1.5% of the weight of commercial cigarette tobacco (Benowitz, 1983; Schmeltz and Hoffmann, 1977). An average tobacco rod contains 10–14 mg of nicotine (Kozlowski et al., 1998), and on average about 1–1.5 mg of nicotine is absorbed systemically during smoking (Benowitz and Jacob III, 1984). Most of the nicotine present in tobacco is mainly the levarotary (s)-isomer, and (R)-nicotine only accounts for 0.1%–0.6% of total nicotine content (Armstrong et al., 1998).

Cotinine is the primary among the metabolites of nicotine in humans, which has been reported to have little to no effect on cardiovascular diseases and cognitive performance in humans, but it may change symptoms and signs of nicotine withdrawal (Benowitz, 1983; Hatsukami et al., 1997; Keenan et al., 1994; Zevin et al., 2000). Nicotine base is well absorbed through the skin. That is the reason for the occupational risk of nicotine poisoning (green tobacco sickness) in tobacco harvesters who are exposed to wet tobacco leaves (McBride et al., 1998). That is also the basis for transdermal delivery technology (Benowitz et al., 1995).

3.3   METABOLIC PATHWAY OF NICOTINE IN THE BODY

Nicotine, a major component in tobacco, is also a major addictive substance in cigarette smoke. It is absorbed through the skin and mucosal lining of the mouth and nose or by inhalation in the lungs by both active and passive smokers. After absorption, nicotine enters the bloodstream; it is about 69% ionized and 31% unionized. Less than 5% of nicotine is cross-linked with plasma proteins (Benowitz et al., 1982). Then, it enters the brain, liver, etc., by the blood delivered. In the liver, nicotine is metabolized to more than 10 kinds of metabolites (Figure 3.1) (Hukkanen et al., 2005). Quantitatively, the most important metabolite of nicotine in most mammalian species is the lactam derivative, cotinine. About 70%–80% of nicotine is converted to cotinine in humans. Nicotine-N-oxide is another primary metabolite of nicotine, although only about 4%–7% of nicotine absorbed by smokers is metabolized via this route (Benowitz et al., 1994). In addition to oxidation of the pyrrolidine ring, nicotine is metabolized by two nonoxidative pathways: methylation of the pyridine nitrogen giving nicotine-N-methylnicotinium ion and glucuronidation. Nicotine glucuronidation leads to form an N-quaternary glucuronide in humans (Benowitz et al., 1994). About 3%–5% of nicotine is transformed into nicotine glucuronide and excreted in urine in humans.

Oxidative N-demethylation is frequently an important pathway in the metabolism of xenobiotics, but this route is, in most species, a minor pathway in the metabolism of nicotine. Conversion of nicotine to nornicotine in humans has been demonstrated (Jacob III and Benowitz, 1991). Nornicotine is a constituent of tobacco leaves. However, most urine nornicotine is derived from metabolism of nicotine with less than 40% coming directly from tobacco, as estimated from the difference in nornicotine excretion in smokers during smoking and transdermal nicotine treatment (0.65% and 0.41%, respectively) (Benowitz et al., 1994). In 2000, Hecht et al. (2000) reported that 2-hydroxylation of nicotine can produce 4-(methylamino)-1-(3-pyridyl)-1-butanone. Then 4-(methylamino)-1-(3-pyridyl)-1-butanone is further metabolized to produce 4-oxo-4-(3-pyridyl)butanoic acid and 4-hydroxy-4-(3-pyridyl)butanoic acid. This new pathway is potentially significant since 4-(methylamino)-1-(3-pyridyl)-1-butanone can be converted to carcinogenic NNK (Hecht et al., 1999).

So far, six primary metabolites of cotinine have been found in humans: 3-hydroxycotinine (Neurath et al., 1987); 5-hydroxycotinine (Neurath, 1994), cotinine-N-oxide; cotinine methonium ion; cotinine glucuronide; and norcotinine.

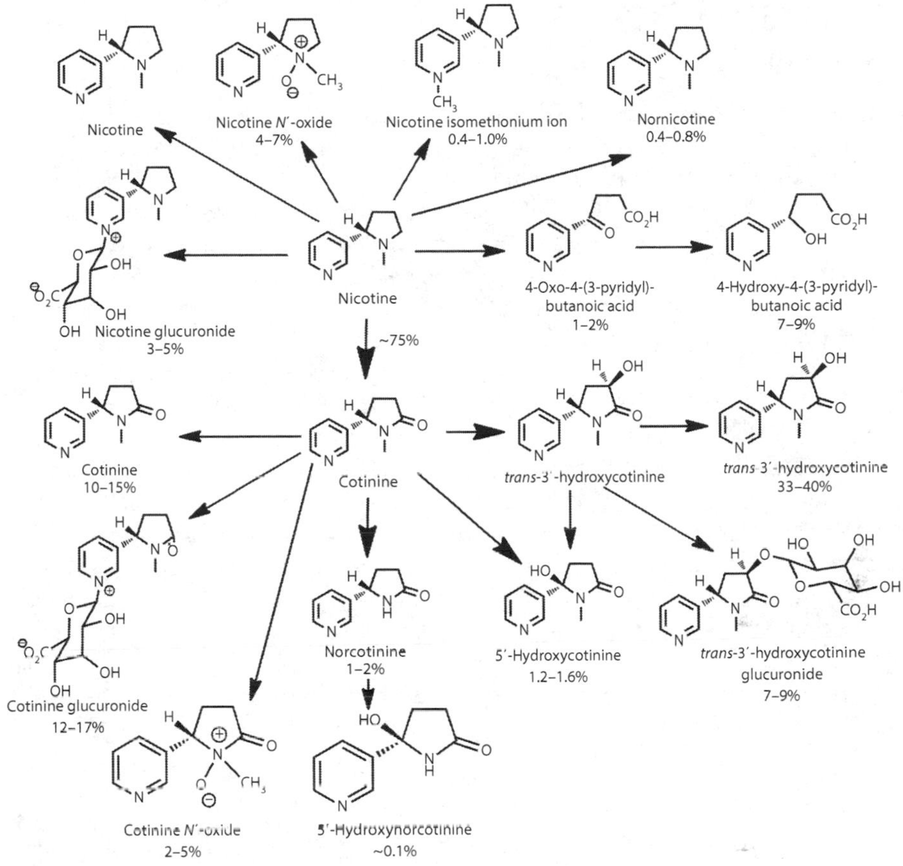

FIGURE 3.1 The metabolic pathways of nicotine in vivo. (From Hukkanen, J. et al., *Pharmacol. Rev.*, 57, 79, 2005.)

3-Hydroxycotinine is the primary metabolite of nicotine measured in urine from smokers. Its glucuronide conjugate is another important metabolite (Benowitz et al., 1994). 3-Hydroxycotinine and its glucuronide conjugate account for 40%–60% of the nicotine dose in urine (Byrd et al., 1992). While nicotine and cotinine conjugates are *N*-glucuronides, *O*-glucuronide is only found in the 3-hydroxycotinine conjugate (Byrd et al., 1994). Approximately 90% of a systemic dose of nicotine was transformed into nicotine and metabolites in urine (Benowitz et al., 1994). The urinary nicotine *N*-oxide accounts for about 4%–7% of systemic nicotine, and nicotine glucuronide accounts for 3%–5% of systemic nicotine (Byrd et al., 1992). A small part of cotinine is unchanged and excreted in urine, which accounts for about 10%–15% of the nicotine and metabolites. The rest is transformed into metabolites, including 33%–40% of trans-3-hydroxycotinine, 12%–17% of cotinine glucuronide, and 7%–9% of trans-3-hydroxycotinine glucuronide.

3.4   SAMPLES USED

3.4.1   URINE

Urine samples have various advantages to use evaluated nicotine and nicotine metabolism biomarkers (e.g., noninjury, convenient sample collection, and simply specimen disposal). Excretion of cotinine and cotinine-N-glucuronide, trans-3'-hydroxycotinine and trans-3'-hydroxycotinine-O-glucuronide, and nicotine-N-glucuronide in 24 h urine was approximately 80%–90% of the systemic nicotine intake from smoking. Nicotine and its metabolism products in urine have been widely used as a biomarker to assess tobacco and smoke exposure (Mendes et al., 2008). Cotinine measured in urine has also been widely used as a biomarker to evaluate tobacco exposure (Neurath et al., 1987). Urinary cotinine concentrations are higher than those obtained from different biological fluids, the reason for this may be that the kidney is the main nicotine-excreting organ. This is corroborated by the results of the Bland–Altman analysis, which showed that urinary cotinine is around five times higher than serum levels. Because the collection of 24 h urine sample is a tedious task for the patient, morning urine is usually used, and urine samples are often adjusted by creatinine for density for assessing, which has a better correlation with serum cotinine (Man et al., 2006). All urinary samples presented significant correlation with serum. However, the variation present in the limits of agreement in unadjusted urine is lower. This finding, and the fact that normalization of urine requires additional costs and work, suggests that urine normalization is unnecessary and urine samples were widely used in evaluated nicotine exposure.

3.4.2   BLOOD

The levels of cotinine in the blood can be used to assess tobacco smoke exposure. Smoking cessation is an effective way to reduce morbidity and mortality caused by smoking in all countries (Neovius et al., 2009). Nicotine metabolite (cotinine and 3'-hydroxycotinine) concentrations in plasma and/or expired carbon monoxide have been examined as validated biomarkers of self-reported cigarette numbers associated with nicotine-metabolizing enzymes cytochrome P450 2A6 (CYP2A6) and CYP2B6 genotypes (Hukkanen et al., 2005). Nicotine contents in blood obtained from smokers is 10 to 50 ng mL^{-1}. Benowitz et al. (1994) found that the mean increase in nicotine was 10.9 ng mL^{-1} in smokers after smoking a cigarette. The peak value of blood nicotine levels decrease rapidly within 20 min after smoking a cigarette, because of tissue distribution. The peak nicotine levels of venous blood from rest tobacco are similar with cigarette smokers (Benowitz et al.,1982). Pipe smokers, particularly those who have previously smoked cigarettes, may have blood and urine levels of nicotine and cotinine as high as cigarette smokers (Jacob III and Benowitz, 1991). Cigar smokers who previously smoked cigarettes may have higher nicotine levels in blood than those who are primary cigar smokers (Wald et al., 1984). The plasma half-life of nicotine after intravenous infusion or cigarette smoking averages about 2 h. However, when half-life is determined using the time course of urinary excretion of nicotine, which is more sensitive in detecting lower levels of nicotine in the body, the terminal half-life averages 11 h (Hecht et al., 2000).

3.4.3 HAIR AND NAILS

The nicotine and cotinine content in hair and nail has been proposed as a way to assess long-term exposure to nicotine from tobacco products (Florescu et al., 2007). The first paper concerning hair nicotine determination was published in 1983 by Ishiyama et al. (1983). Since then, examining hair nicotine content has become a valuable tool facilitating assessment of exposure to tobacco smoke in various groups of people (active and passive smokers or even neonates and fetus) (Koszowski et al., 2008). Florescu et al. analyzed 1746 cases to validate cotinine as a marker for ETS exposure. Their results showed that mean cotinine levels in hair obtained from active nonpregnant and pregnant women smokers were 2.3–3.1 ng/mg and 1.5–1.9 ng/mg, respectively. In the group of passive smokers, mean hair cotinine concentrations were 0.5–0.7 ng/mg for nonpregnant women, 0.04–0.09 ng/mg for pregnant women, and 0.2 ng/mg for pregnant women). A cutoff value of 0.2 ng/mg was accurate in discriminating between exposed and unexposed children (Florescu et al., 2007). Tzatzarakis et al. used hair nicotine/cotinine concentrations as a method of monitoring exposure to tobacco smoke among infants and adults. The results showed the use of hair samples as an effective method for assessing exposure to tobacco, with a high association between nicotine and cotinine especially among infants heavily exposed to SHS (Tzatzarakis et al., 2012).

Stepanov et al. analyzed cotinine, nicotine, and 4-(methylnitrosamino)-1-(3-pyridyl)-1-butanol (NNAL) in toenails and explored their relationship. They found that there was no effect of age or gender on the toenail biomarkers. Toenail nicotine and toenail cotinine correlated significantly with cigarettes smoked per day ($r = 0.24$; $P = 0.015$ and $r = 0.26$; $P = 0.009$, respectively). There is a significant relationship between toenail nicotine and plasma cotinine ($r = 0.45$; $P < 0.001$) and plasma $trans$-3'-hydroxycotinine ($r = 0.30$; $P = 0.008$); and toenail NNAL correlated with urine NNAL ($r = 0.53$; $P = 0.005$) (Stepanov et al., 2007). The results showed that toenail cotinine may be used as a biomarker to investigate the role of chronic tobacco smoke exposure in human cancer.

3.5 ANALYTICAL TECHNOLOGY

Methods published for the quantification of nicotine and its metabolites in different biological specimens include enzyme-linked immunoassay (Matsumoto et al., 2010; Rees et al., 1996), gas chromatography (GC) (Kuo et al., 2002), GC–mass spectrometry (MS) (Culea et al., 2005; Lafay et al., 2010; Man et al., 2009), high-performance liquid chromatography (LC) (Papadoyannis et al., 2002; Petersen et al., 2010), and LC–tandem MS (LC–MS/MS) (Bentley et al., 1999; Lee et al., 2005; Shakleya and Huestis, 2009).

3.5.1 GC-MS[N]

Matsuki et al. (2008) developed a GC–MS method to simultaneously analyze urinary nicotine and its metabolites and studied their half-lives in urine. It was found that the nicotine concentrations in the urine of the subjects exposed to ETS reached a peak at about 2 h after the end of exposure, which was somewhat later than that in active smokers. The half-life times of nicotine, cotinine, and *trans*-3'-hydroxycotinine calculated using theoretical curves were 13.9, 20.0, and 63.0 h, respectively. Lafay et al. (2010) developed a microextraction in packed sorbent (MEPS) and GC–MS method for the determination of cotinine in urine. The assay was simple, rapid, sensitive, and nonconsuming solvent. Compared with the conventional solid-phase extraction/liquid–liquid extraction method, the optimized method had the wide linear range (5–5000 ng/mL) and high sensitivity and allowed application to analysis of urine from smokers as well as nonsmokers susceptible to passive smoking. In general, hair nicotine assay involves alkaline digestion, extraction, and instrumental analysis. In order to simplify extraction procedure and shorten GC analysis time, Man et al. (2009) developed a high-throughput GC–MS assay. The limit of quantitation was 0.04 ng/mg hair, within- and between-assay accuracies and precisions are less than 11.4% and the mean recovery was 92.6%. It was sensitive and applicable for routine analysis of monitoring chronic ETS exposure or tobacco consumption in an epidemiological population.

3.5.2 LC–MS[N]

Due to high sensitivity, LC–MS/MS has been widely used to detect nicotine and its major metabolites in urine or blood. Chadwick and Keevil (2007) developed and validated a method for the measurement of urinary cotinine using reversed phase LC–MS/MS. This technique utilized online ion exchange coupled with an analytical column to eliminate ion suppression effects. Feng et al. (2007) used LC–MS/MS to detect nicotine and its five major metabolites (nicotine-*N*-glucuronide, cotinine, cotinine-*N*-glucuronide, *trans*-3-hydroxycotinine, and *trans*-3-hydroxycotinine-*O*-glucuronide) in 24 h urine samples. They build a first-order elimination pharmacokinetic model to assess the elimination half-life of nicotine equivalents (nicotine and its five major metabolites), the elimination half-life of nicotine equivalents was 19.4 ± 2.6 h, and the nicotine equivalents excretion had reached ~96% of the steady-state levels by day 4. Their results showed that most of the nicotine inhaled from a cigarette is retained (≥98%) in the lungs, and at steady-state, daily urine nicotine equivalent excretion reflects ~90% of the retained nicotine dose from cigarette smoking. However, the aforementioned methods all employed solid-phase extraction for sample preparation. Fan et al. (2008) developed and validated a method based on the direct injection of diluted urine for simultaneous identification and quantification of nicotine and its nine metabolites in human urine. This approach simplifies the process of pretreatment. It can be used to assess smoking-related cotinine exposure.

3.5.3 OTHER ANALYTICAL TECHNOLOGIES

The level of cotinine in biological specimens, such as serum, urine, and saliva, measured by GC or LC is the most validated and reliable indicator of exposure to tobacco smoke. However, chromatographic methods are not always suitable for all types of situations. Matsumoto et al. (2010) developed and validated a commercially available enzyme-linked immunosorbent assay (ELISA) to detect cotinine in urine. The result showed that the levels of IR cotinine in the urine of kindergarten children closely correlated with those of cotinine measured by GC-MS and reflected the smoking behavior of their parents more precisely than cotinine levels determined by GC-MS. Rees et al. (1996) developed a sensitive ELISA method to analyze *trans*-3-hydroxycotinine. Compared with GC/LC methods, no matrix effects were detectable in human saliva, and relatively small matrix effects were found (I-50 for *trans*-3-hydroxycotinine, about 25 ng/mL) in urine. The assay readily detected levels of apparent *trans*-3-hydroxycotinine in urine samples from smoke-exposed mice and rats. ELISA is therefore a sensitive test for the determination of *trans*-3-hydroxycotinine in plasma, saliva, and urine samples from humans and animals and can be used to monitor exposure to tobacco smoke or nicotine.

3.6 METABOLITES OF NICOTINE

3.6.1 COTININE

Approximately 70%–80% of nicotine is metabolized to form cotinine in humans, only 10%–15% of the nicotine absorbed by smokers appears in the urine as unchanged cotinine (Benowitz and Jacob, 1994). A number of cotinine metabolites has been structurally characterized. Six primary metabolites of cotinine have been reported in humans: 3′-hydroxycotinine (Neurath et al., 1987); 5′-hydroxycotinine (also called allohydroxycotinine) (Neurath, 1994), which exists in tautomeric equilibrium with the open-chain derivative 4-oxo-4-(3-pyridyl)-*N*-methylbutanamide (Bowman and McKennis, 1962); cotinine-*N*-oxide (Kyerematen et al., 1990); cotinine methonium ion (McKennis Jr et al., 1963); cotinine glucuronide (Curvall et al., 1991); and norcotinine (also called demethylcotinine) (Kyerematen et al., 1990). 3′-Hydroxycotinine and its glucuronide conjugate in smokers' urine account for 40%–60% of the nicotine dose (Andersson et al., 1997; Byrd et al., 1992). Norcotinine has been detected in smokers' urine (about 1% of total nicotine and metabolites) (Byrd et al., 1995). 5′-Hydroxycotinine has also been measured in urine from smokers, but its level in urine was less than 4% of the 3′-hydroxycotinine (Neurath, 1994). After nicotine administration, 5′-hydroxycotinine *N*-oxide was found in rat urine, but there is no report that it was measured in smokers' urine (Schepers et al., 1999). The procedure (Petersen et al., 2010) involves liquid–liquid extraction, separation on an RP column (Zorbax XDB C8), isocratic pump (0.5 mL/min of water–methanol–sodium acetate [0.1 M]–ACN [50:15:25:10, v/v/v/v], 1.0 mL of citric acid [0.034 M], and 5.0 mL of triethylamine for each liter), and high-pressure LC (HPLC)–UV detection (261 nm). The analytical procedure proved to be sensitive, selective, precise, accurate, and linear ($r > 0.99$) in the range of 5–500.0 ng/mL for cotinine.

3.6.2 3-HYDROXYCOTININE

3-Hydroxycotinine is the main nicotine metabolite detected in smokers' urine. It is also excreted as a glucuronide conjugate (Benowitz and Jacob, 1994; Curvall et al., 1991). 3-Hydroxycotinine and its glucuronide conjugate account for 40%–60% of the nicotine dose in urine (Byrd et al., 1992). The conversion of cotinine to 3-hydroxycotinine in humans is highly stereo selective for the *trans*-isomer, as less than 5% is detected as *cis*-3-hydroxycotinine in urine (Jacob III et al., 1990; Voncken et al., 1990). They had found that concentrations of *trans*-3′-hydroxycotinine smokers' urine exceed cotinine concentrations by two- to threefold (Jacob III et al., 1988). Numerous methods for the determination of cotinine in biological fluids have been reported. GC-MS is the primary technology used to determine the content of cotinine (Gorrod and Jacob III, 1999; Ji et al., 1999). With the development of analytic instruments, HPLC was approximated in progress (Hariharan et al., 1988). However, LC–MS and LC with tandem mass spectrometric detection (Bentley et al., 1999; Bernert et al., 1997; Meger et al., 2002; Xu et al., 2004) are the mostly used techniques. Typically, Bentley and coworkers had validated the method for the simultaneous determination of low-level cotinine and 3-hydroxycotinine in human saliva. In the progress, analytes and deuterated internal standards were extracted from saliva samples using automated solid-phase extraction; the columns containing a hyper-cross-linked styrene–divinylbenzene copolymer sorbent and lower limits of quantitation of 0.05 and 0.10 ng/mL for cotinine and 3-hydroxycotinine, respectively, were achieved (Bentley et al., 1999).

3.6.3 NICOTINE GLUCURONIDE

Nicotine glucuronidation results in an *N*-quaternary glucuronide in humans (Byrd et al., 1992; Curvall et al., 1991) under the action of the uridine diphosphate-glucuronosyl transferase enzyme (Seaton, 1996). About 3%–5% of nicotine is converted to nicotine glucuronide and excreted in urine in humans. Oxidative *N*-demethylation is frequently an important pathway in the metabolism of xenobiotics, but this route is, in most species, a minor pathway in the metabolism of nicotine. Nicotine glucuronidation results in an *N*-quaternary glucuronide in humans (Byrd et al., 1992; Curvall et al., 1991). Ghosheh et al. (2001) reported the identification and the apparent kinetics of formation of nicotine-*N*1-glucuronide in pooled human liver microsomes. The metabolite formed from natural *S*(−)-nicotine was identified by comparison of the HPLC retention time and positive ion electrospray ionization mass spectral characteristics with a synthetic reference standard. The glucuronides of *S*(−)- and *R*(+)-nicotine were formed by one-enzyme kinetics, with Km values of 0.11 and 0.23 mM and Vmax values of 132 and 70 pmol/min/mg of protein, respectively.

3.6.4 COTININE GLUCURONIDE

The major routes of metabolism of nicotine in human involve oxidation and glucuronidation. Cotinine conjugates excreted in smoker's urine were approximately 9%–17% (Byrd et al., 1992; Curvall et al., 1991). Ghosheh and Hawes (2002) conducted research to study the kinetics of the formation of *S*(−)-cotinine *N*1-glucuronide. The direct method for the determination of cotinine glucuronide is LC–MS (Byrd et al., 1994; Caldwell et al., 1992; Nakajima et al., 2002). Miki Nakajima established a highly sensitive system for detecting Nic-glu and Cot-glu with HPLC–UV. The method was successfully applied to determine *N*-glucuronosyltransferase activities of nicotine and cotinine in human liver microsomes (Nakajima et al., 2002).

3.6.5 3-HYDROXYCOTININE GLUCURONIDE

Recently, significant rates of *N*-glucuronidation of 3-hydroxycotinine were detected in human liver microsomes, but this metabolite was not detected in urine (Kuehl and Murphy, 2003). Thus, the *N*-glucuronide of 3-hydroxycotinine might be unstable, or the concentration was too low to detect the *N*-glucuronide by the methodology employed. Hydroxycotinine glucuronide is formed by the sequence in which cotinine is hydroxylated to (3'*R*, 5'*S*)*trans*-hydroxycotinine, which can then be further conjugated (Byrd et al., 1992; Schepers et al., 1992). Isabelle J. Létourneau et al. (2005) firstly chemically synthesized *trans*-hydroxycotinine-*O*-glucuronide and then tested the ability of this compound, nicotine-*N*-glucuronide and cotinine-*N*-glucuronide, to modulate the vesicular transport of several organic anions by MRP1, MRP2, and MRP3 and found that MRP3-mediated transport of 17β-estradiol 17-(β-d-glucuronide) and methotrexate was partially inhibited by *trans*-hydroxycotinine-*O*-glucuronide (300 μM) (by 70% and 50%, respectively).

3.6.6 NORNICOTINE

Neurath et al. (1991) first found that nornicotine can be transformed from nicotine. Nornicotine is an ingredient of tobacco leaves. However, no more than 40% of urinary nornicotine is directly from tobacco; most of urinary nornicotine is from the metabolism of nicotine (Benowitz et al., 1994). The formation of nornicotine from nicotine has been shown to be mediated by the cytochrome P450 system in rabbits (Williams et al., 1990b). Nornicotine has been found in the urine, presenting evidence for the existence of this metabolic step. But only less than 1% of the administered dose of nicotine is excreted as nornicotine, possibly indicating rapid further metabolism (Neurath et al., 1991). Several methods were published, including radioimmunoassay (Knight et al., 1985), enzyme-linked immunoassay (Bjercke et al., 1986), GC (Jaakkola et al., 2003), or GC/MS (Cognard and Staub, 2003; Ji et al., 1999; Pascual et al., 2003).

3.6.7 NORCOTININE

Norcotinine had been reported as a urinary metabolite of nicotine in smokers (Bowman and McKennis, 1962; Gabrielsson and Gumbleton, 1993). Norcotinine occurs as a urinary metabolite of cotinine in dogs, mice, and rats, but could not be detected after administration of cotinine to humans (Bowman and McKennis, 1962). Norcotinine is also formed in vitro from cotinine in hepatic, pulmonary, and renal tissues (Aislaitner et al., 1992). In urine samples from smokers, norcotinine accounts for approximately 1% of total nicotine and its metabolites (Byrd et al., 1992). Demethylation of cotinine or oxidative metabolism of nornicotine is the possible pathway for the formation of norcotinine.

3.6.8 NICOTINE-*N*-OXIDE

Nicotine-*N*-oxide is another primary metabolite of nicotine, although only about 4%–7% of nicotine absorbed by smokers is metabolized via this route

(Benowitz and Jacob, 1994; Byrd et al., 1992). Nicotine-*N*-oxide is stable and rapidly excreted unchanged (Jacob III et al., 1988). Park et al. (1993) carried out the studies of the in vitro and in vivo nicotine-*N*-oxide metabolism of nicotine in humans that are in remarkably good agreement. The data showed that the formation of nicotine-*N*-oxide is a selective functional marker of adult human liver FM03 activity. They concluded that nicotine-*N*-oxide is not back-transformed to nicotine, and this is in good agreement with a previous study (Duan et al., 1991).

3.6.9 COTININE-*N*-OXIDE

Cotinine-*N*-oxide was first found as a new metabolite of nicotine by Ermias Dagne and Neal Castagnoli Jr. (1972). Alexander T. Shulgin et al. presented a method for the isolation and HPLC determination of an additional metabolite, cotinine-*N*-oxide, heretofore not described in humans in 1987 (Shulgin et al., 1987). Cotinine-*N*-oxide is formed by P450 enzymes (Hibberd and Gorrod, 1985). Cotinine-*N*-oxide accounts for 2%–5% of the nicotine and metabolites in smokers' urine (Byrd et al., 1992; Meger et al., 2002). As with nicotine-*N'*-oxide, cotinine-*N*-oxide can be reduced back to the parent amine in vivo as evidenced by a study in rabbits (Yi et al., 1977). To evaluate the nature of nicotine interaction with its derivatives, Riah et al. synthesized cotinine and *N*-oxide cotinine and got the conclusion that nicotine oxidization is an efficient natural mechanism to nicotine detoxification (Riah et al., 1997).

3.7 APPLICATION OF NICOTINE BIOMARKERS ON BIOMONITOR

3.7.1 COTININE

Exposure to nicotine-related alkaloids may also have toxicological significance. Concentrations of nicotine and its metabolites (primarily cotinine) in biological fluids are frequently used to ascertain whether or not a person is using tobacco and to estimate nicotine intake (Benowitz, 1996; Benowitz and Jacob III, 1984; Sepkovic and Haley, 1985). Cotinine, the primary metabolite of nicotine, remains the most commonly used biomarker for nicotine mainly because of its relatively long half-life in biological fluids (16–19 h) and high specificity (Benowitz, 1996; Jarvis et al., 1988). Fustinoni et al. (2013) developed and validated a simple, sensitive LC–MS/MS method to determine unconjugated urinary cotinine (COT-U) levels and to investigate its ability to discriminate active and ETS exposure. The method reliably measured a wide range of COT-U levels. The 30 µg/L cutoff value appropriately classified active tobacco smoke exposure, but the classification of ETS exposure needs further research. Schettgen et al. (2009) investigated spot urine samples from 210 persons (198 male, 12 female; aged 19–80 years, median age of 57.5 years) from the general population with no known occupational exposure to acrylonitrile and/or 1,3-butadiene. According to the results for urinary cotinine as a specific biomarker of cigarette consumption or exposure to ETS and the anamnestic information from the subjects, the study population was divided into four groups: (1) nonsmokers with urinary cotinine levels below 5 µg/L, (2) nonsmokers with slight exposure to ETS and urinary cotinine levels below 10 µg/L, (3) nonsmokers with suspected high exposure to ETS and urinary cotinine levels ranging between 10 and 60 µg/L, (4) and active smokers with cotinine levels ranging from 77 to 4300 µg/L urine. Then the influence of smoking on urinary *N*-acetyl-*S*-2-cyanoethylcysteine (CEMA) derived from smoking acrylonitrile and *N*-acetyl-*S*-(3,4-dihydroxybutyl)cysteine (DHBMA) and MHBMA (an isomeric mixture of *N*-acetyl-*S*-((1-hydroxymethyl)-2-propenyl) cysteine and *N*-acetyl-*S*-((2-hydroxymethyl)-3-propenyl)-cysteine) derived from 1,3-butadiene was studied. Their results showed that there is a strong background exposure to acrylonitrile for the general population, and that is mainly effected by individual exposure to cigarette smoke. Urinary MHBMA from smokers had a significantly higher excretion than in nonsmokers, whereas the difference of DHBMA in urine was not significant.

3.7.2 3-HYDROXYCOTININE

The concentration of 3-hydroxycotinine in urine is the highest (Neurath et al., 1987); it may be a more credible biomarker than cotinine to assess nicotine intake (Benowitz and Jacob III, 1997), particularly at low levels of exposure. This could be especially good for studying the effect of lower ETS exposure levels on adverse health consequences (Matt et al., 2006). Determining both levels of 3-hydroxycotinine metabolites may yield another potential advantage. Because 3-hydroxycotinine has been metabolized from cotinine, the ratio of 3-hydroxycotinine to cotinine (3-hydroxycotinine/ cotinine) can shed light on the efficiency of the metabolic processes converting cotinine to 3-hydroxycotinine (Dempsey et al., 2004). This is a good answer to the question about the effect of interindividual variances in nicotine metabolism and the degree to which such variances may confuse cotinine and 3-hydroxycotinine as ETS exposure markers. Benowitz et al. (Matt et al., 2006) used urinary *trans*-3-hydroxy cotinine (3HC) as a biomarker of ETS exposure, and 3HC was investigated in comparison with urinary cotinine (COT), the sum (3HC/COT), and the ratio of the two nicotine metabolites (3HC/COT). They found that COT, 3HC, and 3HC/ COT are approximately equivalent and equally strong biomarkers of ETS exposure in children. Moreover, 3HC/COT may provide a useful indicator to investigate age- and race-related differences in the metabolism of COT and 3HC. Strasser et al. (2011) used the ratio of the metabolites cotinine and 3-hydroxycotinine to assess the variability in smoking behavior. Their results indicated that faster nicotine metabolizers (third and fourth quartiles vs. first quartile) based on the ratio of 3HC/COT exhibited significantly greater total puff volume and total NNAL; the total puff volume by daily cigarette consumption interaction was a significant predictor of total NNAL level.

3.7.3 NICOTINE GLUCURONIDE

Nicotine, the major addictive agent in tobacco and tobacco smoke, undergoes a complex metabolic pathway, with ~22% of nicotine urinary metabolites in the form of phase II *N*-glucuronidated compounds. In most individuals, nicotine 5′-oxidation is the major pathway of nicotine metabolism. However, nicotine is also metabolized to its N-glucuronide conjugate (Kuehl and Murphy, 2003). Blacks excreted significantly less nicotine as nicotine-N-glucuronide and less cotinine as cotinine-N-glucuronide than whites, but there was no difference in the excretion of 3'-hydroxycotinine-O-glucuronide. Nicotine and cotinine glucuronidation appeared to be polymorphic, with evidence of slow and fast N-glucuronide formers among blacks but was unimodal with fast conjugators only among whites. Seaton et al. (1993) reported that a new HPLC assay was adapted for radiometric detection of nicotine metabolites in rat bile. In their study of nicotine glucuronide and 3-hydroxycotinine glucuronide, cotinine was detected in bile after administration to rats of a single subcutaneous dose of (–)-S-nicotine (0.2 or 1.0 mg/kg) that contained a tracer dose of rac-[pyrrolidine-2'-14C]nicotine (20 μCi). They found that dose dependency of nicotine metabolism occurred: less nicotine glucuronide was excreted at the low dose than at the high dose.

REFERENCES

Aislaitner, G., Li, Y., Gorrod, J. In vitro metabolic studies on(-)-(S)-nornicotine. *Medical Science Research*. 1992, 20, 897–889.

Andersson, G., Vala, E. K., Curvall, M. The influence of cigarette consumption and smoking machine yields of tar and nicotine on the nicotine uptake and oral mucosal lesions in smokers. *Journal of Oral Pathology & Medicine*. 1997, 26, 117–123.

Armstrong, D. W., Wang, X., Ercal, N. Enantiomeric composition of nicotine in smokeless tobacco, medicinal products, and commercial reagents. *Chirality*. 1998, 10, 587–591.

Benowitz, N. L. Cotinine as a biomarker of environmental tobacco smoke exposure. *Epidemiologic Reviews*. 1996, 18, 188–204.

Benowitz, N. L. The role of nicotine in smoking-related cardiovascular disease. *Preventive Medicine*. 1997, 26, 412–417.

Benowitz, N. L. The use of biologic fluid samples in assessing tobacco smoke consumption. *NIDA Research Monographs*. 1983, 48, 6–26.

Benowitz, N. L., Jacob III, P. Daily intake of nicotine during cigarette smoking. *Clinical Pharmacology & Therapeutics*. 1984, 35, 499–504.

Benowitz, N. L., Jacob, P. Metabolism of nicotine to cotinine studied by a dual stable isotope method. *Clinical Pharmacology & Therapeutics*. 1994, 56, 483–493.

Benowitz, N. L., Jacob III, P. Individual differences in nicotine kinetics and metabolism in humans. *NIDA Research Monographs*. 1997, 173, 48–64.

Benowitz, N. L., Jacob, P., Fong, I., Gupta, S. Nicotine metabolic profile in man: Comparison of cigarette smoking and transdermal nicotine. *Journal of Pharmacology and Experimental Therapeutics*. 1994, 268, 296–303.

Benowitz, N. L., Jacob, P., Jones, R. T., Rosenberg, J. Interindividual variability in the metabolism and cardiovascular effects of nicotine in man. *Journal of Pharmacology and Experimental Therapeutics*. 1982, 221, 368–372.

Benowitz, N. L., Jacob III, P., Sachs, D. P. Clinical pharmacology rounds deficient C-oxidation of nicotine. *Clinical Pharmacology & Therapeutics*. 1995, 57, 590–594.

Bentley, M. C., Abrar, M., Kelk, M., Cook, J., Phillips, K. Validation of an assay for the determination of cotinine and 3-hydroxycotinine in human saliva using automated solid-phase extraction and liquid chromatography with tandem mass spectrometric detection. *Journal of Chromatography B: Biomedical Sciences and Applications*. 1999, 723, 185–194.

Bernert, J. T., Turner, W. E., Pirkle, J. L., Sosnoff, C. S., Akins, J. R., Waldrep, M. K., Ann, Q., Covey, T. R., Whitfield, W. E., Gunter, E. W. Development and validation of sensitive method for determination of serum cotinine in smokers and nonsmokers by liquid chromatography/atmospheric pressure ionization tandem mass spectrometry. *Clinical Chemistry*. 1997, 43, 2281–2291.

Bjercke, R. J., Cook, G., Rychlik, N., Gjika, H. B., Van Vunakis, H., Langone, J. J. Stereospecific monoclonal antibodies to nicotine and cotinine and their use in enzyme-linked immunosorbent assays. *Journal of Immunological Methods*. 1986, 90, 203–213.

Bowman, E. R., McKennis, H. Studies on the metabolism of (-)-cotinine in the human. *Journal of Pharmacology and Experimental Therapeutics*. 1962, 135, 306–311.

Bramer, S. L., Kallungal, B. A. Clinical considerations in study designs that use cotinine as a biomarker. *Biomarkers*. 2003, 8, 187–203.

Byrd, G., Robinson, J., Caldwell, W., de Bethizy, J. Comparison of measured and FTC-predicted nicotine uptake in smokers. *Psychopharmacology*. 1995, 122, 95–103.

Byrd, G., Uhrig, M., Debethizy, J., Caldwell, W., Crooks, P., Ravard, A., Riggs, R. Direct determination of cotinine-N-glucuronide in urine using thermospray liquid chromatography/mass spectrometry. *Biological Mass Spectrometry*. 1994, 23, 103–107.

Byrd, G. D., Chang, K., Greene, J. M. Evidence for urinary excretion of glucuronide conjugates of nicotine, cotinine, and trans-3'-hydroxycotinine in smokers. *Drug Metabolism and Disposition*. 1992, 20, 192–197.

Caldwell, W., Greene, J., Byrd, G., Chang, K., Uhrig, M., DeBethizy, J., Crooks, P., Bhatti, B., Riggs, R. Characterization of the glucuronide conjugate of cotinine: A previously unidentified major metabolite of nicotine in smokers' urine. *Chemical Research in Toxicology*. 1992, 5, 280–285.

Chadwick, C. A., Keevil, B. Measurement of cotinine in urine by liquid chromatography tandem mass spectrometry. *Annals of Clinical Biochemistry*. 2007, 44, 455–462.

Cognard, E., Staub, C. Determination of nicotine and its major metabolite cotinine in plasma or serum by gas chromatography-mass spectrometry using ion-trap detection. *Clinical Chemistry and Laboratory Medicine*. 2003, 41, 1599–1607.

Culca, M., Cozar, O., Nicoara, S., Podea, R. Exposure assessment of nicotine and cotinine by GC-MS. *Indoor and Built Environment*. 2005, 14, 293–299.

Curvall, M., Vala, E. K., Englund, G. Conjugation pathways in nicotine metabolism. In: *Effects of Nicotine on Biological Systems*, F. Adlkofer and K. Thurau (eds.), 1991, Birkhauser Verlag, Basel, Sweden, pp. 69–75.

Dagne, E., Castagnoli, N. Cotinine N-oxide, a new metabolite of nicotine. *Journal of Medicinal Chemistry*. 1972, 15, 840–841.

Dempsey, D., Tutka, P., Jacob, P., Allen, F., Schoedel, K., Tyndale, R. F., Benowitz, N. L. Nicotine metabolite ratio as an index of cytochrome P450 2A6 metabolic activity. *Clinical Pharmacology & Therapeutics*. 2004, 76, 64–72.

Duan, M., Yu, L., Savanapridi, C., Jacob, P., Benowitz, N. L. Disposition kinetics and metabolism of nicotine-1'-N-oxide in rabbits. *Drug Metabolism and Disposition*. 1991, 19, 667–672.

Fan, Z., Xie, F., Xia, Q., Wang, S., Ding, L., Liu, H. Simultaneous determination of nicotine and its nine metabolites in human urine by LC–MS–MS. *Chromatographia*. 2008, 68, 623–627.

Feng, S., Kapur, S., Sarkar, M., Muhammad, R., Mendes, P., Newland, K., Roethig, H. J. Respiratory retention of nicotine and urinary excretion of nicotine and its five major metabolites in adult male smokers. *Toxicology Letters*. 2007, 173, 101–106.

Florescu, A., Ferrence, R., Einarson, T. R., Selby, P., Kramer, M., Woodruff, S., Grossman, L., Rankin, A., Jacqz-Aigrain, E., Koren, G. Reference values for hair cotinine as a biomarker of active and passive smoking in women of reproductive age, pregnant women, children, and neonates: Systematic review and meta-analysis. *Therapeutic Drug Monitoring*. 2007, 29, 437–446.

Fustinoni, S., Campo, L., Polledri, E., Mercadante, R., Erspamer, L., Badea, V. A validated method for urinary cotinine quantification used to classify active and environmental tobacco smoke exposure. *Current Analytical Chemistry*. 2013, 9, 447–456.

Gabrielsson, J., Gumbleton, M. Kinetics of nicotine and its metabolites in animals. In: *Nicotine and Related Alkaloids,* J. W. Gorrod and J. Wahren (eds.), 1993, Springer, the Netherlands, pp. 181–195.

Ghosheh, O., Hawes, E. M. N-glucuronidation of nicotine and cotinine in human: Formation of cotinine glucuronide in liver microsomes and lack of catalysis by 10 examined UDP-glucuronosyltransferases. *Drug Metabolism and Disposition*. 2002, 30, 991–996.

Ghosheh, O., Vashishtha, S. C., Hawes, E. M. Formation of the quaternary ammonium-linked glucuronide of nicotine in human liver microsomes: Identification and stereoselectivity in the kinetics. *Drug Metabolism and Disposition*. 2001, 29, 1525–1528.

Gorrod, J. W., Jacob III, P. *Analytical Determination of Nicotine and Related Compounds and their Metabolites,* J. W. Gorrod and Jacob (eds.), 1999, Elsevier.

Hariharan, M., Van Noord, T., Greden, J. F. A high-performance liquid-chromatographic method for routine simultaneous determination of nicotine and cotinine in plasma. *Clinical Chemistry,* J. W. Gorrod and Jacob (eds.), 1988, 34, 724–729.

Hatsukami, D. K., Grillo, M., Pentel, P. R., Oncken, C., Bliss, R. Safety of cotinine in humans: Physiologic, subjective, and cognitive effects. *Pharmacology Biochemistry and Behavior*. 1997, 57, 643–650.

Hecht, S. S., Carmella, S. G., Chen, M., Koch, J. D., Miller, A. T., Murphy, S. E., Jensen, J. A., Zimmerman, C. L., Hatsukami, D. K. Quantitation of urinary metabolites of a tobacco-specific lung carcinogen after smoking cessation. *Cancer Research*. 1999, 59, 590–596.

Hecht, S. S., Hochalter, J. B., Villalta, P. W., Murphy, S. E. 2'-Hydroxylation of nicotine by cytochrome P450 2A6 and human liver microsomes: Formation of a lung carcinogen precursor. *Proceedings of the National Academy of Sciences*. 2000, 97, 12493–12497.

Hibberd, A., Gorrod, J. Comparative N-oxidation of nicotine and cotinine by hepatic microsomes. In: *Biological Oxidation of Nitrogen in Organic Molecules,* A. Hibberd and J. W. Gorrod, VCH-Verlag, Chichester, U.K. 1985, pp. 246–250.

Hukkanen, J., Jacob, P., Benowitz, N. L. Metabolism and disposition kinetics of nicotine. *Pharmacological Reviews*. 2005, 57, 79–115.

Ishiyama, I., Nagai, T., Toshida, S. Detection of basic drugs (methamphetamine, antidepressants, and nicotine) from human hair. *Journal of Forensic Sciences*. 1983, 28, 380–385.

Jaakkola, M. S., Ma, J., Yang, G., Chin, M.-F., Benowitz, N. L., Ceraso, M., Samet, J. M. Determinants of salivary cotinine concentrations in Chinese male smokers. *Preventive Medicine*. 2003, 36, 282–290.

Jacob III, P., Benowitz, N. L. Oxidative metabolism of nicotine in vivo. In: *Effects of Nicotine on Biological Systems,* F. Adlkofer and K. Thurau (eds.), 1991, Birkhauser Verlag, Basel, Sweden, pp. 35–44.

Jacob III, P., Benowitz, N. L., Shulgin, A. T. Recent studies of nicotine metabolism in humans. *Pharmacology Biochemistry and Behavior*. 1988, 30, 249–253.

Jacob III, P., Shulgin, A. T., Benowitz, N. L. Synthesis of (3'R, 5'S)-trans-3'-hydroxycotinine, a major metabolite of nicotine. Metabolic formation of 3'-hydroxycotinine in humans is highly stereoselective. *Journal of Medicinal Chemistry*. 1990, 33, 1888–1891.

Jarvis, M. J., Russell, M., Benowitz, N., Feyerabend, C. Elimination of cotinine from body fluids: Implications for noninvasive measurement of tobacco smoke exposure. *American Journal of Public Health*. 1988, 78, 696–698.

Ji, A. J., Lawson, G. M., Anderson, R., Dale, L. C., Croghan, I. T., Hurt, R. D. A new gas chromatography–mass spectrometry method for simultaneous determination of total and free trans-3'-hydroxycotinine and cotinine in the urine of subjects receiving transdermal nicotine. *Clinical Chemistry*. 1999, 45, 85–91.

Kataoka, H., Inoue, R., Yagi, K., Saito, K. Determination of nicotine, cotinine, and related alkaloids in human urine and saliva by automated in-tube solid-phase microextraction coupled with liquid chromatography–mass spectrometry. *Journal of Pharmaceutical and Biomedical Analysis*. 2009, 49, 108–114.

Keenan, R. M., Hatsukami, D. K., Pentel, P. R., Thompson, T. N., Grillo, M. A. Pharmacodynamic effects of cotinine in abstinent cigarette smokers. *Clinical Pharmacology & Therapeutics*. 1994, 55, 581–590.

Knight, G., Wylie, P., Holman, M., Haddow, J. Improved 125I radioimmunoassay for cotinine by selective removal of bridge antibodies. *Clinical Chemistry*. 1985, 31, 118–121.

Koszowski, B., Czogala, J., Goniewicz, M. L., Sobczak, A., Kolasinska, E., Kosmider, L., Kuma, T. Use of hair nicotine as a tool to assess tobacco smoke exposure. *Przegląd lekarski*. 2008, 65, 696–699.

Kozlowski, L. T., Goldberg, M. E., Yost, B. A., White, E. L., Sweeney, C. T., Pillitteri, J. L. Smokers' misperceptions of light and ultra-light cigarettes may keep them smoking. *American Journal of Preventive Medicine*. 1998, 15, 9–16.

Kuehl, G. E., Murphy, S. E. N-glucuronidation of nicotine and cotinine by human liver microsomes and heterologously expressed UDP-glucuronosyltransferases. *Drug Metabolism and Disposition*. 2003, 31, 1361–1368.

Kuo, H. W., Yang, J. S., Chiu, M. C. Determination of urinary and salivary cotinine using gas and liquid chromatography and enzyme-linked immunosorbent assay. *Journal of Chromatography B-Analytical Technologies in the Biomedical and Life Sciences*. 2002, 768, 297–303.

Kyerematen, G. A., Morgan, M. L., Chattopadhyay, B., Donald deBethizy, J., Vesell, E. S. Disposition of nicotine and eight metabolites in smokers and nonsmokers: Identification in smokers of two metabolites that are longer lived than cotinine. *Clinical Pharmacology & Therapeutics*. 1990, 48, 641–651.

Lafay, F., Vulliet, E., Flament-Waton, M.-M. Contribution of microextraction in packed sorbent for the analysis of cotinine in human urine by GC-MS. *Analytical and Bioanalytical Chemistry*. 2010, 396, 937–941.

Lee, D., Ryu, H., Nam, M., Kong, S. Simultaneous and sensitive measurement of nicotine and cotinine in small amount of human hair by liquid chromatography-tandem mass spectrometry. *Clinical Chemistry*. 2005, 51, A67.

Létourneau, I. J., Bowers, R. J., Deeley, R. G., Cole, S. P. C. Limited modulation of the transport activity of the human multidrug resistance proteins MRP1, MRP2 and MRP3 by nicotine glucuronide metabolites. *Toxicology Letters*. 2005, 157, 9–19.

Man, C. N., Gam, L.-H., Ismail, S., Lajis, R., Awang, R. Simple, rapid and sensitive assay method for simultaneous quantification of urinary nicotine and cotinine using gas chromatography–mass spectrometry. *Journal of Chromatography B*. 2006, 844, 322–327.

Man, C. N., Ismail, S., Harn, G. L., Lajis, R., Awang, R. Determination of hair nicotine by gas chromatography-mass spectrometry. *Journal of Chromatography B-Analytical Technologies in the Biomedical and Life Sciences*. 2009, 877, 339–342.

Matsuki, H., Hashimoto, K., Arashidani, K., Akiyama, Y., Amagai, T., Ishizu, Y., Matsushita, H. Studies on a simultaneous analytical method of urinary nicotine and its metabolites, and their half-lives in urine. *Journal of UOEH*. 2008, 30, 235–252.

Matsumoto, A., Ino, T., Ohta, M., Otani, T., Hanada, S., Sakuraoka, A., Matsumoto, A., Ichiba, M., Hara, M. Enzyme-linked immunosorbent assay of nicotine metabolites. *Environmental Health* and *Preventive Medicine*. 2010, 15, 211–216.

Matt, G., Quintana, P., Liles, S., Hovell, M., Zakarian, J., Jacob III, P., Benowitz, N. Evaluation of urinary trans-3'-hydroxycotinine as a biomarker of children's environmental tobacco smoke exposure. *Biomarkers*. 2006, 11, 507–523.

McBride, J. S., Altman, D. G., Klein, M., White, W. Green tobacco sickness. *Tobacco Control.* 1998, 7, 294–298.

McKennis, Jr., H., Turnbull, L. B., Bowman, E. R., Tamaki, E. The synthesis of hydroxycotinine and studies on its structure. *Journal of Organic Chemistry.* 1963, 28, 383–387.

Meger, M., Meger-Kossien, I., Schuler-Metz, A., Janket, D., Scherer, G. Simultaneous determination of nicotine and eight nicotine metabolites in urine of smokers using liquid chromatography–tandem mass spectrometry. *Journal of Chromatography B.* 2002, 778, 251–261.

Mendes, P., Kapur, S., Wang, J., Feng, S., Roethig, H. A randomized, controlled exposure study in adult smokers of full flavor Marlboro cigarettes switching to Marlboro Lights or Marlboro Ultra Lights cigarettes. *Regulatory Toxicology and Pharmacology.* 2008, 51, 295–305.

Nakajima, M., Kwon, J.-T., Tanaka, E., Yokoi, T. High-performance liquid chromatographic assay for N-glucuronidation of nicotine and cotinine in human liver microsomes. *Analytical Biochemistry.* 2002, 302, 131–135.

Neovius, M., Sundström, J., Rasmussen, F. Combined effects of overweight and smoking in late adolescence on subsequent mortality: Nationwide cohort study. *BMJ: British Medical Journal.* 2009, 338, b496.

Neurath, G. Aspects of the oxidative metabolism of nicotine. *The Clinical Investigator.* 1994, 72, 190–195.

Neurath, G., Orth, D., Pein, F. Detection of nornicotine in human urine after infusion of nicotine. In: *Effects of Nicotine on Biological Systems*, F. Adlkofer and K. Thurau (eds.), 1991, Birkhauser Verlag, Basel, Sweden, pp. 45–49.

Neurath, G. B., Dünger, M., Orth, D., Pein, F. G. Trans-3′-hydroxycotinine as a main metabolite in urine of smokers. *International Archives of Occupational and Environmental Health.* 1987, 59, 199–201.

Papadoyannis, I. N., Samanidou, V. F., Stefanidou, P. G. Clinical assay of nicotine and its metabolite, cotinine, in body fluids by HPLC following solid phase extraction. *Journal of Liquid Chromatography & Related Technologies.* 2002, 25, 2315–2335.

Park, S. B., Jacob III, P., Benowitz, N. L., Cashman, J. R. Stereoselective metabolism of (S)-(-)-nicotine in humans: Formation of trans-(S)-(-)-nicotine N-1′-oxide. *Chemical Research in Toxicology.* 1993, 6, 880–888.

Pascual, J. A., Diaz, D., Segura, J., Garcia-Algar, Ó., Vall, O., Zuccaro, P., Pacifici, R., Pichini, S. A simple and reliable method for the determination of nicotine and cotinine in teeth by gas chromatography/mass spectrometry. *Rapid Communications in Mass Spectrometry.* 2003, 17, 2853–2855.

Petersen, G. O., Leite, C. E., Chatkin, J. M., Thiesen, F. V. Cotinine as a biomarker of tobacco exposure: Development of a HPLC method and comparison of matrices. *Journal of Separation Science.* 2010, 33, 516–521.

Rees, W. A., Kwiatkowski, S., Stanley, S. D., Granstrom, D. E., Yang, J. M., Gairola, C. G., Drake, D., Glowczyk, J., Woods, W. E., Tobin, T. Development and characterization of an ELISA for trans-3-hydroxycotinine, a biomarker for mainstream and sidestream smoke exposure. In: Bengtson D. A., Henshel, D. S. (Eds.), *Environmental Toxicology and Risk Assessment: Biomarkers and Risk Assessment—Fifth Volume,* ASTM, West Conshohocken, PA. 1996, Vol. 1306, pp. 149–162.

Riah, O., Dousset, J.-C., Courriere, P., Genevièvebaziard-Mouysset, Renéecalle. Synthesis of cotinine and cotinine N-oxide: Evaluation of their interaction with nicotine in the insecticidal activity. *Natural Product Letters.* 1997, 11, 37–45.

Schepers, G., Demetriou, D., Rustemeier, K., Voncken, P., Diehl, B. Nicotine phase 2 metabolites in human urine-structure of metabolically formed trans-3′-hydroxycotinine glucuronide. *Medical Science Research.* 1992, 20, 863–863.

Schepers, G., Demetriou, D., Stabbert, R., Diehl, B., Seeman, J. 5'-Hydroxycotinine-N-oxide, a new nicotine metabolite isolated from rat urine. *Xenobiotica.* 1999, 29, 793–801.

Schettgen, T., Musiol, A., Alt, A., Ochsmann, E., Kraus, T. A method for the quantification of biomarkers of exposure to acrylonitrile and 1,3-butadiene in human urine by column-switching liquid chromatography-tandem mass spectrometry. *Analytical and Bioanalytical Chemistry.* 2009, 393, 969–981.

Schmeltz, I., Hoffmann, D. Nitrogen-containing compounds in tobacco and tobacco smoke. *Chemical Reviews.* 1977, 77, 295–311.

Seaton, M. Interpolations of Rosseland-mean opacities for variable X and Z. *Monthly Notices of the Royal Astronomical Society.* 1996, 279, 95–100.

Seaton, M., Vesell, E. Variables affecting nicotine metabolism. *Pharmacology & Therapeutics.* 1993, 60, 461–500.

Seaton, M. J., Kyerematen, G. A., Vesell, E. S. Rates of excretion of cotinine, nicotine glucuronide, and 3-hydroxycotinine glucuronide in rat bile. *Drug Metabolism and Disposition.* 1993, 21, 927–932.

Sepkovic, D., Haley, N. Biomedical applications of cotinine quantitation in smoking related research. *American Journal of Public Health.* 1985, 75, 663–665.

Shakleya, D. M., Huestis, M. A. Optimization and validation of a liquid chromatography-tandem mass spectrometry method for the simultaneous quantification of nicotine, cotinine, trans-3'-hydroxycotinine and norcotinine in human oral fluid. *Analytical and Bioanalytical Chemistry.* 2009, 395, 2349–2357.

Shulgin, A. T., Jacob, P., Benowitz, N. L., Lau, D. Identification and quantitative analysis of cotinine-N-oxide in human urine. *Journal of Chromatography B: Biomedical Sciences and Applications.* 1987, 423, 365–372.

Stepanov, I., Hecht, S. S., Lindgren, B., Jacob, P., Wilson, M., Benowitz, N. L. Relationship of human toenail nicotine, cotinine, and 4-(M ethylnitrosamino)-1-(3-Pyridyl)-1-butanol to levels of these biomarkers in plasma and urine. *Cancer Epidemiology Biomarkers & Prevention.* 2007, 16, 1382–1386.

Strasser, A. A., Benowitz, N. L., Pinto, A. G., Tang, K. Z., Hecht, S. S., Carmella, S. G., Tyndale, R. F., Lerman, C. E. Nicotine metabolite ratio predicts smoking topography and carcinogen biomarker level. *Cancer Epidemiology Biomarkers & Prevention.* 2011, 20, 234–238.

Tricker, A. R. Biomarkers derived from nicotine and its metabolites: A review. *Beiträge zur Tabakforschung International.* 2006, 22, 147–175.

Tzatzarakis, M. N., Vardavas, C. I., Terzi, I., Kavalakis, M., Kokkinakis, M., Liesivuori, J., Tsatsakis, A. M. Hair nicotine/cotinine concentrations as a method of monitoring exposure to tobacco smoke among infants and adults. *Human & Experimental Toxicology.* 2012, 31, 258–265.

Voncken, P., Rustemeier, K., Schepers, G. Identification of cis-3'-hydroxycotinine as a urinary nicotine metabolite. *Xenobiotica.* 1990, 20, 1353–1356.

Wald, N. J., Idle, M., Boreham, J., Bailey, A., Van Vunakis, H. Urinary nicotine concentrations in cigarette and pipe smokers. *Thorax.* 1984, 39, 365–368.

Xu, X., Iba, M. M., Weisel, C. P. Simultaneous and sensitive measurement of anabasine, nicotine, and nicotine metabolites in human urine by liquid chromatography–tandem mass spectrometry. *Clinical Chemistry.* 2004, 50, 2323–2330.

Yi, J., Sprouse, C., Bowman, E. R., McKennis, H. The interrelationship between the metabolism of (S)-cotinine-N-oxide and (S)-cotinine. *Drug Metabolism and Disposition.* 1977, 5, 355–362.

Zevin, S., Jacob, P., Geppetti, P., Benowitz, N. L. Clinical pharmacology of oral cotinine. *Drug and Alcohol Dependence.* 2000, 60, 13–18.

4 Biomarkers of Tobacco-Specific Nitrosamines

4.1 INTRODUCTION

Tobacco and tobacco smoke contain three types of nitrosamines, which include volatile nitrosamines, tobacco-specific nitrosamines (TSNAs), and nitrosamines derived from residues of agricultural chemicals on tobacco (Spiegelhalder and Bartsch, 1996). TSNAs have arisen as a leading class of carcinogens in tobacco products (Hecht and Hoffmann, 1988). Extensive research has indicated the role of TSNAs in cancer induction (Hecht, 1998). So far, seven TSNAs have been identified in cigarettes (Hoffmann et al., 1994). TSNAs include 4-(methylnitrosamino)-1-(3-pyridyl)-1-butanone (NNK), N-nitrosonornicotine (NNN), N-nitrosoanabasine (NAB) and N-nitrosoanatabine (NAT) 4-(methylnitrosamino)-1-(3-pyridyl)-1-butanol (NNAL), iso-NNAL, and iso-NNAC. Because their origin is specific to tobacco, the study of TSNA uptake will provide very useful insight into the mechanistic and epidemiologic role of these compounds in human cancer. A major challenge in this area is the ability to quantitatively measure the amounts of these compounds and their metabolites in complex biological matrices, as a measure of exposure to these tobacco-specific carcinogens. A significant amount of research investigating in vivo disposition and excretion of TSNAs in laboratory animals and humans has been carried out (Figure 4.1).

4.1.1 TOXICOLOGICAL EVALUATION

TSNAs play an important role in cancer induction by tobacco products. Smoking is by far the major cause of lung cancer, a disease that was expected to cause 1.37 million deaths in the world in 2008. It is estimated that 156,940 lung cancer deaths will have occurred in the United States in 2011 (Siegel et al., 2012). NNK is a significantly potent inducer of adenocarcinoma of the lung in rodents. Adenocarcinoma is now the leading lung cancer type in the United States, and a recent study demonstrates that this change in histology is due to the changing cigarette and the methods of diagnosis are not improved (Thun et al., 1997). Smoking is a major cause of induced formation of pancreatic cancer and esophageal cancer, it is estimated that 25% of pancreatic cancer death and 70–80% of esophageal cancer death are related to smoking in the United States (Fuchs et al., 1996; Shopland, 1995). NNK and its major metabolite NNAL are pancreatic carcinogens in cigarette smoke (Hoffmann and Hecht, 1990). NNK, NNAL, and NNN certainly play a significant role in cancer induction. The fact that NNK and NNN have a significant role in oral cancer induced by smokeless tobacco has been generally accepted, despite its mechanism being still unclear (Bartsch and Spiegelhalder, 1996; Magee, 1989). TSNAs require undergoing

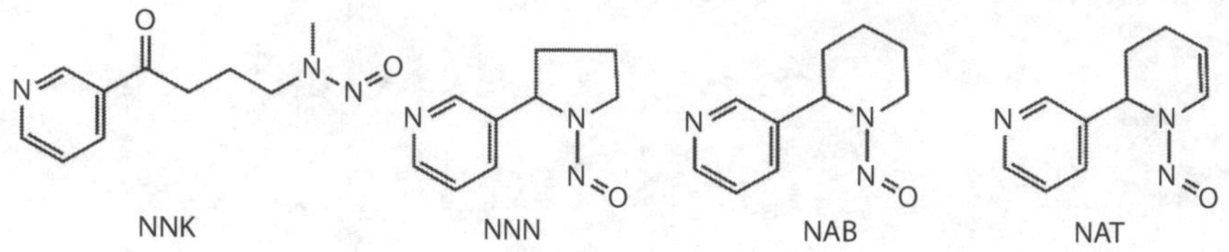

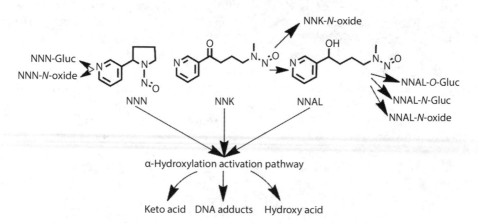

FIGURE 4.1   The structures of TSNAs.

metabolic activation to release their toxic and carcinogenic effects. NNK, NNAL, and NNN metabolism, adduct formation, and detoxification have been quite well understood. Some metabolism reactions occur in rodents and humans, though there are quantitative differences. DNA adducts of NNK and NNN, hemoglobin adducts, and urinary metabolites of NNK have been quantified in humans. So, the overall pathway resulting in the initiation of cancers has been demonstrated for TSNAs in humans. These data taken together with the epidemiologic evidence that tobacco products cause cancer in humans directly point to the relevance of these compounds in human cancer etiology.

4.1.2   METABOLIC PATHWAY IN THE BODY

Of all the TSNAs identified, NNK and NNN are the most prevalent strong carcinogens in tobacco products (Hecht and Hoffmann, 1988; Hoffmann et al., 1995). Moreover, NNK and NNN have been classified as the only TSNAs carcinogenic to humans by the International Agency for Research on Cancer (Humans, 2007). Figure 4.2 shows an overview of NNK and NNN metabolism. It may be argued that measurement of urinary keto and hydroxy acids is more appropriate because these are the end products of the DNA adduct–forming α-hydroxylation pathway of NNK and NNN metabolism. However, as suggested by Hecht et al. (2002), these compounds cannot be used to measure the extent of α-hydroxylation because these are

FIGURE 4.2   Metabolic scheme for NNN and NNK. (From Shah, K.A. and Karnes, H.T., *Crit. Rev. Toxicol.*, 40(4), 305, 2010.)

formed from nicotine as well, which is a primary constituent of tobacco products. Recently, nevertheless, Stepanov et al. (2008) have suggested the use of [pyridine-d4]NNK as a biomarker to measure the deuterium-labeled keto and hydroxyl acid specifically formed from NNK. Biomarker strategies using stable isotope-labeled compounds can however be challenging.

As depicted in Figure 4.2, conversion of NNK to NNAL is an important metabolic pathway in humans, with NNAL having similar carcinogenicity as NNK (Castonguay et al., 1983). A characteristic feature of NNAL metabolism is the formation of NNAL-Gluc, which is the most important detoxification product of the NNK–NNAL metabolic pathway in humans and animals (Hecht et al., 1993c; Morse et al., 1990). Additionally, NNAL and its glucuronide have a longer half-life (at least 10–15 days in smokers and about 40–45 days in users of oral tobacco) when compared with most other urinary metabolites (Carmella et al., 2009; Hecht et al., 2002). Considering all of this, urinary NNAL and NNAL-Gluc have evolved as the most prominently studied TSNA biomarkers.

4.2 NNK

NNK is a TSNA and a potent and selective lung carcinogen in the rat, independent of the route of administration. Thus, adenoma and adenocarcinoma of the lung are the major tumor types induced by NNK in the rat, whether it is administered by subcutaneous injection, in the drinking water, by oral swabbing, or by instillation in the bladder (Hecht and Hoffmann, 1988). NNK is believed to play a significant role, along with polynuclear aromatic hydrocarbons, in the induction of lung cancer in smokers. The metabolism of NNK is summarized in Figure 4.3 (Hecht, 1996). The presence of the pyridine ring and carbonyl group leads to a somewhat more complex series of reactions than observed for simple dialkyl nitrosamines. The major pathway of NNK metabolism in laboratory animals and humans is reduction of the carbonyl group, catalyzed by carbonyl reductase enzymes such as 1-hydroxysteroid dehydrogenase. The product, NNAL, is also a potent pulmonary carcinogen and appears to be the major transport form of NNK (Castonguay et al., 1983; Rivenson et al., 1988). Glucuronidation of NNAL produces NNAL-Gluc, which is excreted in the urine of laboratory animals and humans and is believed to be a detoxification product of NNK (Carmella et al., 1995; Hecht et al., 1993c). Other detoxification products are NNK-N-oxide and NNAL-N-oxide, resulting from cytochrome P450–catalyzed oxidation of the pyridine ring. An interesting series of products is formed in vitro, by substitution of NNK or NNAL for nicotinamide in NADPH, catalyzed by NAD glycohydrolase (Peterson et al., 1994). The a-hydroxylation pathways of NNK, catalyzed by cytochromes P450, result in intermediates that methylate and pyridyloxobutylate DNA. The strong pulmonary tumorigenicity of NNK is associated with the formation and persistence of these adducts in the Clara and type II cells of the rat lung (Belinsky et al., 1990).

4.2.1 NNAL

In humans, NNK is almost entirely reduced to NNAL, which is also carcinogenic and excreted in the urine (Hecht, 1998) (Figure 4.1). Higher urine NNAL was also

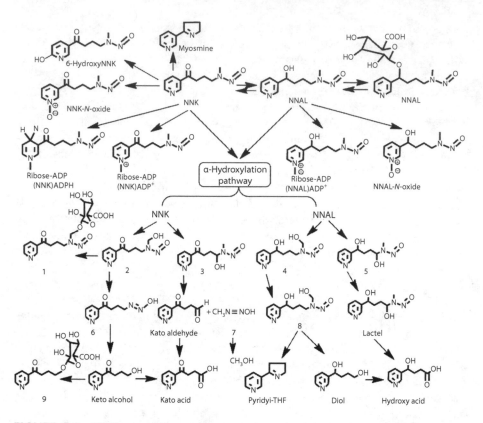

FIGURE 4.3 NNK metabolism pathways as determined by studies in lab animals and humans. (From Hecht, S.S., Approaches to cancer prevention based on an understanding of *N*-nitrosamine carcinogenesis, *Proceedings of the Society for Experimental Biology and Medicine*, Society for Experimental Biology and Medicine, New York, Royal Society of Medicine, 1997.)

related to greater dyspnea, poorer physical health status, and more restricted activity (Eisner et al., 2009). The urine NNAL-to-creatinine ratio (per interquartile increment) was associated with greater chronic obstructive pulmonary disease severity. A study indicated there was a direct association between the NNK of a 24 h mouth-level exposure from cigarette smoking and the NNAL concentration of its primary metabolite in the urine of smokers (Ashley et al., 2010). An important method to determine individual and collective risk from exposure to tobacco products is evaluation of tobacco carcinogen biomarkers (Hatsukami et al., 2006; Hecht, 2002). Urinary metabolites have arose as highly practical biomarkers for determining uptake of specific carcinogens and toxicants in tobacco smoke and are likely to have more utility in predicting tobacco-associated harm than machine measurements of smoke constituents. But the urine matrix was very complex, where the presence of many potentially interfering substances in concentrations far greater than those of the NNAL set demands for high selectivity and low limit of quantification (LOQ) for unequivocal identification and quantification. The analytical procedures for urinary

NNAL are challenging indeed. Recently, liquid chromatography/tandem mass spectrometry (LC–MS/MS) has been developed for the analysis of urinary NNAL (Bhat et al., 2011; Jacob III et al., 2008; Shah et al., 2009, 2011; Xia et al., 2005; Xia and Bernert, 2008; Yang et al., 2010). In order to get good recoveries and improve method's accuracy and precision, complex sample cleanup is often required including liquid–liquid or solid-phase extraction (SPE). However, traditional liquid–liquid extractions need relatively large solvent consumption, and the process is complex. SPE has many advantages compared to traditional liquid–liquid extraction methods. Bhat et al. (Shah et al., 2009) described the use of LC–MS combined with a novel sample cleanup method using SPE on a WCX column developed specifically for extracting NNAL from urine samples. The method made it possible to analyze free NNAL in only 0.25 mL urine. But the prepreparation was very complex; it involved derivatization, liquid–liquid extraction, and SPE. Recently, a molecularly imprinted polymer (MIP) column was applied to prepare for SPE of urinary NNAL. Xia et al. (2005) described the use of LC/APCI–MS/MS combined with a novel sample cleanup method using SPE for extracting NNAL from urine samples. However, the matrix suppression effect was strong. In 2009, Shah et al. (2009) improved the assay, by changing the liquid chromatography conditions, the response for this method was enhanced approximately 25-fold through avoidance of ionization suppression, and the lower LOQ for the assay was 20 pg/mL. The combination of MIP column extraction and LC/ESI–MS/MS can provide a sensitive and relatively simple analytical method.

We developed and validated a method for the determination of total NNAL in human urine by extraction on an MIP column and LC/ESI–MS/MS. Higher sensitivity and good recovery were achieved. The limit of detection was 0.30 pg/mL and the analysis time was 6 min. The validated method was applied to quantify urinary NNAL levels of 207 smokers and 36 nonsmokers, and then the relationship between the urinary concentrations of total NNAL and Chinese Virginia cigarette smoke analytes of NNK and tar was assessed. NNAL was found to be significantly higher in the urine of smokers compared with nonsmokers. Compared with smokers with blended cigarettes, Chinese Virginia cigarette smokers had low urinary NNAL levels. There was a direct association between the 24 h mouth-level exposure of carcinogen NNK from cigarette smoking and the concentration of NNAL in the urine of smokers. However, there was not a positive correlation between urinary total NNAL levels in 24 h and tar. Total urinary NNAL is a valuable biomarker for monitoring exposure to carcinogenic NNK in smokers and in nonsmokers. A prediction model of cigarette smoke NNK and urinary average NNAL levels in 24 h was established ($y = 2.8987x - 245.38$, $r^2 = 0.9952$, $n = 204$) (Hou et al., 2012).

4.2.2   4-(Methylnitrosamino)-1-(3-Pyridyl)- 1-Butanol Glucuronide (NNAL-Gluc)

An important advance was the identification of NNAL-Gluc as a urinary metabolite of NNK in mice and rats. The amount of the NNK dose excreted as NNAL-Gluc in the urine reached a maximum of 22% at relatively high NNK doses but was substantially lower at NNK doses closer to human exposure levels. Research on the patas monkey indicated that there are two diastereomers of NNAL-Gluc in urine, in contrast to the rat and mouse, which had mainly one diastereomer (Hecht et al., 1993c). These two NNAL-Gluc diastereomers are about 15%–20% of the NNK dose. In Cramella's study, they found that concentrations of NNAL-Gluc in urine were 0.16 to 19.0 pmol/mg creatinine, and levels of NNAL in urine were 0.08 to 4.89 pmol/mg creatinine (Carmella et al., 1995). These levels are consistent with the amounts of NNK reported in mainstream cigarette smoke and support the proposal that uptake of NNK by a smoker results in a lifetime dose similar to that which can induce lung tumors in laboratory animals. Levels of NNAL plus NNAL-Gluc in urine are correlated with cotinine in urine. The ratio of NNAL-Gluc to NNAL is a good indicator of NNK detoxification because the carcinogenic of NNAL-Gluc is less than NNK and NNAL (although this has not been tested to date). NNAL-Gluc/NNAL ratios appear to segregate into two phenotypes, as illustrated in Figure 4.2 (Carmella et al., 1995). To some extent, the higher the ratio in urine from smokers, the higher the possibility of cancer. The ratio phenotype has been applied as a useful biomarker in molecular epidemiology studies. In addition, NNAL and NNAL-Gluc have been detected in the urine obtained from nonsmokers who were exposed to environmental tobacco smoke (Hecht et al., 1993b). This study established the presence of a lung carcinogen NNAL in the urine of nonsmokers exposed to environmental tobacco smoke, demonstrating that nonsmokers take up and metabolize NNK. The results provided experimental support for the proposal that environmental tobacco smoke can cause lung cancer. In another investigation, NNAL and NNAL-Gluc levels were quantified in the urine of people from Sudan who used an oral tobacco product—toombak—known to contain unusually high levels of NNK (Murphy et al., 1994).

4.3 *N*-NITROSONORNICOTINE

N'-nitrosonornicotine (NNN) is an important tobacco-specific carcinogen that may be involved in the development of tobacco-related cancers (Hecht et al., 1979). NNN content in tobacco and tobacco smoke is at relatively high concentrations; convincing evidence indicates its carcinogenic activity in mice, rats, and hamsters (Boyland et al., 1964; Hoffmann et al., 1981). Mammalian metabolism of NNN leads to a variety of products resulting from α-hydroxylation, β-hydroxylation, and pyridine *N*-oxidation (Chi-Hong et al., 1978; Hecht et al., 1980). These metabolites are most effectively quantified by high-pressure liquid chromatography (HPLC). In a previous report, an HPLC method for determining the in vitro rates of α-hydroxylation of NNN was described (Chi-Hong et al., 1979). Hecht et al. (Stephen et al., 1981) developed a HPLC method to detect metabolites of NNN in urine of F-344 rats. The percentage excretion of the principal urinary metabolites was determined over a dose range of 3–300 mg/kg in the F-344 rat, as follows: 4-hydroxy-4-(3-pyridyl) butyric acid (37.1%–53.3%, respectively, of the dose), 4-oxo-4-(3-pyridyl)butyric acid (31.1%–12.8%), *N'*-nitrosonomicotine-l-*N*-oxide (6.7%–10.7%), *N'*-nitrosonornicotine (3.3%–5.2%), and norcotinine (3.2%–5.1%). The urinary metabolites in the strain mouse and Syrian golden hamster were qualitatively similar to those observed in the F-344 rat (Figure 4.4).

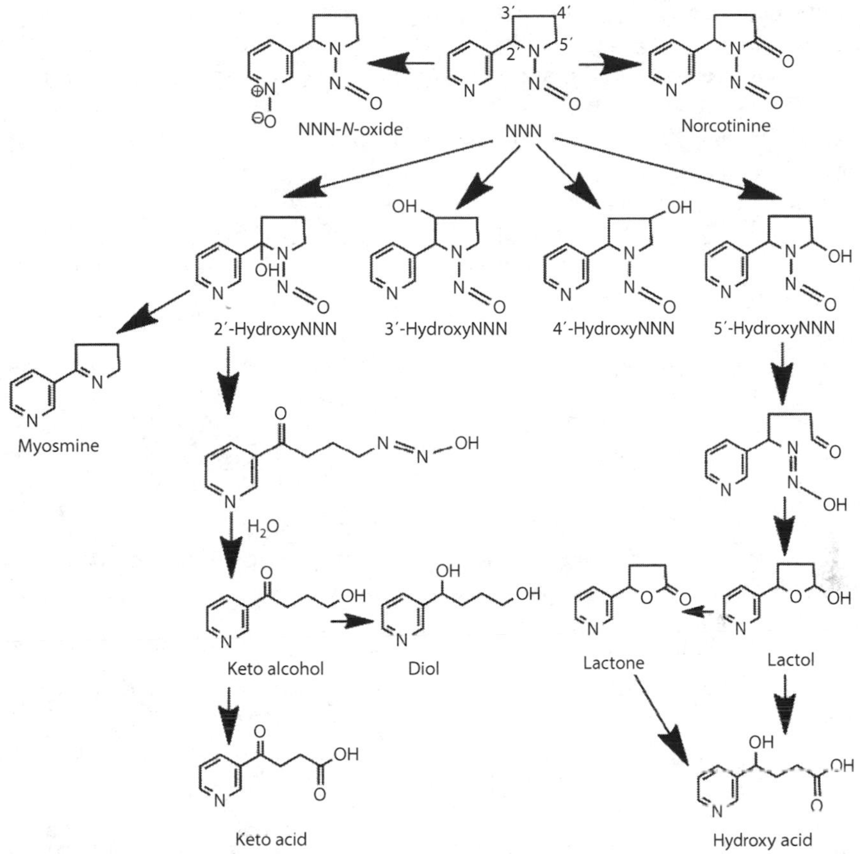

FIGURE 4.4 NNN metabolism pathways in vivo. (From Hecht, S.S., *Chem. Res. Toxicol.*, 11(6), 559, 1998.)

4.4 TSNA BIOMARKERS IN OTHERS

4.4.1 DNA ADDUCTS DERIVED FROM TSNAS IN TISSUES

Metabolic activation of NNK by methylene hydroxylation or methyl hydroxylation results in methylation or pyridyloxobutylation of DNA. The presence of methyl and pyridyloxobutyl adducts in DNA of tissues from animals treated with NNK has been established conclusively (Castonguay et al., 1984; Hecht et al., 1986, 1988). Methyl adducts derived from NNK include 7-mG, O^4-mT O^6-mG, and O^6-mG O^4-mT, which can be detected by GC–MS or HPLC-MS/MS method. Pyridyloxobutyl adducts are often quantified by the measurement of 4-hydroxy-1-(3-pyridyl)-1-butanone (HPB) released by either acid or neutral thermal hydrolysis. Then the HPB can be quantified by radiochromatography and GC–MS. Because HPB-releasing adducts are unstable at the nucleoside level, and under conventional analytical conditions, the structures of HPB-releasing adducts under the condition of acid hydrolysis have not been identified. However, the lifetime of HPB-releasing adducts in DNA is very long. Persisting research in rat lung and liver indicated that it presents up to weeks after a single injection of NNK (Peterson et al., 1991).

Levels of methyl adducts and pyridyloxobutyl adducts have been compared in rat lung, liver, and nasal mucosa as well as in mouse lung. In rats, the ratio of 7-mG to HPB released varied from 7 to 25 (lung) and from 13 to 49 (liver), with increasing dose (Murphy et al., 1990). The concentrations of HPB released from the DNA of rat lung and liver were equal to or higher than that of O^6-mG. However, concentrations of O^6-mG were approximately five to seven times higher than that of HPB released in the nasal mucosa of rats treated chronically with NNK (Trushin et al., 1994). In A/J mice treated with NNK, O^6-mG levels were persistent in lung for up to 15 day, and levels were consistently greater than those of HPB released (Peterson and Hecht, 1991). The two adducts were formed and removed at different rates, consistent with differing cytochrome P450–mediated catalysis of the two metabolic activation pathways.

Single-strand breaks have been observed in the DNA of rat and hamster liver after treatment with NNK, as well as in hepatocytes incubated with NNK (Jorquera et al., 1994). Aldehydes generated in NNK metabolism via the α-hydroxylation pathway are partially involved in this mode of DNA damage (Carmella et al., 1995). Oxidative damage, as quantified by 8-oxodeoxyguanosine, has been detected in the lungs of rats and mice treated with NNK (Peterson and Hecht, 1991).

4.4.2 PROTEIN ADDUCTS AND HEMOGLOBIN ADDUCTS FROM TSNAs

HPB-releasing hemoglobin adducts have been characterized in rats and mice treated with NNK (Peterson et al., 1990). One adduct is an ester, probably with aspartate or glutamate, which is hydrolyzed under mild basic conditions to release HPB (Carmella et al., 1992). The ester adduct comprises 20%–40% of the total binding to hemoglobin; the remainder does not release HPB upon mild base hydrolysis and has not been fully characterized, although it is known that cysteine adducts are not formed to a significant extent (Carmella et al., 1990b). As in DNA, the released HPB can be quantified by radiochromatography or GC–MS. Levels of HPB-releasing adducts in hemoglobin correlate with levels in DNA, although the relationship is not linear (Belinsky et al., 1990). Following metabolic activation of NNK in rat hepatocytes, the α-hydroxynitrosamine or its transport form can migrate out of the hepatocyte and into the red blood cell, where it reacts with hemoglobin. Hemoglobin itself also can activate NNK by methylene hydroxylation, resulting in methylation of the protein, but not in the formation of HPB-releasing adducts (Murphy and Coletta, 1993). The HPB-releasing adduct has proven to be a useful indicator of extents of metabolic activation of NNK in studies with laboratory animals.

The quantitation of hemoglobin adducts thus presents an attractive alternative because hemoglobin is readily available. A GC–MS method for quantifying HPB-releasing adducts from human hemoglobin has been standardized (Carmella et al., 1990a; Hecht et al., 1994). HPB-releasing adducts have been quantified in smokers, snuff dippers, and nonsmokers (Hecht et al., 1993a). In two recent studies, levels of HPB-releasing hemoglobin adducts were quantified in nasal snuff users from Germany and in Sudanese toombak users (Falter et al., 1994). Both groups showed elevated levels of hemoglobin adducts, similar to the earlier finding of elevated adduct levels in snuff dippers from the United States. In smokers, HPB-releasing adducts are elevated above background in only a subset of individuals and appear to be generally lower than in snuff users. This may relate to patterns of enzyme induction or inhibition due to constituents of tobacco smoke, but this requires further investigation.

REFERENCES

Ashley, D. L. et al. Effect of differing levels of tobacco-specific nitrosamines in cigarette smoke on the levels of biomarkers in smokers. *Cancer Epidemiology Biomarkers & Prevention*. 2010, 19(6), 1389–1398.

Bartsch, H., Spiegelhalder, B. Environmental exposure to N-nitroso compounds (NNOC) and precursors: An overview. *European Journal of Cancer Prevention*. 1996, 5, 11–17.

Belinsky, S. A. et al. Dose-response relationship between O6-methylguanine formation in Clara cells and induction of pulmonary neoplasia in the rat by 4-(methylnitrosamino)-1-(3-pyridyl)-1-butanone. *Cancer Research*. 1990, 50(12), 3772–3780.

Bhat, S. H. et al. A new liquid chromatography/mass spectrometry method for 4-(methylnitrosamino)-1-(3-pyridyl)-1-butanol (NNAL) in urine. *Rapid Communications in Mass Spectrometry*. 2011, 25(1), 115–121.

Boyland, E., Roe, F., Gorrod, J. Induction of pulmonary tumours in mice by nitrosonornicotine, a possible constituent of tobacco smoke. *Nature*, 1964, 202, 1126.

Carmella, S. G. et al. Evaluation of cysteine adduct formation in rat hemoglobin by 4-(methylnitrosamino)-1-(3-pyridyl)-1-butanone and related compounds. *Cancer Research*. 1990a, 50(17), 5453–5459.

Carmella, S. G. et al. Mass spectrometric analysis of tobacco-specific nitrosamine hemoglobin adducts in snuff dippers, smokers, and nonsmokers. *Cancer Research*. 1990b, 50(17), 5438–5445.

Carmella, S. G. et al. Evidence that a hemoglobin adduct of 4-(methylnitrosamino)-1-(3-pyridyl)-1-butanone is a 4-(3-pyridyl)-4-oxobutyl carboxylic acid ester. *Chemical Research in Toxicology*. 1992, 5(1), 76–80.

Carmella, S. G. et al. Intraindividual and interindividual differences in metabolites of the tobacco-specific lung carcinogen 4-(methylnitrosamino)-1-(3-pyridyl)-1-butanone (NNK) in smokers' urine. *Cancer Epidemiology Biomarkers & Prevention*. 1995, 4(6), 635–642.

Carmella, S. G. et al. Effects of smoking cessation on eight urinary tobacco carcinogen and toxicant biomarkers. *Chemical Research in Toxicology*. 2009, 22(4), 734–741.

Castonguay, A. et al. Comparative carcinogenicity in A/J mice and metabolism by cultured mouse peripheral lung of N'-nitrosonornicotine, 4-(methylnitrosamino)-1-(3-pyridyl)-1-butanone, and their analogues. *Cancer Research*. 1983, 43(3), 1223–1229.

Castonguay, A. et al. Kinetics of DNA methylation by the tobacco-specific carcinogen 4-(methylnitrosamino)-1-(3-pyridyl)-1-butanone in F344 rats. *IARC Scientific Publications*. 1984, (57), 805.

Chi-hong, B. C. et al. Metabolic α-hydroxylation of the tobacco-specific carcinogen, N'-nitrosonornicotine. *Cancer Research*. 1978, 38(11 Part 1), 3639–3645.

Chi-hong, B. C. et al. Assay for microsomal α-hydroxylation of N'-nitrosonornicotine and determination of the deuterium isotope effect for α-hydroxylation. *Cancer Research*. 1979, 39(12), 5057–5062.

Eisner, M. D. et al. Longer term exposure to secondhand smoke and health outcomes in COPD: Impact of urine 4-(methylnitrosamino)-1-(3-pyridyl)-1-butanol. *Nicotine & Tobacco Research.* 2009, 11(8), 945–953.

Falter, B. et al. Biomonitoring of hemoglobin adducts: Aromatic amines and tobacco-specific nitrosamines. *The Clinical Investigator.* 1994, 72(5), 364–371.

Fuchs, C. S. et al. A prospective study of cigarette smoking and the risk of pancreatic cancer. *Archives of Internal Medicine.* 1996, 156(19), 2255.

Hatsukami, D. K. et al. Biomarkers to assess the utility of potential reduced exposure tobacco products. *Nicotine & Tobacco Research.* 2006, 8(2), 169–191.

Hecht, S. S. Recent studies on mechanisms of bioactivation and detoxification of 4-(methylnitrosamino)-1-(3-pyridyl)-1-butanone (NNK), a tobacco-specific lung carcinogen. *CRC Critical Reviews in Toxicology.* 1996, 26(2), 163–181.

Hecht, S. S. Approaches to cancer prevention based on an understanding of N-nitrosamine carcinogenesis. *Proceedings of the Society for Experimental Biology and Medicine.* Society for Experimental Biology and Medicine, New York, Royal Society of Medicine. 1997.

Hecht, S. S. Biochemistry, biology, and carcinogenicity of tobacco-specific N-nitrosamines. *Chemical Research in Toxicology.* 1998, 11(6), 559–603.

Hecht, S. S. Human urinary carcinogen metabolites: Biomarkers for investigating tobacco and cancer. *Carcinogenesis.* 2002, 23(6), 907–922.

Hecht, S. S., Hoffmann, D. Tobacco-specific nitrosamines, an important group of carcinogens in tobacco and tobacco smoke. *Carcinogenesis.* 1988, 9(6), 875.

Hecht, S. S., Lin, D., Chen, C. B. Comprehensive analysis of urinary metabolites of N'-nitrosonornicotine. *Carcinogenerfs.* 1981, 2(9), 833–838.

Hecht, S. S. et al. Tobacco-specific nitrosamines: Occurrence, formation, carcinogenicity, and metabolism. *Accounts of Chemical Research.* 1979, 12(3), 92–98.

Hecht, S. S. et al. Metabolism in the F344 rat of 4-(N-methyl-N-nitrosamino)-1-(3-pyridyl)-1-butanone, a tobacco-specific carcinogen. *Cancer Research.* 1980, 40(11), 4144–4150.

Hecht, S. S. et al. Comparative tumorigenicity and DNA methylation in F344 rats by 4-(methylnitrosamino)-1-(3-pyridyl)-1-butanone and N-nitrosodimethylamine. *Cancer Research.* 1986, 46(2), 498–502.

Hecht, S. S. et al. Evidence for 4-(3-pyridyl)-4-oxobutylation of DNA in F344 rats treated with the tobacco-specific nitrosamines 4-(methylnitrosamino)-1-(3-pyridyl)-1-butanone and N'-nitrosonornicotine. *Carcinogenesis.* 1988, 9(1), 161–165.

Hecht, S. S. et al. A tobacco-specific lung carcinogen in the urine of men exposed to cigarette smoke. *New England Journal of Medicine.* 1993a, 329(21), 1543–1546.

Hecht, S. S. et al. Metabolism of the tobacco-specific nitrosamine 4-(methylnitrosamino)-1-(3-pyridyl)-1-butanone in the patas monkey: Pharmacokinetics and characterization of glucuronide metabolites. *Carcinogenesis.* 1993b, 14(2), 229–236.

Hecht, S. S. et al. Tobacco-specific nitrosamine adducts: Studies in laboratory animals and humans. *Environmental Health Perspectives.* 1993c, 99, 57.

Hecht, S. S. et al. Tobacco-specific nitrosamine-hemoglobin adducts. *Methods in Enzymology.* 1994, 231, 657–667.

Hecht, S. S. et al. Quantitation of metabolites of 4-(methylnitrosamino)-1-(3-pyridyl)-1-butanone after cessation of smokeless tobacco use. *Cancer Research.* 2002, 62(1), 129–134.

Hoffmann, D., Hecht, S. Advances in tobacco carcinogenesis. In: *Chemical Carcinogenesis and Mutagenesis I.* C. S. Cooper and P. L. Grover (eds.). Springer, London, 1990, pp. 63–102.

Hoffmann, D. et al. Comparative carcinogenicity and metabolism of 4-(methylnitrosamino)-1-(3-pyridyl)-1-butanone and N'-nitrosonornicotine in Syrian golden hamsters. *Cancer Research.* 1981, 41(6), 2386–2393.

Hoffmann, D. et al. Tobacco-specific N-nitrosamines and ARECA-derived N-nitrosamines: Chemistry, biochemistry, carcinogenicity, and relevance to humans. *Journal of Toxicology and Environmental Health, Part A Current Issues.* 1994, 41(1), 1–52.

Hoffmann, D. et al. Five leading US commercial brands of moist snuff in 1994: Assessment of carcinogenic N-nitrosamines. *Journal of the National Cancer Institute.* 1995, 87(24), 1862–1869.

Hou, H. et al. Development of a method for the determination of 4-(methylnitrosamino)-1-(3-pyridyl)-1-butanol in urine of nonsmokers and smokers using liquid chromatography/tandem mass spectrometry. *Journal of Pharmaceutical and Biomedical Analysis.* 2012, 63, 17–22.

IARC Working Group on the Evaluation of Carcinogenic Risks to Humans. *IARC Monographs on the Evaluation of Carcinogenic Risks to Humans.* World Health Organization, Lyon, France, 2007.

Jacob III, P. et al. Subpicogram per milliliter determination of the tobacco-specific carcinogen metabolite 4-(methylnitrosamino)-1-(3-pyridyl)-1-butanol in human urine using liquid chromatography–tandem mass spectrometry. *Analytical Chemistry.* 2008, 80(21), 8115–8121.

Jorquera, R. et al. DNA single-strand breaks and toxicity induced by 4–(methyl-nitrosamino)–1–(3-pyridyl)-1-butanone or N-nitrosodimethylamine in hamster and rat liver. *Carcinogenesis.* 1994, 15(2), 389–394.

Magee, P. The experimental basis for the role of nitroso compounds in human cancer. *Cancer Surveys.* 1989, 8(2), 207.

Morse, M. A. et al. Characterization of a glucuronide metabolite of 4-(methylnitrosamino)-1-(3-pyridyl)-1-butanone (NNK) and its dose-dependent excretion in the urine of mice and rats. *Carcinogenesis.* 1990, 11(10), 1819–1823.

Murphy, S. E., Coletta, K. A. Two types of 4-(methylnitrosamino)-1-(3-pyridyl)-1-butanone hemoglobin adducts, from metabolites which migrate into or are formed in red blood cells. *Cancer Research.* 1993, 53(4), 777–783.

Murphy, S. E., Palomino, A., Hecht, S. S., Hoffmann, D. Dose-response study of DNA and hemoglobin adduct formation by 4-(methylnitrosamino)-1-(3-pyridyl)-1-butanone in F344 rats. *Cancer Research.* 1990, 50, 5446.

Murphy, S. E. et al. Uptake and metabolism of carcinogenic levels of tobacco-specific nitrosamines by Sudanese snuff dippers. *Cancer Epidemiology Biomarkers & Prevention.* 1994, 3(5), 423–428.

Peterson, L. A., Hecht, S. S. O6-methylguanine is a critical determinant of 4-(methylnitrosamino)-1-(3-pyridyl)-1-butanone tumorigenesis in A/J mouse lung. *Cancer Research.* 1991, 51(20), 5557–5564.

Peterson, L. A. et al. Investigations of metabolic precursors to hemoglobin and DNA adducts of 4-(methylnitrosamino)-1-(3-pyridyl)-1-butanone. *Carcinogenesis.* 1990, 11(8), 1329–1333.

Peterson, L. A. et al. In vivo and in vitro persistence of pyridyloxobutyl DNA adducts from 4-(methylnitrosamino)-1-(3-pyridyl)-1-butanone. *Carcinogenesis.* 1991, 12(11), 2069–2072.

Peterson, L. A. et al. Formation of NADP (H) analogs of tobacco-specific nitrosamines in rat liver and pancreatic microsomes. *Chemical Research in Toxicology.* 1994, 7(5), 599–608.

Rivenson, A. et al. Induction of lung and exocrine pancreas tumors in F344 rats by tobacco-specific and areca-derived N-nitrosamines. *Cancer Research.* 1988, 48(23), 6912–6917.

Shah, K. A., Karnes, H. T. A review of the analysis of tobacco-specific nitrosamines in biological matrices. *Critical Reviews in Toxicology.* 2010, 40(4), 305–327.

Shah, K. A. et al. A modified method for the determination of tobacco specific nitrosamine 4-(methylnitrosamino)-1-(3-pyridyl)-1-butanol in human urine by solid phase extraction using a molecularly imprinted polymer and liquid chromatography tandem mass spectrometry. *Journal of Chromatography B.* 2009, 877(14), 1575–1582.

Shah, K. A. et al. Microfluidic direct injection method for analysis of urinary 4-(methylnitrosamino)-1-(3-pyridyl)-1-butanol (NNAL) using molecularly imprinted polymers coupled on-line with LC–MS/MS. *Journal of Pharmaceutical and Biomedical Analysis.* 2011, 54(2), 368–378.

Shopland, D. R. Tobacco use and its contribution to early cancer mortality with a special emphasis on cigarette smoking. *Environmental Health Perspectives.* 1995, 103(Suppl. 8), 131.

Siegel, R. et al. Cancer statistics, 2012. *CA: A Cancer Journal for Clinicians* 2012, 62(1), 10–29.

Spiegelhalder, B., Bartsch, H. Tobacco-specific nitrosamines. *European Journal of Cancer Prevention.* 1996, 5, 33–38.

Stepanov, I. et al. Extensive metabolic activation of the tobacco-specific carcinogen 4-(methylnitrosamino)-1-(3-pyridyl)-1-butanone in smokers. *Cancer Epidemiology Biomarkers & Prevention.* 2008, 17(7), 1764–1773.

Thun, M. J. et al. Cigarette smoking and changes in the histopathology of lung cancer. *Journal of the National Cancer Institute.* 1997, 89(21), 1580–1586.

Trushin, N. et al. Evidence supporting the role of DNA pyridyloxobutylation in rat nasal carcinogenesis by tobacco-specific nitrosamines. *Cancer Research.* 1994, 54(5), 1205–1211.

Xia, Y., Bernert, J. Quantitation of the tobacco-specific nitrosamine 4-(methylnitrosamino)-1-(3-pyridyl)-1-butanol (NNAL) in urine by LC tandem mass spectrometry. *Epidemiology.* 2008, 19(6), S230.

Xia, Y. et al. Analysis of the tobacco-specific nitrosamine 4-(methylnitrosamino)-1-(3-pyridyl)-1-butanol in urine by extraction on a molecularly imprinted polymer column and liquid chromatography/atmospheric pressure ionization tandem mass spectrometry. *Analytical Chemistry.* 2005, 77(23), 7639–7645.

Yang, Y. et al. On-line concentration and determination of tobacco-specific N-nitrosamines by cation-selective exhaustive injection–sweeping–micellar electrokinetic chromatography. *Talanta.* 2010, 82(5), 1797–1801.

5 Biomarkers of Polycyclic Aromatic Hydrocarbons

5.1 INTRODUCTION

Polycyclic aromatic hydrocarbons (PAHs) are products of incomplete burning of organic substances (Angerer et al., 2007; Hagedorn et al., 2009; Jacob and Seidel, 2002; Xue and Warshawsky, 2005), which can be formed in the environments related to production, refining, and application of coal, mineral oil, tobacco, and oil shale. Some PAHs are potent carcinogenics, and complex mixtures of PAHs have been shown to be carcinogenic to humans. The International Agency for Research on Cancer (IARC) described the metabolism of several of the PAH compounds (Hansen et al., 2008), but only a limited number of PAH metabolites are commercially available. The reliable and specific PAH metabolites, such as 1-hydroxypyren (1-OHP), hydroxyphenanthrenes, and 1,2-hydroxynaphthalene, etc., in urine are listed in Table 5.1. Smoking is a major source for PAH exposure for subjects not occupationally exposed to PAHs (Hansen et al., 2008), and urinary 1-OHP is used in the assessment of PAH exposures. The smokers had significantly higher amounts of 1-OHP than nonsmokers, and the number of cigarettes smoked correlated rather well with urinary 1-OHP concentrations. Urinary 1-OHP was used as a tool for the risk assessment of exposure to PAHs for the reason that 1-OHP excretion increases linearly with airborne concentrations of pyrene.

5.2 PAHS AND PAH EXPOSURE AND CARCINOGENESIS

PAHs are readily absorbed into the body through the skin, lungs, and gastrointestinal tract. Exposure to PAHs in occupational settings can lead to lung, skin, or other human cancers (Strickland and Kang, 1999). Epidemiological studies have consistently demonstrated that exposure to PAHs is associated with increases in mortality and/or morbidity from respiratory diseases, cardiovascular diseases, and cancer (Taioli et al., 2007). PAHs are planar aromatic compounds, which have various potencies of carcinogenicity due to their individual structures (Siddens et al., 2012). Different types of combustion result in different compositions of PAHs both in relative amounts and individual PAHs present. The relationship between exposure to PAHs and skin cancers is illustrated by epidemiological studies (Mensing et al., 2005). Benzo[a]pyrene (B[a]P), the most extensively studied carcinogenic PAH, is classified by IARC as a Group 1 or known human carcinogen. Four of the top ten priority pollutants, defined by the Agency for Toxic Substances and Disease Registry (ATSDR) in 2011, are single PAH or PAH mixtures (PAHs, B[a]P, benzo[b]fluoranthene, and dibenzo[a,h]anthracene). PAHs are carcinogenic in a number of animal

TABLE 5.1

Biomarkers of PAH

Metabolites	Matrix
1-,3-,9-Hydroxybenz[a]anthracene	Urine
1-,2-,3-Hydroxybenzo[c]phenanthrene	Urine
3-Hydroxyfluoranthene	Urine
2-,3-,9-Hydroxyfluorene	Urine
1-,2-,3-,4-,9-Hydroxyphenanthrene	Urine
1-OHP	Urine
3-OH-B[a]P	Urine
1-,2-Hydroxynapthalene	Urine

models with multiple objectives, including the skin, liver, lung, breast, ovaries, and hematopoietic tissue.

PAHs required bioactivation through metabolism in order to be mutagenic, carcinogenic, or teratogenic to target cellular macromolecules (IARC, 2010). The most well-characterized bioactivation pathways have been cytochrome P450 (CYP)–dependent epoxygenation (Jain, 2010; Snyder, 2000; Stoilov, 2011; Zanger and Schwab, 2013), hydrolysis by epoxide hydrolase, and a second CYP epoxygenation to the B[a]P-7,8-dihydrodiol-9,10-epoxide (BPDE) in the case of B[a]P and to the 11,12-dihydrodiol-13,14 epoxide (DBCDE) in the case of DBC (Xue and Warshawsky, 2005).

5.3 ENZYMES OF METABOLIC ACTIVATION AND DETOXICATION OF PAHS

Generally, oxidation of PAHs by CYP enzymes is an initial and important step in the activation process to produce the polar biochemically reactive electrophilic substances (the carcinogenic metabolites), which can interact with cellular macromolecules, particularly nucleic acids and proteins (Xue and Warshawsky, 2005). The identification and characterization of the electrophilic metabolites is important for the illumination of the mechanism of the metabolic activation of PAHs. PAHs are metabolized in the liver and other tissues by the CYP enzyme system into reactive electrophiles (Rybicki et al., 2006). These reactive species are then detoxified and made water soluble through phase II conjugation with glutathione (GSH) or glucuronic acid. However, the reactive electrophilic PAH species can also bind to cellular macromolecules such as DNA, forming the PAH–DNA adducts. B[a]P is considered as a lead compound and one of the most intensively studied carcinogenic PAHs (Chung et al., 2010). The ultimate carcinogens derived from B[a]P metabolism are the isomeric B[a]P-diol-epoxides (BPDEs) that bind to DNA, proteins, and other macromolecules to form adducts. Adducts of the BPDEs accumulate in blood and can be used as biomarkers of exposure to B[a]P and other

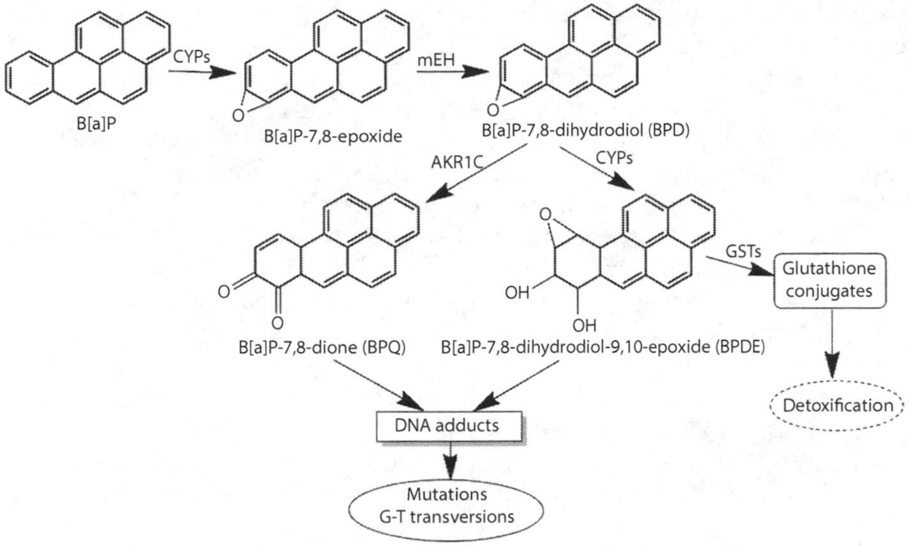

FIGURE 5.1 Enzymatic conversion pathways of B[a]P catalyzed by CYPs and the aldo-keto reductase 1C (AKR1C).

PAHs from the same source (Bartsch, 2000). B[a]P is initially oxidized to an epoxide (e.g., B[a]P-7,8-epoxide) by CYP enzymes CYP1A1 or CYP1B1 (Figure 5.1). After the epoxide (B[a]P-7,8 epoxide) is formed, it may be further metabolized by microsomal epoxide hydrolase (mEH), to form a dihydrodiol (B[a]P-7,8-dihydrodiol [BPD]). This dihydrodiol has two critical subpathways, both of which may create metabolites that can damage DNA. The most common route involves metabolism of the dihydrodiol (BPD) by CYP1A1, CYP1A2, or CYP1B1 to form a diol epoxide (BPDE), which may covalently bind to DNA forming a BPDE–DNA adduct. Of the four possible diastereomers of BPDE, the (C)-anti-BPDE is the most abundant, mutagenic, and tumorigenic.

5.3.1 ENZYMES OF METABOLIC ACTIVATION

Presently, there are three principal metabolic pathways for PAHs that have been proposed and established with experimental evidence (Xue and Warshawsky, 2005). Pathway 1 is the theory of bay-region dihydrodiol epoxide. It was widely accepted by scientists in this field as the dominant mechanism of chemical carcinogenesis of PAH (Islam et al., 1999). This metabolic activation pathway of carcinogenic PAHs has been well established through studies over the past half century. It involves three enzyme-mediated reactions: first, oxidation of a double bond catalyzed by CYP enzymes to unstable arene oxides; second, hydrolysis of the arene oxides by mEH to *trans*-dihydrodiols (diols and all diols mentioned in this review are *trans* unless stated otherwise); finally, a second CYP-catalyzed oxidation at the double bond adjacent to the diol function to generate a vicinal diol epoxide. This pathway can lead to sterically hindered bay- or fjord-region diol epoxides. The bay (fjord) region diol epoxides are electrophiles, which can bind to DNA, while some of the diol epoxide stereoisomers of PAHs are found to be ultimate carcinogens.

Pathway 2 is the formation of radical cations of PAH through one-electron oxidation catalyzed by CYP peroxidase. It is known that the radical cation of PAH is formed by the removal of one electron from the k electron system of the molecule through one-electron oxidation. Radical cations are electrophilic in nature, capable of interacting with nucleophilic centers in cellular macromolecules. Therefore, the formation of radical cation in metabolic oxidation process catalyzed by CYP peroxidase was proposed to be one of the pathways of PAH activation, shown in Figure 5.2.

Pathway 3 is the formation of o-quinones catalyzed by dihydrodiol dehydrogenases (DDs) in the process of PAH activation. The formation of o-quinones by DD-catalyzed oxidation was proposed as the third major metabolic activation pathway of PAH (shown in Figure 5.3). Under physiological conditions, the DD enzyme competes with CYP to oxidize the non-K-region diol, a proximate carcinogenic metabolite of PAH, but not the K-region diol. The mechanism of this NADP+-dependent oxidation involves the initial formation of a ketol that spontaneously rearranges to form a catechol by the action of DD followed by autoxidation of the unstable catechol to o-quinone. These o-quinones of PAH exhibited strong reactivity toward the cellular thiols, GSH, and cysteine, leading to formation of water-soluble conjugates. PAH-o-quinones were also excellent substrates for the human placental NADP-linked 15-hydroxyprostaglandine dehydrogenase enzymes. Lately, more evidence has revealed that PAH-o-quinones are highly reactive Michael acceptors, which can form both stable and depurinating DNA adducts. The redox-active PAH-o-quinone can be reduced to reform catechol by a nonenzymatic two-electron reduction or to reform the semiquinone anion radical (SQ) via a one-electron enzymatic reduction. In the metabolic process, reactive oxygen species (ROS) are generated.

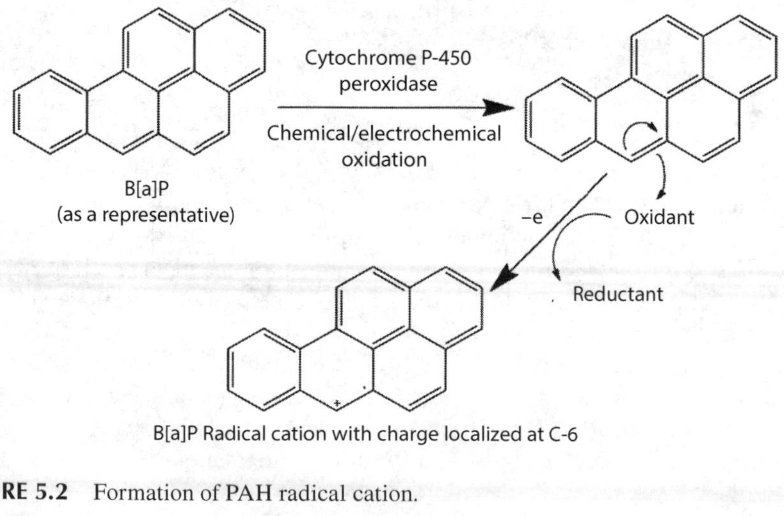

FIGURE 5.2 Formation of PAH radical cation.

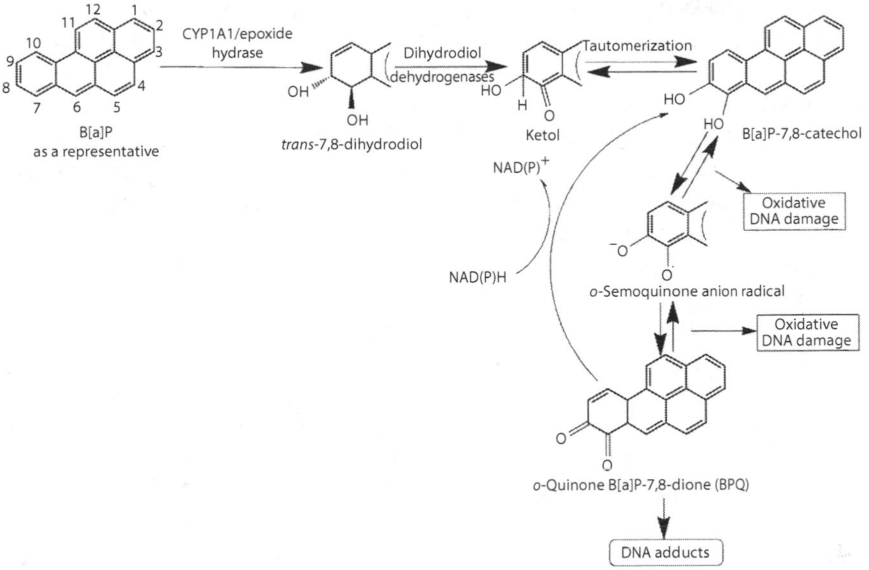

FIGURE 5.3 Metabolic activation pathway of PAH via *o*-quinone.

Therefore, *o*-quinone formation is now considered to be a metabolic activation pathway rather than detoxification.

5.3.2 DETOXICATION OF PAH

Following PAHs' metabolic activation to diastereomeric bay- and fjord-region diol epoxides, the mutagenic and carcinogenic activities appear to be linked to the covalent binding of these reactive intermediates to the exocyclic amino group of deoxyguanosine (dG) and deoxyadenosine (dA) in DNA (Sundberg et al., 1998). Available information indicates that fjord-region diol epoxides in general are more biologically possible than bay-region diol epoxides. The former intermediates demonstrate a higher preference for adduct formation with dA than bay-region diol epoxides, which in part may explain the increased potency. Furthermore, whereas the (+)-anti-enantiomers of the bay-region diol epoxides with R,S,S,R-absolute configuration seem to be the principal tumor initiators among the four possible stereoisomers, both the anti- and syn-stereoisomers of the fjord-region diol epoxides demonstrate high mutagenic and/or tumorigenic activity. The diequatorial orientation of the hydroxyl groups in all fjord-region diol epoxides due to the nonplanarity and the steric hindrance in the molecule is likely part of the explanation. The most important intracellular system preventing the formation of diol epoxide–DNA adducts is glutathione *S*-transferase (GST)-catalyzed conjugation of the intermediates with GSH. The Pi class of GSTs seems to be of particular importance, and recently it was demonstrated that GSTPi gene–deleted mice exhibit an increased susceptibility to PAH-induced tumors. In humans, an allelic variant with Val105 (GSTP1-1/V-105) rather than Ile105 (GSTP1-1/I-105) has been associated with higher tumor susceptibility in organs exposed to PAH. The frequency of GSTP1*A (encoding GSTP1-1/V-105) in individuals with squamous carcinomas of the upper aerodigestive tract was lower than in control individuals. The individuals with colorectal or lung cancer were found to have no significant associations between a particular GSTP1-1 variant and their tumor. Thus, the results to date on GSTP1-1 polymorphism and cancer susceptibility are variable, and further evaluation in larger populations seems to be needed. Previous studies have shown that the GSTP1-1/I-105 and GSTP1-1/V-105 exhibit different catalytic efficiencies toward bay-region diol epoxides. With most compounds examined, GSTP1-1/V-105 was more efficient than GSTP1-1/I-105. In organs in which tumor formation is initiated by DNA adducts derived from bay-region diol epoxides, this difference may be a significant factor in tumor susceptibility. Fjord-region diol epoxides are considerably more carcinogenic in experimental animals and may pose a higher risk as tumor initiators in humans. Therefore, it is important to evaluate the role of their GST-catalyzed detoxication. The catalytic efficiencies of the two naturally occurring GSTP1-1 allelic variants, GSTP1-1/I-105 and GSTP1-1/V-105, toward several of these compounds have been determined. The metabolically most relevant (3)-anti- and (+)-syn-diol epoxide stereoisomers from benzo[c]phenanthrene (B[c]Ph), benzo[c]chrysene (B[c]C), benzo[g]chrysene (B[g]C), and dibenzo[a,l]pyrene (DB[a,l]P), the most carcinogenic PAH yet identified, have been investigated. As part of the active site of GSTP1-1, the side chain of the amino acid residue in position 105 appears to influence catalytic efficiency. To gain insight into the steric requirements for efficient catalysis of fjord-region diol epoxides conjugation, mutants with Ala^{105} (GSTP1/A-105) or Trp^{105} (GSTP1-1/W-105) at this position have also been studied.

5.3.2.1   UDP-Glucuronosyltransferases

The phenols and polyphenols of B[a]P can be converted to radicals, semiquinones, and quinones, the latter undergoing quinone/quinol redox cycles that generate semiquinones and ROS (Gschaidmeier et al., 1995). The extent of detoxification of PAH quinones is largely determined by the balance between phase I and phase II drug metabolism. Previous studies have shown that mutagenicity of B[a]P and B[a]P-3,6-quinone in the Ames test was markedly reduced upon addition of uridine diphosphate (UDP)-glucuronic acid, thereby emphasizing the inactivating role of glucuronidation. After reduction of quinones by quinone reductase, glucuronidation of the resulting quinols appears to be mainly catalyzed by 3-methylcholanthreneinducible UDP-glucuronosyltransferases (UGTs). Similar results have been obtained with structurally related 3,6-dihydroxychrysene as substrate. Diglucuronides of B[a]P-3,6-quinol have been found as major biliary metabolites after intratracheal instillation of B[a]P.

5.3.2.2   Glutathione S-Transferases and GSH Conjugates

Among the well-known PAH biomarkers, the phase I enzyme CYP1A is responsible for the biotransformation of xenobiotics (Nahrgang et al., 2009). The induction of CYP1A can be triggered via the cytoplasmic aryl hydrocarbon receptor 2 (AhR2). The AhR2 was found to play a role in the regulation of xenobiotic-metabolizing enzymes in case of exposure to the AhR ligands such as PAHs. The phase II enzymes, GSTs, provide cellular protection against the toxic effects of a variety of endogenous and environmental chemicals (Hecht et al., 2010; Obolenskaya et al., 2010). Their main function is to catalyze the conjugation of GSH with xenobiotics, such as PAHs, to facilitate the excretion of the formed complexes (Binkova et al., 2007; Rihs et al., 2005; Singh et al., 2007; Thier et al., 2003). Furthermore, CYP1A is directly involved in the generation of oxidative stress, via the production and accumulation of ROS beyond the capacity of an organism to scavenge them (Santella et al., 1995; Sundberg et al., 1998). The evaluation of oxidative stress is commonly used in monitoring programs based on measurements of catalase (CAT), GSH peroxidase (GPX), and superoxide dismutase (SOD) activities.

5.4 BIOMONITORING PAHs AND THEIR METABOLITES IN HUMAN URINE

The obtained data on urine concentrations of PAH metabolites reflect a more accurate estimation of the quantity of the actual PAH intake of an individual compared to ambient air measurements because it estimates the internal dose from exposure through several routes including both skin absorption and respiratory uptake (Jacob and Seidel., 2002).

Urinary metabolites of PAH were measured as the environmental carcinogen exposure markers for humans (Elovaara et al., 2006; Ichiba et al., 2006; Rihs et al., 2005; Sobus et al., 2009). The most common metabolite is urinary 1-OHP, a metabolite of pyrene; urinary 2-naphthol (2-NP), a metabolite of naphthalene (Nap), has also been measured as a PAH exposure marker. Their metabolite measurements were useful for exposure monitoring, because their urinary concentrations are relatively high and easier to measure compared to those of carcinogenic PAHs such as B[a]P. Pyrene and Nap did not used to be thought of as carcinogens. Nap was found to show clear evidence for carcinogenic activity by the National Toxicology Program 2000. Nap was therefore reclassified by the IARC, Deutsche Forschungsgemeinschaft (DFG), and the EPA as being possibly carcinogenic to humans since 2001. Consequently, the importance of measuring Nap metabolite levels has increased.

5.4.1 B[a]P AND 3-OH-B[a]P

The past few years have seen increased study of urinary biomarkers of PAH exposure other than 1-OH-Pyr, which may provide direct evidence of exposure to PAHs of greater toxicological relevance than pyrene. Among them, 3-hydroxybenzo[a] pyrene (3-OH-B[a]P), the main primary metabolite of B[a]P present in urine, has received particular attention (Leroyer et al., 2010). During the validation of a high-performance liquid chromatography fluorometric (HPLC-fluo.) analytical method, in the urine of 10 nonexposed children, while 1-OH-Pyr was quantifiable in the urine of all the children, 3-OH-B[a]P was detected in only 1 urine sample but was not quantifiable (limit of detection [LOD]: 0.19 nmol/L; limit of quantification [LOQ]: 1.12 nmol/L). The use of urinary 3-OH-B[a]P as a biomarker of exposure to B[a]P in the general population is in the early stages of development. In an early study,

TABLE 5.2
Urinary Excretion of 3-OH-B[a]P by Nonexposed
and Occupationally PAH-Exposed Subjects

Workplace	No. of Subjects	3-OH-B[a]P (ng/L)
Nonexposed	48	6
Coke plant workers	40	280
Road pavers	10	19
Fireproof material plant	19	5–356

a method for the detection of urinary 3-OH-B[a]P using reduction by hydrogen iodide found levels of 0.12 mg/L in the urine of exposed workers. Later, an HPLC method for the detection of 3-OH-B[a]P reported recoveries of 43% and a detection limit of 1 ng/24 h urine. In a subsequent study, a detection limit of 1 mg/L (4 nmol/L) was reported. However, urinary levels of 3-OH-B[a]P in workers of a coal tar distillation plant were found to be below the detection limit, and the method appears to be less suitable for biomonitoring. The method that improved the sensitivity and reproducibility of the HPLC was the addition of ascorbic acid to the eluent. More recently, an HPLC method with fluorescence detection using an enriching precolumn consisting of silica is modified with copper phthalocyanine as developed, which allows the simultaneous determination of 3-OH-B[a]P and 3-hydroxybenz[a]anthracene in human urine with detection limits of 6 and 8 ng/L, respectively. The chromatograms presented, however, evidenced a poor resolution of the analytes and co-elution of interfering compounds. Nevertheless, concentrations of 3–198 ng 3-OH-B[a]P/g creatinine and 15–1871 ng 3-hydroxybenz[a]anthracene/g creatinine were found for the postshift urine of workers ($n = 519$) of a fireproof material plant. Selected data on the urinary content of 3-OH-B[a]P are presented in Table 5.2.

The detection limit for 3-OH-B[a]P using HPLC with conventional fluorescence detection is too high to allow its accurate analysis in urine samples from nonexposed persons. Therefore, other more sensitive methods such as HPLC with laser-induced fluorescence (LIF) detection along with an improved sample cleanup or g-cyclodextrin-modified micellar electrokinetic chromatography with LIF are necessary for low-level (0.5–8 ng/L) measurements of 3-OH-B[a]P in urine.

5.4.2   PAH AND 1-OH-PHENANTHRENE

The use of 1-OHP, a urinary metabolite of pyrene, has been proposed as biological marker of occupational exposure to PAHs since pyrene is always present in PAH mixtures and good correlations have been found between 1-OHP and both airborne pyrene and total airborne PAHs (Campo et al., 2006). However, due to the variable composition of PAH mixtures from different sources, pyrene may not be representative of all exposure conditions. Moreover, pyrene is not carcinogenic, and the measurement of the metabolites of a carcinogenic compound could be more appropriate for biological monitoring. For these reasons, the measurement of other PAHs hydroxy derivatives has been proposed, with a particular attention to Nap and phenanthrene (Phe) for their high volatility and abundance in PAHs mixtures, and a particular attention to B[a]P for its toxicity. Urinary levels of unmetabolized Nap and Phe, and of hydroxylated Nap and Phe metabolites (1- and 2-hydroxynaphthalene and 1-, 2-, 3-, 4-, and 9-hydroxyphenanthrene), have been shown to reflect recent exposures to asphalt emissions (Sobus et al., 2009). The ranges of biomarker concentrations in urine (nmol/L) were 1-naphthol, 14–159; 2-naphthol, 9–166; 1-plus 2-naphthol, 35–269; 1-hydroxyphenanthrene (1-OH-Phe), 6–56; 2-plus 3-OH-Phe, 6–70; 4-OH-Phe, 1–6; 9-OH-Phe 1–7; the sum of phenanthrols, 15–135; and 1-OHP, 2.2–67 (Elovaara et al., 2006).

In conclusion, determination of the complete urinary profile of Phe metabolites (1-OH-Phe, 4-OH-Phe, etc.) in addition to the well-established measurement of 1-OHP appears to be more sensitive, because human exposure to Phe is in most cases higher than to pyrene. Nonsmokers as well as smokers exposed to diesel exhaust show an increase in the excretion of Phe metabolites as compared to nonexposed controls indicating an induction of CYP enzymes such as 1A1 competent for PAH metabolism (Seidel et al., 2002). OH-Phe was additionally measured, which turned out to be stronger correlated with ambient PAH assessed as Phe and less affected by confounders such as smoking and particular type of industry. Thus, OH-Phe can be considered as a suitable biomarker of occupational PAH exposure (Rihs et al., 2005). Phe 1 is metabolized by CYP to three different arene oxides 2, 3, and 4 from which the *trans*-dihydrodiols 5, 8, and 13 are formed by mEH. The isomeric phenols 6, 7, 10, 11, and 12 of Phe are formed by nonenzymatic rearrangement reactions of the arene oxides. However, the arene oxides 2, 3, and 4 also give rise to react with GSH catalyzed by GST to produce the GSH conjugates 9, 14, and 15 (Figure 5.4). The phenols and *trans*-dihydrodiols are predominantly excreted in urine as sulfates and glucuronides, although excretion of mercapturic acids derived from GSH conjugates such as 9, 14, and 15 has also been observed.

5.5 PAH BIOMARKERS IN OTHERS

Several metabolites in urine have been used to measure PAH exposure. 1-OHP is the most widely used urinary PAH metabolite and has been applied to subjects with various PAH exposures (Kang et al., 2005). The assay for the glucuronide conjugate of 1-OHP (1-OHPG) has been developed and successfully applied to population with various PAH exposures. 1-OHPG is more sensitive than 1-OHP, since 1-OHPG is three to five times more fluorescent than 1-OHP. On the other hand, the 1-OHPG assay requires more urine than the 1-OHP assay. Urinary 2-naphthol, a stable PAH metabolite, reflects more specifically ambient PAH exposure, whereas 1-OHP levels can be affected by diet and smoking. Thus, urinary 2-naphthol is suggested as a specific marker for exposure to airborne particulates, whereas urinary 1-OHP has been used as a marker for exposure to PAHs by nonspecific exposure routes. When absorption of PAH leads to a reaction in the body, which is similar for all PAH, this reaction can be used to develop a biomarker (Jongeneelen, 1997). A small proportion of the dose of PAH may react with nucleophilic sites in DNA, resulting in PAH–DNA adducts. PAH may also react with GSH and undergo further metabolism,

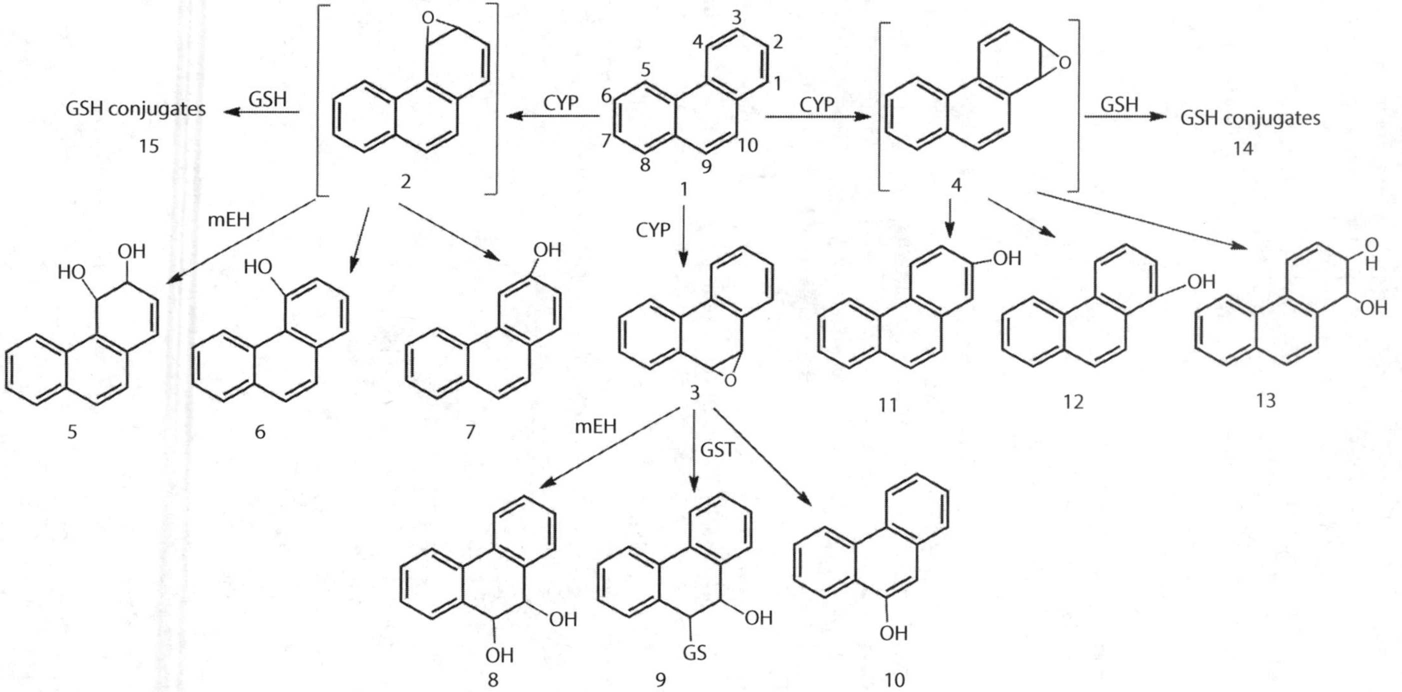

FIGURE 5.4 Partial metabolism of Phe 1 is illustrated based on excreted metabolites in urine of mammals. Phe 1 is oxidized by CYP to three isomeric arene oxides 2, 3, and 4, which as unstable intermediates are further converted to (i) three *trans*-dihydrodiols 5, 8, and 13 by mEH, to (ii) five isomeric phenols 6, 7, 10, 11, and 12 by nonenzymatic rearrangement, or to (iii) GSH conjugates 9, 14, and 15 by reaction with GSH catalyzed by GST. Phenols and dihydrodiols are predominantly excreted in urine as glucuronide conjugates, whereas the GSH conjugates are transformed into mercapturic acids before urinary excretion.

finally resulting in excreting of sulfur-containing compounds, known as thioethers. Another part of the metabolites may be excreted as mutagenic metabolites, which can be detected in the *salmonella/microsome* mutagenicity assay. Mutagenicity in urine is also a biomarker of exposure to a mixture of PAH.

5.5.1 Aromatic DNA Adducts

A large research effort has been made to use the extent of binding of PAH to DNA as a biomarker of PAH exposure. DNA of white blood cells (WBCs) is usually taken as surrogate target DNA. Agent-specific immunoassays and non-agent-specific 32P-past-labeling assays have been suggested. The 32P-past-labeling assay detects the total of aromatic DNA adducts and is more or less an indicator of exposure to the total PAH mixture. The results of DNA adduct studies in workers are not very clear; prior studies have produced conflicting results on adducts in WBC of smokers. In some studies, an enhanced DNA adduct level in highly exposed workers is not found; other studies, however, report on a positive relationship. PAH–DNA adducts in lung tissue of lung cancer patients appeared not to be correlated with adducts in lymphocytes; in fact, an inverse relationship was found. Much research remains to be done to overcome the problems encountered in measuring and interpreting DNA adducts. At the moment, the present technical complexity of the method makes it more convenient for research applications than for routine applications in occupational health practice.

5.5.2 Urinary Thioethers

The variation of the baseline excretion of thioethers is high, and tobacco smoking is a very strong interfering confounding factor. Occupational exposure to PAH in the range of 1 ng/m^3 to 10 µg/m^3 of PAH in workroom air does not lead to an increased level of thioethers. Therefore, monitoring of urinary thioethers as biomarker of exposure to mixtures of PAH lacks sensitivity. The method is not suitable for routine monitoring of workers at the presently found concentrations in the occupational environment nor for the even lower environmental exposure level.

5.5.3 Urinary Mutagenicity

Several studies showed that the urinary mutagenicity is hardly increased at the occupational exposure level of 1 ng/m^3 to 10 µg/m^3 of PAH. Analysis of urinary mutagenicity as a biological monitoring method of exposure to PAH lacks sensitivity. Moreover, smoking is a very strong interfering confounding factor, and results have a large variability. The method is suitable for routine application of monitoring of PAH neither in the occupational environment nor in the environmental setting.

5.5.4 Metabolites in Urine of the PAH Mixture

The determination of various PAHs in urine has been suggested as a biomarker, either after *reversed metabolism* to parent PAH or by determining various hydroxylated

PAH. Analytical shortcomings have been shown in the total PAH assay, especially when a *reversed metabolism* step is introduced. The *reversed metabolism* methodology has large analytical drawbacks and cannot be used repeatedly for application of monitoring of the exposure levels of PAH. The determination of urinary hydroxylated PAH in urine may be executed in separated steps for each metabolite or in one run for all urinary hydroxylated PAH. The determination of all hydroxylated metabolites in one run seems to be an informative method, but the experience is yet limited.

5.5.5 PAH DNA Adducts in Human Tissues

Occupational PAH exposure is associated with PAH–DNA adduct formation in lymphocytes, and notably the epidemiologic evidence for a link between PAH and prostate carcinogenesis comes mainly through occupational studies (Rybicki et al., 2006). Recent work with prostate tissue that has shown the presence of the metabolic enzyme activity necessary for PAH activation and the formation of DNA adducts upon exposure to PAH supports the biologic feasibility of PAH carcinogenesis. The detection and quantification of PAH–DNA adduct formation in the PAH prostate carcinogenesis pathway is key, since it substantiates a well-established mechanism of environmental genotoxicity through DNA damage that can lead to carcinogenic mutations. Demonstration of PAH–DNA adducts in prostate cancer cases supports the notion that PAH carcinogenesis in the prostate is initiated through DNA adduct formation, but it fails to establish a temporal relationship between adduct formation and cancer onset.

B[a]P is predominantly metabolized to the (+)-enantiomer of anti-B[a]P 7,8-diol-9,10-epoxide (BPDE) (Mensing et al., 2005). BPDE is known to show the highest tumor-inducing activity of all biological reactive B[a]P metabolites and was found covalently bound to dG and dA residues in DNA of a variety of cells and organs of mammalian species that have been exposed to B[a]P. Unless removed by DNA repair processes, the resulting BPDE adducts may give rise to mutations during DNA replication. Determination of BPDE–DNA adduct levels in human tissue is of importance in risk assessment of individuals exposed to PAH. BPDE–DNA adducts can be detected by various techniques including 32P-postlabeling, immunoassay, and special mass spectrometric methods. Another more indirect method is HPLC with fluorescence detection that is applied to determine B[a]P-tetrol as the hydrolysis product of BPDE–DNA adducts. The latter method has been used in recent years in several field studies to determine BPDE–DNA adducts as a biomarker of PAH exposure. In most cases, the formation of PAH adducts in humans was studied in DNA isolated from lymphocytes and WBCs. An exposure-related correlation between DNA adducts in the lung or larynx and in peripheral blood cells has been observed, and the potential use of DNA adducts isolated from WBC as surrogates for DNA adducts from lung tissue has been previously demonstrated. Although there is already a large number of studies concerning measurements of B[a]P–DNA adduct levels in WBC, only a few studies measured ambient B[a]P concentration at the workplace to examine the relation between B[a]P and BPDE–DNA adducts. The individual measurements of BPDE–DNA adducts by HPLC-FLD and of B[a]P exposure can be useful for biological monitoring of exposed workers from different occupational settings. At present, the relative contribution of dermal and respiratory absorption of B[a]P to the formation of BPDE–DNA adducts at the workplace is unknown. However, the results mentioned earlier support the idea that inhalation—at least in refractory workers—is only of minor importance for the uptake of B[a]P at the workplace, because BPDE–DNA adduct concentrations were independent on B[a]P levels in the air and even smoking did not increase BPDE–DNA adduct rates. Therefore, BPDE–DNA adducts in refractory workers are more likely caused by environmental exposure (e.g., diet) than by exposure at the workplace. The contribution of other routes of uptake of PAH or B[a]P, for example, skin absorption and diet as well as the role of genetic polymorphisms and DNA repair should be investigated in more detail in future human studies. In addition, more data are needed on different occupational settings (such as refractory workers, coke-oven workers, converter workers, workers producing fireproofed material), because they have different working tasks and production procedures, which in turn result in different exposure and risk.

5.5.6 PAH PROTEIN ADDUCTS (HEMOGLOBIN AND SERUM ALBUMIN ADDUCTS)

Mutagenic and carcinogenic substances bind to macromolecules, especially to proteins. That reactive electrophilic intermediates of mutagenic substances bind to nucleophilic sites of proteins is the underlying principle. The preferred sites are the sulfhydryl group of cysteine and nitrogen of histidine and N-terminal valine because the pKa values are in the range of the pH of blood (pH in the narrow range of 7.35 to 7.45). Hemoglobin (Hb) and serum albumin (SA) are the preferred monitor molecules because they are accessible in large amounts. They are chemically stable and are not prone to repair mechanisms like DNA adducts. Because of the long life span of Hb (120 days) and the long half-life of SA (20 days), these adducts cumulate in the human body, making them a very sensitive parameter for human biomonitoring.

Protein adducts as surrogates for DNA adducts are considered to be potentially valuable markers of internal dose (Boysen and Hecht, 2003). Some research showed that PAH forms a stable ester with one or more carboxylic acid groups in human Hb. Similar studies showed that diol epoxides of PAH bind predominantly to the histidine[146] or lysine[137] in subdomain IB of SA. It has been demonstrated that (7S,8R)-dihydroxy-(9R,10S)-epoxy-7,8,9,10-tetrahydrobenzo[a]pyrene (with opposite absolute configuration to BPDE) forms adducts with histidine[146], while BPDE forms relatively unstable ester adducts with aspartate[187] or glutamate[188] in human SA. HPLC–FD and gas chromatography–mass spectrometry (GC–MS) have been used in a number of studies to quantify BPDE adducts with Hb and SA. Some results appear to be influenced by environmental exposures. For example, nonsmoking newspaper vendors working in high traffic density areas had approximately threefold higher BPDE–Hb adduct levels, and the frequency of detected adducts was twice (60% vs. 29%) as high, when compared with nonsmoking vendors working in low traffic density areas. No effect of the newsstand location was observed for vendors who smoked. Levels of adducts in smokers and nonsmokers were not significantly different in that study. It was found that significantly higher levels of BPDE–Hb adducts in smokers than nonsmokers, and higher levels of BPDE–SA adducts in smokers than in nonsmokers. Levels of BPDE–Hb adducts were also higher in smokers than in nonsmokers, but the difference was not significant.

Hb adducts are the preferably monitored molecules because cumulation is still greater than that of SA. Moreover, reactive intermediates have to cross a cell membrane showing this way that they are sufficiently stable to reach the DNA in the critical organ. Because of these reasons, the scientific community takes Hb adducts as surrogates of DNA adducts, which are thought to be the initial step of carcinogenicity. Hb adduct levels in blood enable the estimation of internal exposure as well as biochemical effects. Hb adducts seem to be better estimates for cancer risk than measuring the genotoxic substances or their metabolites in human body fluids. For the genotoxic substances like PAH and tobacco-specific nitrosamines, the determination of Hb adducts has been described. In these cases, the protein chain has to be cleaved to yield these adducts. Such procedures lead to multiple products, which have to be separated in tedious chromatographic procedures. In spite of all obstacles, however, protein adduct monitoring should be pushed to get a better means of risk estimation of carcinogenic substances.

5.5.7 BIOMONITORING OF PAH IN HAIR

Although the biomonitoring of human exposure to PAH is generally achieved through the determination of urinary metabolites or the detection of PAH–DNA adducts in WBCs, hair has also been investigated for the detection of parent molecules as well as for their monohydroxy metabolites (Appenzeller and Tsatsakis, 2012). Hair analysis is particularly suitable for the biomonitoring of chronic exposures (Appenzeller et al., 2012). This approach, providing information on the average level of exposure to PAHs, is not affected by intra- and interday extreme variations in the exposure that complicate the interpretation of results obtained from urine analysis. Moreover, unlike urine, which only provides information on very recent exposure (hours), detection windows accessible by hair analysis may reach several months, which is more relevant in the context of long-term exposure-associated diseases. The highest concentrations were observed for Nap (range: 136–1,370 pg/mg), Phe (28.9–255 pg/mg), and fluorene (range: 6.7–42.9 pg/mg) and their respective metabolites (3,499 and 14,198 pg/mg for 1- and 2-naphthols, respectively; 1,998 pg/mg for 9-OH-Phe; and 679 pg/mg for 9-OH-fluorene), and significantly higher concentrations were detected in smokers' hair for anthracene, chrysene, and benzo[k]fluoranthene (Appenzeller and Tsatsakis, 2012). Table 5.3 shows the PAHs detected in human hair, and B(b þ k)F, B[a]P, IcdP, DBA, and BghiP were not detected in hair samples (Wang et al., 2013). Nap (29.5–294, median 152 ng/g) and Phe (51.8–431, median 136 ng/g) were the most dominant congeners detected in human hair samples. This was consistent with previous study that Phe and Nap were dominant PAHs detected in human adipose tissue of HK. Limited research is available on PAHs in human hair. Pyr and Flu were also important congeners detected in hair, with a median of 60.9 and 77.3 ng/g, respectively.

The concentrations of hydroxy-PAHs detected in hair specimens were highly varying according to both the subject and the molecule considered

TABLE 5.3

PAHs Concentrations in Human Hair Samples (ng/g)

Congener	Nap	Acy	Ace	Fl	Phe	Ant	Flu	Pyr	BaA	Chr
Min	29.5	0.16	1.00	ND	51.8	ND	5.08	ND	1.69	1.64
Max	294	5.96	3.84	35.6	431	13.7	265	325	22.0	44.5
Mean	153	1.86	2.28	10.8	174	3.86	82.8	72.7	5.77	12.0
SD	77.2	1.89	0.91	9.56	121	3.22	59.9	69.0	4.92	10.4
Median	152	0.82	2.30	7.18	136	2.74	77.3	60.9	3.79	7.07

(Appenzeller et al., 2012). In 31 samples, no OH-PAH was detected. For the remaining samples, the most common situation was samples containing 1 and 2 different OH-PAHs: respectively, 42 and 26 samples. Only 1 sample contained 4 different OH-PAHs, 2 samples tested positive for 5 OH-PAHs, and 1 sample contained 10 different OH-PAHs. Only five samples had an OH-PAHs concentration above 1 nmol/g. The mean concentration of each metabolite ranged from 60 pmol/g for 2-OH-Flu up to 4627 pmol/g for 1-OH-Naph. The most abundant metabolites detected were naphthols (1-OH-Naph and 2-OH-Naph). 2-OH-Naph was by far the most frequently detected molecule: 42 out of 61 smokers (69%) and 22 out of 44 nonsmokers (50%). The nine remaining PAH metabolites investigated here were all detected in less than 10% of the specimens tested, with no significant difference between the groups.

5.6 DISCUSSION

PAH-exposed people undergo inhalation exposure, dermal exposure, and oral exposure to a certain extent. Any exposure to genotoxic carcinogens as represented by PAH mixtures is assumed to pose a certain excess risk of cancer.

The concentration of 1-OH-PAH in urine can be used as a biological indicator of recent doses of PAH. The presence of 1-OH-PAH in the urine represents the sum of resorption in the airways and resorption in the gastrointestinal tract due to the swallowing of coarse particles and via the dermal route. The urine test is specific for exposure to PAH. At present, 1-OH-PAH in urine is often used as a biomarker of the dosage of PAH. PAH mixtures contain many other genotoxic PAH, and the relative proportion of individual PAH in the PAH mixtures from different exposed level may vary. This factor is an important confounder of the dose–response relationship of the long-term average of urinary 1-OH-PAH versus excess cancer in exposed people.

Only when results from epidemiological studies of persons with long-term exposure to PAH, with urinary 1-OH-PAH as a dose indicator, will a more reliable estimate of the average 1-OH-PAH level versus excess tumor incidence be made. Only with these data will an accurate and precise estimate of the health-based limit value of urinary 1-OH-PAH be made. A guideline based on epidemiological data for exposed persons will be superior to the three-level benchmark guideline presented here, and such a standard will supersede the current proposal when such studies are available.

5.7 CONCLUSION

Many individual PAHs are genotoxic carcinogens. Humans are exposed to PAHs from tobacco and tobacco smoke. PAH metabolites in human urine and other tissues can be used as biomarkers of internal dose to assess recent exposure to PAHs. PAH metabolites that have been detected in human urine and tissues include 1-OHP, 1-hydroxypyrene-O-glucuronide (1-OHP-gluc), 3-hydroxybenzo[a]pyrene, 7,8,9,10-tetrahydroxy-7,8,9,10-tetrahydrobenzo[a]pyrene, and a number of other hydroxylated PAHs. The most widely used of these is l-OHP-gluc, the major form of 1-OHP in human urine and tissues, by virtue of its relatively high concentration and prevalence in body and its ease of measurement. This metabolite of pyrene can be measured as 1-OHP after deconjugation of the glucuronide with P-glucuronidase or directly as 1-OHP-gluc without deconjugation. Elevated levels of 1-OHP or 1-OHP-gluc have been demonstrated in smokers (vs. nonsmokers). Such biomarkers provide an alternative approach to improve assessment of PAH exposure in experimental and epidemiological studies.

REFERENCES

Angerer, J., Ewers, U., Wilhelm, M. Human biomonitoring: State of the art. *International Journal of Hygiene and Environmental Health*. 2007, 210, 201–228.

Appenzeller, B. M. et al. Simultaneous determination of nicotine and PAH metabolites in human hair specimen: A potential methodology to assess tobacco smoke contribution in PAH exposure. *Toxicology Letters*. 2012, 210, 211–219.

Appenzeller, B. M., Tsatsakis, A. M. Hair analysis for biomonitoring of environmental and occupational exposure to organic pollutants: State of the art, critical review and future needs. *Toxicology Letters*. 2012, 210, 119–140.

Bartsch, H. Studies on biomarkers in cancer etiology and prevention a summary and challenge of 20 years of interdisciplinary research. *Mutation Research*. 2000, 462, 255–279.

Binkova, B. et al. PAH-DNA adducts in environmentally exposed population in relation to metabolic and DNA repair gene polymorphisms. *Mutation Research*. 2007, 620, 49–61.

Boysen, G., Hecht, S. S. Analysis of DNA and protein adducts of benzo[a]pyrene in human tissues using structure-specific methods: A review. *Mutation Research*. 2003, 543, 17–30.

Campo, L. et al. Biological monitoring of exposure to polycyclic aromatic hydrocarbons by determination of unmetabolized compounds in urine. *Toxicology Letters*. 2006, 162, 132–138.

Chung, M. K. et al. A sandwich enzyme-linked immunosorbent assay for adducts of polycyclic aromatic hydrocarbons with human serum albumin. *Analytical Biochemistry*. 2010, 400, 123–129.

Elovaara, E. et al. Assessment of soil remediation workers' exposure to polycyclic aromatic hydrocarbons (PAH): Biomonitoring of naphthols, phenanthrols, and 1-hydroxypyrene in urine. *Toxicology Letters*. 2006, 162, 158–163.

Gschaidmeier, H. et al. Formation of mono- and diglucuronides and other glycosides of benzo(a) pyrene-3,6-quinol by V79 cell-expressed human phenol UDP-glucuronosyltransferases of the UGT1 gene complex. *Biochemical Pharmacology*. 1995, 49, 1601–1606.

Hagedorn, H.-W. et al. Urinary excretion of phenolic polycyclic aromatic hydrocarbons (OH-PAH) in nonsmokers and in smokers of cigarettes with different ISO tar yields. *Journal of Analytical Toxicology*. 2009, 33, 301–309.

Hansen, A. M. et al. Urinary 1-hydroxypyrene (1-HP) in environmental and occupational studies—A review. *International Journal of Hygiene and Environmental Health*. 2008, 211, 471–503.

Hecht, S. S., Yuan, J.-M., Hatsukami, D. Human placental glutathione S-transferase activity and polycyclic aromatic regulation and cancer prevention. *Chemical Research in Toxicology*. 2010, 23, 1001–1008.

Ichiba, M. et al. Decreasing urinary PAH metabolites and 7-methylguanine after smoking cessation. *International Archives of Occupational and Environmental Health*. 2006, 79, 545–549.

Islam, G. A. et al. HPLC analysis of benzo[a]pyrene-albumin adducts in benzo[a]pyrene exposed rats. Detection of cis-tetrols arising from hydrolysis of adducts of anti- and syn-BPDE-III with proteins. *Chemico-Biological Interactions*. 1999, 123, 133–148.

Jacob, J., Seidel, A. Biomonitoring of polycyclic aromatic hydrocarbons in human urine. *Journal of Chromatography B*. 2002, 778, 31–47.

Jain, K. K. *The Handbook of Biomarkers*. 2010, Springer Science, New York.

Jongeneelen, F. J. Methods for routine biological monitoring of carcinogenic PAH-mixture. *The Science of the Total Environment*. 1997, 199, 141–149.

Kang, D. et al. Design issues in cross-sectional biomarkers studies: Urinary biomarkers of PAH exposure and oxidative stress. *Mutation Research*. 2005, 592, 138–146.

Leroyer, A. et al. 1-Hydroxypyrene and 3-hydroxybenzo[a]pyrene as biomarkers of exposure to PAH in various environmental exposure situations. *Science of the Total Environment*. 2010, 408, 1166–1173.

Mensing, T. et al. DNA adduct formation of benzo[a]pyrene in white blood cells of workers exposed to polycyclic aromatic hydrocarbons. *International Journal of Hygiene and Environmental Health*. 2005, 208, 173–178.

Nahrgang, J. et al. PAH biomarker responses in polar cod (*Boreogadus saida*) exposed to benzo(a)pyrene. *Aquatic Toxicology*. 2009, 94, 309–319.

Obolenskaya, M. Y. et al. Human placental glutathione S-transferase activity and polycyclic aromatic hydrocarbon DNA adducts as biomarkers for environmental oxidative stress in placentas from pregnant women living in radioactivity- and chemically-polluted regions. *Toxicology Letters*. 2010, 196, 80–86.

Rihs, H. P. et al. Occupational exposure to polycyclic aromatic hydrocarbons in German industries: Association between exogenous exposure and urinary metabolites and its modulation by enzyme polymorphisms. *Toxicology Letters*. 2005, 157, 241–255.

Rybicki, B. A. et al. Polycyclic aromatic hydrocarbon-DNA adduct formation in prostate carcinogenesis. *Cancer Letters*. 2006, 239, 157–167.

Santella, R. M. et al. Polycyclic aromatic hydrocarbon-DNA and protein adducts in coal tar treated patients and controls and their relationship to glutathione S-transferase genotype. *Mutation Research*. 1995, 334, 117–124.

Seidel, A. et al. Biomonitoring of polycyclic aromatic compounds in the urine of mining workers occupationally exposed to diesel exhaust. *International Journal of Hygiene and Environmental Health*. 2002, 204, 333–338.

Siddens, L. K. et al. Polycyclic aromatic hydrocarbons as skin carcinogens: Comparison of benzo[a]pyrene, dibenzo[def,p]chrysene and three environmental mixtures in the FVB/N mouse. *Toxicology and Applied Pharmacology*. 2012, 264, 377–386.

Singh, R. et al. The relationship between biomarkers of oxidative DNA damage, polycyclic aromatic hydrocarbon DNA adducts, antioxidant status and genetic susceptibility following exposure to environmental air pollution in humans. *Mutation Research*. 2007, 620, 83–92.

Snyder, M. J. Cytochrome P450 enzymes in aquatic invertebrates recent advances and future directions. *Aquatic Toxicology*. 2000, 48, 529–547.

Sobus, J. R. et al. Investigation of PAH biomarkers in the urine of workers exposed to hot asphalt. *The Annals of Occupational Hygiene*. 2009, 53, 551–560.

Stoilov, I. Cytochrome P450s: Coupling development and environment. *Trends in Genetics*. 2011, 17, 629–632.

Strickland, P., Kang, D. Urinary 1-hydroxypyrene and other PAH metabolites as biomarkers of exposure to environmental PAH in air particulate matter. *Toxicology Letters*. 1999, 108, 191–199.

Sundberg, K. et al. Detoxication of carcinogenic fjord-region diol epoxides of polycyclic aromatic hydrocarbons by glutathione transferase P1-1 variants and glutathione. *FEBS Letters*. 1998, 438, 206–210.

Taioli, E. et al. Biomarkers of exposure to carcinogenic PAHs and their relationship with environmental factors. *Mutation Research*. 2007, 620, 16–21.

Thier, R. et al. Markers of genetic susceptibility in human environmental hygiene and toxicology: The role of selected CYP, NAT and GST genes. *International Journal of Hygiene and Environmental Health*. 2003, 206, 149–171.

Wang, W. et al. Risk assessment of non-dietary exposure to polycyclic aromatic hydrocarbons (PAHs) via house PM2.5, TSP and dust and the implications from human hair. *Atmospheric Environment*. 2013, 73, 204–213.

Xue, W., Warshawsky, D. Metabolic activation of polycyclic and heterocyclic aromatic hydrocarbons and DNA damage: A review. *Toxicology and Applied Pharmacology*. 2005, 206, 73–93.

Zanger, U. M., Schwab, M. Cytochrome P450 enzymes in drug metabolism: Regulation of gene expression, enzyme activities, and impact of genetic variation. *Pharmacology and Therapeutics*. 2013, 138, 103–141.

6 Biomarkers of Volatile Organic Compounds

6.1 INTRODUCTION

The combustion of a cigarette creates an aerosol containing numerous chemical compounds (Stedman, 1968). Several specific classes of hazardous compounds pose a great concern among the thousands of reported chemicals generated during smoking (Hoffmann et al., 1997; Matsushima et al., 1979; Rustemeier et al., 2002). According to their carcinogenicity on humans, individual chemicals and chemical mixtures have been classified as known, probable, and possibly carcinogenic by the International Agency for Research on Cancer (IARC). So far, 69 smoke constituents have been classified as carcinogens by the IARC, and cigarette smoke as a mixture has been classified as a human carcinogen (Pánková, 1986); many of the individual volatile organic compounds (VOCs) present in whole smoke, such as benzene, ethylbenzene, and styrene, are known or potential human carcinogens. Although VOCs comprise only a small fraction (by weight) of mainstream cigarette smoke, smoking is a primary exposure source for many toxic volatile compounds, and this fraction has been proposed as the most hazardous fraction of mainstream smoke (Fowles and Dybing, 2003).

6.2 TOXICOLOGICAL EVALUATION

Acrolein, 1,3-butadiene, and benzene are volatile emissions in the priority list of the Framework Convention on Tobacco Control (FCTC) (2008). Levels of acrylonitrile and crotonaldehyde in mainstream smoke should also be the measurements according to Health Canada's regulations (FCTC, 2010; WHO, 2005). Those toxic constituents are cytotoxic, mutagenic, and possibly carcinogenic to humans. The IARC has classified benzene as a Group 1 carcinogen (IARC, 1982), 1,3-butadiene as Group 2A carcinogen (IARC, 1992), acrylonitrile as Group 2B (IARC, 1999), and acrolein and crotonaldehyde as Group 3 carcinogen (IARC, 1995).

6.3 METABOLIC PATHWAY IN THE BODY

Acrolein, 1,3-butadiene, benzene, acrylonitrile, and crotonaldehyde can react with glutathione (GSH) and finally convert to mercapturic acid metabolites in urine. Urinary mercapturic acid metabolites have emerged as highly practical biomarkers for determining uptake of specific carcinogens and toxicants in tobacco smoke. The relationship of mercapturic acids in urine can standardize the metabolite way in the body and further help us to know the interaction of tobacco toxic constituents.

N-acetyl-*S*-(2-cyanoethyl)-cysteine (CEMA) is one of the major urinary mercapturic acid metabolites after exposure to acrylonitrile (Sumner et al., 1997). In previous study, urinary excretion of CEMA was found to be higher in smokers compared with nonsmokers and strongly correlated with the smoking dosage (Hou et al., 2012; Scherer et al., 2007a). 3-Hydroxypropylmercapturic acid (3-HPMA) (Parent et al., 1998), has been identified as a main urinary metabolite of acrolein in rats (IARC, 1992). It increases after smoking or experimentally high exposure to environmental tobacco smoke (ETS) (Hou et al., 2012). The major 1,3-butadiene-derived mercapturic acids are monohydroxybutenyl-mercapturic acid (MHBMA, also termed MII) and *N*-acetyl-S-(3,4-dihydroxybutyl)cysteine (DHBMA) (Boogaard et al., 2001; van Sittert et al., 2000). *S*-phenylmercapturic acid (SPMA) is a specific urinary metabolite of benzene and is recommended for the biological monitoring of benzene by the BEI committee (ACGIH, 2005). 3-Hydroxy-1-methylpropylmercapturic acid (HMPMA, major metabolite) and 2-carboxy-1-methylethylmercapturic acid (CMEMA, minor metabolite) are identified in the urine of rats subcutaneously injected with crotonaldehyde. Scherer et al. (2007b) reported that smoking cessation or switching from smoking conventional cigarettes to experimental cigarettes with lower crotonaldehyde delivery led to significant reductions of urinary HMPMA excretion, but not CMEMA excretion.

6.3.1 ALDEHYDES

Aldehydes are highly reactive molecules, which are intermediary or final products of metabolism involved in biochemical, physiological, and pharmacological processes. Endogenous aldehydes may be derived from the metabolism of amino acids, carbohydrates, lipids, biogenic amines, vitamins, and steroids. Exogenous aldehydes mainly generate from the biotransformation of a large number of environmental agents and drugs (Voulgaridou et al., 2011). Aldehydes interact with phospholipids, proteins, and DNA, and if not well repaired, it may lead to physiological and homeostatic to cytotoxic, mutagenic, or carcinogenic (Vasiliou et al., 2000, 2004).

6.3.2 FORMALDEHYDE

Formaldehyde has been classified as a human and animal carcinogen (cat.1) by the IARC (Humans, 2006). The most important sources of human exposure to formaldehyde are cigarette smoke, particle board or plywood furniture containing formaldehyde-based resins, water-based paints, fabrics, household cleaning agents, disinfectants, pesticide formulations, and other building materials (Sarigiannis et al., 2011). At the same time, formaldehyde is endogenously produced from glycine, methionine, serine, and choline as well as being generated from the metabolism of foods, drugs, and chemicals by demethylation. The endogenous concentration of formaldehyde in the blood of human subjects is about 0.1 mM/L (Heck et al., 1985). Previous research of test systems has demonstrated that formaldehyde is genotoxic, causing mutations in multiple genes (Li et al., 2010; Speit et al., 2008, 2010). In vivo, formaldehyde can react with proteins and DNA to form corresponding protein adducts (Lu et al., 2008; Metz et al., 2004), DNA adducts (McGhee and Von Hippel, 1975),

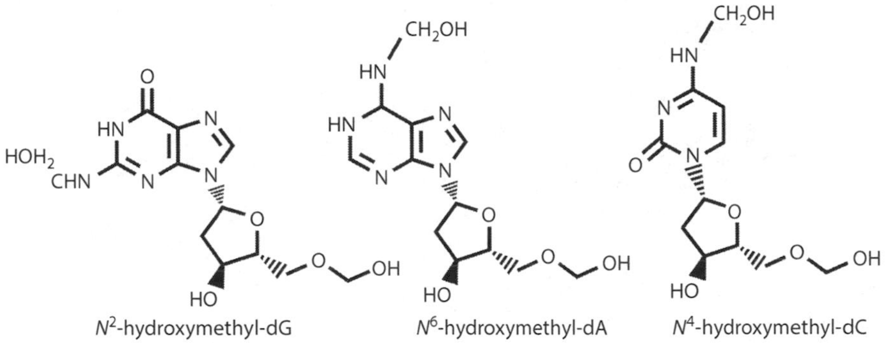

FIGURE 6.1   The structure of formaldehyde-derived DNA adducts.

and DNA–protein cross-links (Heck and Casanova, 1987; Lu et al., 2009; Merk and Speit, 1999; Quievryn and Zhitkovich, 2000). DNA adducts play an important role in mutagenesis and carcinogenesis. The structures of aldehyde-induced DNA adducts in vitro have been known for decades (McGhee and Von Hippel, 1975). Formaldehyde can typically result in N^6-hydroxymethyl-deoxyadenosine (N^6-hydroxymethyl-dA), N^2-hydroxymethyl-deoxyguanosine (N^2-hydroxymethyl-dG), and N^4-hydroxymethyl-deoxycytosine (N^4-hydroxymethyl-dC) in vitro (Figure 6.1) (Zhong and Hee, 2004; Zhong and Que Hee, 2004). Hydroxymethyl DNA adducts derived from formaldehyde are not stable and need to be reduced to their methyl forms for robust quantitation. Lu et al. (2012) used isotope labeled compounds coupled with highly sensitive mass spectrometry to differentiate between endogenous and exogenous hydroxymethyl and methyl DNA adducts. Their results demonstrated that N^2-hydroxymethyl-dG was the main DNA adduct derived from formaldehyde exposure in cells. At the same time, they demonstrated that methyl adducts induced by alkylating agents were at N^2-dG and N^6-dA positions. Wang et al. (2009) used liquid chromatography–electrospray ionization–tandem mass spectrometry (LC–ESI–MS/MS) to quantify the formaldehyde–DNA adduct N^6-hydroxymethyldeoxyadenosine (N^6-HOMe-dAdo) in leukocyte DNA samples from 32 smokers of 10 cigarettes/day and 30 nonsmokers. Clear peaks coeluting with the internal standard in two different systems were seen in samples from smokers but rarely in nonsmokers. N^6-HOMe-dAdo was detected in 29 of 32 smoker samples (mean F SD, 179 F 205 fmol/Mmol dAdo). In contrast, it was detected in only 7 of 30 nonsmoker samples (15.5 F 33.8 fmol/Mmol dAdo; $P < 0.001$). Their study results showed remarkable differences between smokers and nonsmokers in levels of a leukocyte formaldehyde–DNA adduct, suggesting a potentially important and previously unrecognized role for formaldehyde as a cause of cancer induced by cigarette smoking.

6.3.3   ACETALDEHYDE

Acetaldehyde, which is found widely in the human environment, is genotoxic and carcinogenic. It causes mutations, sister chromatid exchanges, micronuclei, and aneuploidy in cultured mammalian cells and gene mutations in bacteria (Cancer, 1985).

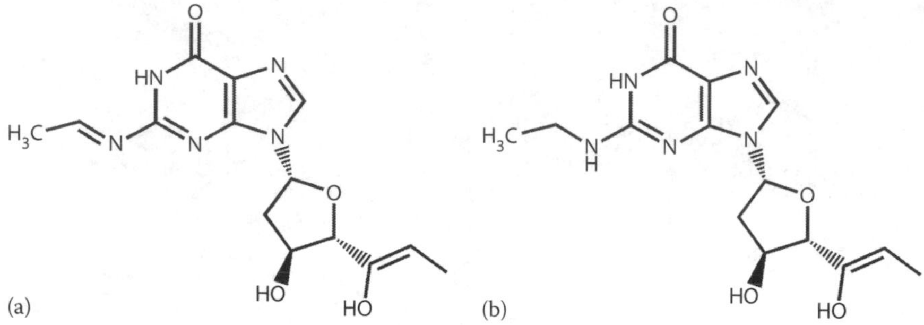

FIGURE 6.2 The structure of N^2-ethylidene-deoxyguanosine (a) and N^2-ethyl-deoxyguanosine (b).

Inhalation of acetaldehyde produces adenocarcinoma and squamous cell carcinoma of the nasal mucosa in rats and laryngeal carcinoma in hamsters (Cancer, 1999). Acetaldehyde is "reasonably anticipated to be a human carcinogen" by the U.S. Department of Health and Human Services and it has been classified as "possibly carcinogenic in humans" by the IARC (Cancer, 1999). The major DNA adduct of acetaldehyde reacted with DNA is N^2-ethylidene-dGuo (Figure 6.2); several minor adducts as well as an interstrand cross-link have also been identified (Wang et al., 2000). NaBH$_3$CN treatment of DNA can convert adduct N^2-ethylidene-dGuo derived from acetaldehyde to N^2-ethyl-dGuo. Chen et al. (2007) developed and validated a LC-MS/MS method for the measurement of N^2-ethylidene-dGuo in human leukocyte DNA. The mean concentrations of N^2-ethyl-dGuo in leukocytes DNA obtained from the smokers were 1310 ± 1720 and 1120 ± 1140 fmol/μmol dGuo at the two baseline points and 705 ± 438 fmol/μmol dGuo after 4 weeks of cessation. The concentrations of N^2-ethyl-dGuo reduced markedly by 28% after quitting smoking ($P = 0.02$). These results indicated that cigarette smoking had an important influence on the levels of N^2-ethyl-dGuo in leukocytes DNA. Zhang et al. (2006) developed a sensitive and specific LC–ESI–MS/MS method to detect acetaldehyde-derived 1,N^2-propanodeoxyguanosine adducts in DNA from human liver, lungs, and blood. The Cro-dGuo adducts were detected more frequently in human lung DNA than in liver DNA but were not detected in DNA from blood. The results of this study provide quantified data on Cro-dGuo adducts in human tissues. The higher frequency of Cro-dGuo in lung DNA than in the other tissues investigated is potentially important and deserves further study.

6.3.4 ACROLEIN

Acrolein is an α,β-unsaturated aldehyde that can originate from endogenous sources (carbohydrate consumption, and LPO) or exogenous exposure to tobacco smoking and automobile exhaust (Esterbauer et al., 1991). The general population is exposed to acrolein via smoking, secondhand smoke, and exposure to wood and plastic smoke. Endogenous acrolein is produced by peroxidation of amino acids (methionine and threonine), polyamines, and lipids and also by the metabolism of cyclophosphamide,

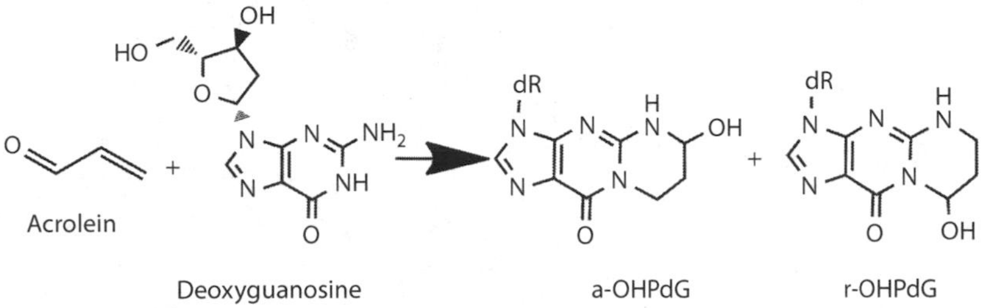

FIGURE 6.3 Structures of DNA adducts induced by acrolein. (From Voulgaridou, G.-P. et al., *Mut. Res. Fund. Mol. Mech. Mutagen.*, 711, 13, 2011.)

a drug administered for the treatment of cancer (Singh et al., 2010; Stevens and Maier, 2008). Which has high reactivity and can react with cysteine, lysine, and histidine residues of proteins by blocking their sulfhydryl groups, and resulting in toxicity, blockage of intermediary metabolism, and cell proliferation transitions (Kaminskas et al., 2004; Uchida, 1999). At low exposure levels, acrolein vapor may cause eye, nasal, and respiratory tract irritations. There is a dose–effect relationship between the degree of severity and the exposure dosage. Acrolein leads to respiratory, gastrointestinal, and ocular irritations by decreasing the release of peptides in the nerve terminals innervating these systems. Exposures of 22 and 249 ppm acrolein for 10 min lead to a significant decrease in substance P (a short-chain polypeptide that functions as a neurotransmitter or neuromodulator) (Faroon et al., 2008).

Acrolein can preferentially react with guanine nucleotides and specifically in 5-CpG-3 regions to form DNA adducts, this may be related to the methylation of the cytosine at site (Feng et al., 2006). Common mutations caused by acrolein mainly lead to G → C (Stevens and Maier, 2008) and GC → TA transversions and GC → AT transitions (Feng et al., 2006). Acr-dG DNA adducts (Figure 6.3) have been detected in animal and human tissues as common DNA lesions (Nath and Chung, 1994). Yin et al. (2013) developed a sensitive and accurate method for simultaneous quantification of α-Acr-dG (two stereoisomers), γ-Acr-dG, α-Acr-dC, and α-Acr-dA in human cells using stable isotope dilution UHPLC-MS/MS, which only requires 3 μg of genomic DNA for each analysis. The limits of detection (S/N = 3) are estimated to be about 40–80 amol.

6.3.4.1 3-HPMA

The main pathway for elimination of acrolein is conjugation with GSH in the liver, followed by enzymatic cleavage of the c-glutamic acid and glycine residues, respectively, in the liver and in the kidney (Jösch et al., 2003; Lieberman et al., 1995), and N-acetylation of the resultant cysteine conjugate to form *S*-(3-oxopropyl)-*N*-acetylcysteine (OPMA) in the kidney. Reduction of this aldehyde yields *S*-(3-hydroxy-propyl)-*N*-acetylcysteine (3-HPMA), the main metabolite of acrolein found in urine (Kaye, 1973).

3-HPMA is the primary metabolite of acrolein in urine. Oral administration of acrolein to rats leads to 79% excretion as 3-HPMA in the 0–24-h urine (Sanduja et al., 1989). In another study, the sum of 3-HPMA and CEMA excreted in the

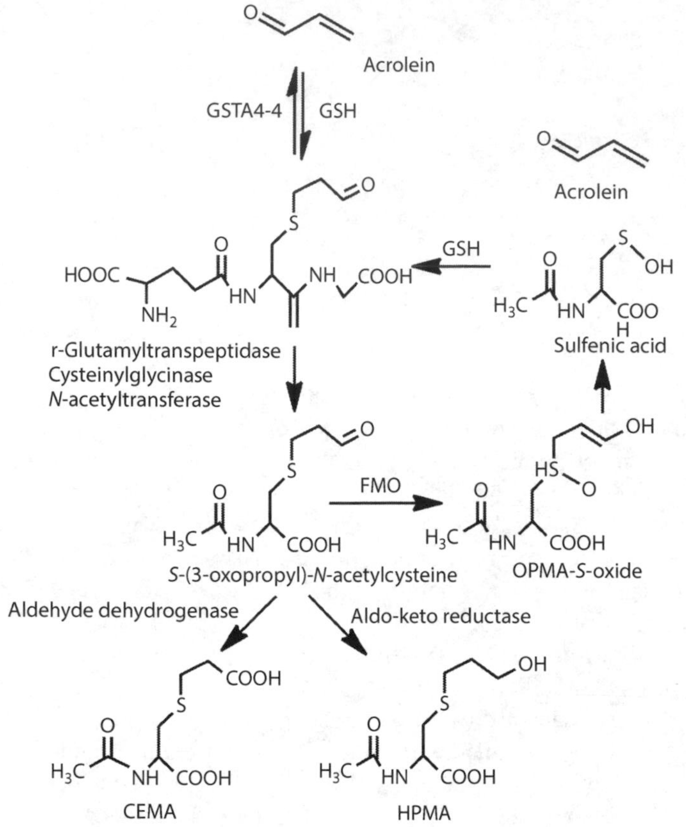

FIGURE 6.4 The metabolic pathways of acrolein in vivo. (From Stevens, J.F. and Maier, C.S., *Mol. Nutr. Food Res.*, 52, 7, 2008.)

urine during the first 24 h after intraperitoneal administration of acrolein to rats (0.5–2.0 mg/kg) accounted for 29.2% ± 6.5% of the dose (Figure 6.4). CEMA was estimated to represent less than 10% of the total amount of mercapturic acids recovered in the urine (Linhart et al., 1996). In the same study, 3-HPMA and CEMA were also found in the urine of rats exposed to acrolein by inhalation, but at lower recovery rates (Linhart et al., 1996). Acrylic acid derived from acrolein was found in the liver tissue preparations from rats, but it did not detect in the lung tissue preparations from rats, which indicated that the metabolism of acrolein in the lung and in the liver was different (Patel et al., 1980).

In humans, the mean levels of 3-HPMA in first-void morning urine are about 4.0 nmol/mg creatinine for smokers and 0.7 nmol/mg creatinine for nonsmokers (Carmella et al., 2007). The average concentration of 3-HPMA in the urine of the smoker subjects was 1095 ng/mL urine, which translates into HPMA excretion of about 1.7–2 mg/day (Carmella et al., 2007) (= 7.7–9 µmol). For comparison, the exposure to acrolein from one cigarette is about 60 µg or 1.1 µmol.

6.3.5 1,3-BUTADIENE

1,3-Butadiene (CAS No. 106-99-0) is a colorless, noncorrosive gas that forms explosive peroxides when exposed to air. It has been classified as a known human carcinogen by the Deutsche Forschungsgemeinschaft (DFG) (Group 1) (DFG, 2007) and as a probable human carcinogen (Group 2A) by the IARC. Occupational exposure to 1,3-butadiene has been linked with leukemia and lymph hematopoietic cancers. For the general population, the main exposure pathway to 1,3-butadiene is engine exhausts (Tang et al., 2007) and cigarette smoke (Brunnemann et al., 1990). Cigarette smoke was reported to contain approximately 30–40 µg 1,3-butadiene per cigarette (Brunnemann et al., 1990).

6.3.5.1 Monohydroxybutenyl-Mercapturic Acid, Dihydroxybutenyl-Mercapturic Acid, and Trihydroxybutyl Mercapturic Acid

1,3-Butadiene metabolites are excreted in urine as mercapturic acids. GSH conjugates of EB are processed via the mercapturic acid pathway to yield 2-(N-acetyl-L-cystein-S-yl)-1-hydroxybut-3-ene and 1-(N-acetyl-L-cystein-S-yl)-2-hydroxybut-3-ene, collectively called MHBMA (Figure 6.5) (Boogaard et al., 2001; van Sittert et al., 2000). Hydroxymethyl vinyl ketone is further metabolized to form N-acetyl-S-(3,4-dihydroxybutyl)cysteine (DHBMA) under the role of glutathione conjugation (Carmella et al., 2009). 3,4-epoxy-1,2-diol goes through similar glutathione conjugation and metabolic transformation to form 4-(N-acetyl-L-cystein-S-yl)-1,2,3-trihydroxybutane (THBMA) (Richardson et al., 1999). N-Acetylcysteine conjugates of EB (MHBMA in Scheme 1), HMVK (DHBMA), and EBD (THBMA) can serve as useful indicators of 1,3-butadiene exposure and metabolic activation to ultimate genotoxic species in smokers and occupationally exposed individuals, allowing for human biomonitoring and risk assessment from 1,3-butadiene exposure (Roethig et al., 2009; Sapkota et al., 2006). MHBMA, DHBMA, and THBMA have been measured in urine from smokers and nonsmokers. In 2009, Schettgen et al. (2009a) developed and validated a fast and specific automated multidimensional LC-MS/MS method for the simultaneous quantification of MHBMA and DHBMA in urine from no occupational exposure population. The results indicated that MHBMA concentrations in urine from smokers were significantly higher than those from nonsmokers, but the differences between DHBMA in smokers and nonsmokers were not significant. THBMA was first found in urine obtained from rats and mice exposed to 200 ppm of radiolabeled ^{14}C-1,3-butadiene for 6 h. It is estimated that THBMA was approximately 4.1% and 6.7% of the total 1,3-butadiene dose in the rat and in mice, respectively (Richardson et al., 1999).

Kotapati et al. (2011) developed and employed an isotope dilution HPLC–ESI–MS/MS methodology to quantify THBMA in urine of known smokers and nonsmokers (1927 per group). The new method has excellent sensitivity (LOQ, 1 ng/mL urine) and achieves accurate quantitation using a small sample volume (100 µL). Mean urinary THBMA concentrations in smokers and nonsmokers were found to be 21.6 and 13.7 ng/mg creatinine, respectively, suggesting that there are sources of THBMA other than exposure to tobacco smoke in humans, as is also the case for DHBMA. However, THBMA concentrations are significantly greater in urine of

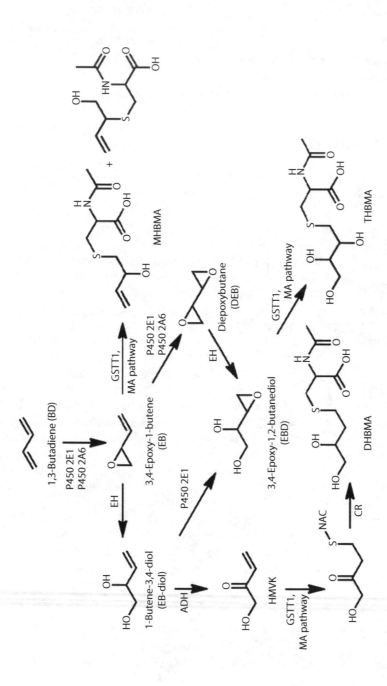

FIGURE 6.5 The metabolism of 1,3-butadiene to MHBMA, DHBMA, and THBMA *in vivo*. (From Kotapati, S. et al., *Chem. Res. Toxicol.*, 24, 1516, 2011.)

smokers than that of nonsmokers ($P < 0.01$). Furthermore, THBMA amounts in human urine declined 25%–50% following smoking cessation, suggesting that smoking is an important source of this metabolite in humans.

6.3.5.2   1,3-Butadiene Adducts in Blood and Tissue

Several hemoglobin adducts can be formed from 1,3-butadiene or its metabolites in vivo. N-(2-hydroxy-3-butenyl)-valine (HBVal) adducts were induced by 3,4-epoxy-1-butene and the previous research indicated that its concentration in mice was greater than that in rats by 1,3-butadiene inhalation (Osterman-Golkar et al., 1998). The species difference exhibited some concentration dependence, yielding approximately fourfold greater concentrations in mice than rats when exposed to 100 ppm 1,3-butadiene for 4 weeks, but dropping to an approximate twofold difference at concentrations of 10 ppm. The N-(2,3,4-trihydroxybutyl)-valine (THBVal) hemoglobin adducts, although potentially formed by either EBD or 1,2:3,4-diepoxybutane (DEB), have been shown to be produced almost entirely by the EBD (Boysen et al., 2004; Koivisto et al., 1999). The THBVal adducts are greater in concentrations in mice than in rats, which demonstrates that 3,4-epoxy-1,2-butanediol is the most abundant electrophilic metabolite of 1,3-butadiene metabolism in vivo in both mice and rats (Pérez et al., 1997; Swenberg et al., 2000). More recent studies have measured HBVal and THBVal adduct concentrations at even lower external exposure concentrations and, importantly, have also measured the critical DEB-specific hemoglobin adduct, N,N-(2,3-dihydroxy-1,4-butadiyl)-valine (pyrVal), providing new insights into species and exposure differences in 1,3-butadiene metabolism (Boysen et al., 2004). The formation of pyrVal adducts has been studied in male and female mice and rats exposed to 1.0 ppm by inhalation for 6 h/day for 4 weeks (Swenberg et al., 2007). At this low exposure concentration, although clearly detectable, the adduct concentrations for male and female rats were only 0.9 ± 0.1 and 0.7 ± 0.1 pmol/g, respectively—more than 30-fold lower than the corresponding values in mice. However, data from Boysen et al. (2004) suggest that at high concentrations, and shorter exposure durations (10 days), the difference between species becomes smaller (approximately 12-fold at 3 ppm and only 3-fold at 62.5 ppm). These observations suggest that 1,3-butadiene is primarily metabolized via the B-diol pathway in both mice and rats, but that mice are much more efficient in producing EB and DEB, especially DEB, at low 1,3-butadiene exposure levels. These results indicated that 1,3-butadiene is mainly metabolized by the B-diol pathway in both mice and rats. The comparison of pyrVal hemoglobin adduct between animal (mice and rats) and human was conducted by Swenberg et al. (2007). pyrVal levels showed strong species differences, with approximately 30 times higher levels found in mice than in rats at low 1,3-butadiene exposures (1 ppm). pyrVal adducts were not quantifiable in exposed humans using one-half the detection limit (0–3 pmol/g), and normalizing the data for differences in cumulative exposures, human levels were estimated to be approximately 4- to 9-fold lower than found in rats and approximately 120- to 320-fold lower than found in mice; recent unpublished studies from the same laboratory with a lower detection limit indicate that the pyrVal levels in humans were approximately threefold lower than detected in rats. These findings are in accord with and extend the direct metabolic studies in rodents.

6.3.6 BENZENE

Benzene is a ubiquitous indoor and outdoor pollutant causing occupational and public health concerns because of its hematotoxic, genotoxic, and carcinogenic properties (Danzon et al., 2000). In exhaled breath of smokers, mean benzene concentration can reach 522 g/m³ (Gordon et al., 2002). For smokers, benzene in cigarette smoke accounts for approximately 90% of their benzene exposure (Johnson et al., 2007). It is estimated that about one-tenth to one-half of smoking-induced total leukemia mortality and up to three-fifths of smoking-related acute myeloid leukemia mortality are related to benzene (Korte et al., 2000).

6.3.6.1 Benzene in Blood and Urine

Routine exposures to benzene can increase the background body burden, so differences in benzene levels in blood, breath, and urine can help to discriminate between exposed and nonexposed populations particularly for occupational exposures and for smokers versus nonsmokers. For example, benzene levels in blood distinguished three groups of occupationally exposed workers in Mexico: service station attendants (median 0.1 ppm), street vendors (0.02 ppm), and office workers (0.013 ppm), with nonsmokers having lower benzene blood levels than smokers (Cardinali et al., 1995). Individual benzene blood levels were not highly matched to the paired air concentration, because benzene concentration in air changed over time, while the blood concentrations were reflective of the exposure during the last minutes prior to the blood collection. Thus, the air sample and blood sample would represent different exposure concentrations. Benzene blood levels measured as part of the National Health and Nutritional Examination Survey (NHANES) (Arif and Shah, 2008; Symanski et al., 2009) have been compared to benzene air concentrations to determine their applicability as a biomarker and have been used to try to distinguish between exposed and nonexposed workers on a population basis. A significant difference was found in both the geometric mean and 75th percentile benzene blood concentrations of smokers compared to nonsmokers (factor of 3) than across the range of air concentrations for these two groups. For the NHANES data, the adjusted R^2 were stronger for smokers than nonsmokers, and the overall association in a generalized linear regression model between benzene blood and benzene air concentration was influenced by smoking, exposure–smoking interactions, gender, age, and body mass index (Lin et al., 2007b). For individuals, benzene concentrations in blood were not highly related to the paired air concentration, the reason for this may be the benzene concentration in air were average values over time, but its concentrations in blood were reflective of the exposure during the last minutes before the blood collection. Blood benzene levels in occupational benzene exposure workers who smoke was higher than the nonsmokers in the same industry; however, benzene concentrations in urine from smokers were not always higher than those in nonsmokers. When the exposure range was 0.2–100 ppm, benzene levels in urine were linearly related to work shift air concentrations (Ghittori et al., 1995; Kim et al., 2006). Individuals with paired urinary benzene concentration pre- and postshift samples do not always show increases in the postshift samples, which may reflect benzene exposure that occurred during the previous work shift and from the environment. Blood and breath benzene levels have biological half-lives of seconds to minutes, while urinary benzene levels reflect exposures since the previous one to two voids for single exposure. Individuals who are routinely exposed to benzene will have elevated background benzene in these biological fluids compared to a nonexposed population, though their peak benzene levels will be within minutes of the end of the exposure.

6.3.6.2   Urinary Metabolites

Urinary t,tMA has been shown to increase with benzene exposures from <0.1 to 20 ppm across a variety of occupational settings as well as between smokers and nonsmokers. Urinary metabolite levels represent exposures of the previous several hours for brief exposures, though these can be elevated for days for routinely exposed individuals after the exposure has stopped. These results indicate that t,tMA is a useful and a valid biomarker for low-level exposures (sub-ppm). But concentration of urinary t,tMA is not always related to the levels of benzene in air (Sanguinetti et al., 2001). One possible reason is the influence of the diet. Sorbic acid is a common food additive, whose metabolite in vivo is also t,tMA, which lead to increases in urinary t,tMA even without the presence of benzene exposure (Hoet et al., 2009). The mean t,tMA concentrations in urine from people who were occupationally exposed to benzene at tens to several hundreds ppbs were 2–33 times higher than those in urine from the controls (Wiwanitkit et al., 2005). Urinary sPMA has been proposed as a better biomarker than t,tMA for benzene exposure below 1 ppm (Farmer et al., 2005). Urinary SPMA has been demonstrated to increase with benzene exposure.

6.3.6.3   Benzene Adducts in Blood and Tissue

Benzene adducts have been proposed as biomarkers of longer-term benzene exposure since several benzene metabolites include reactive electrophiles: benzene oxides, 1,2 and 1,4 benzoquinone, muconaldehydes, and benzene diol epoxide, which have the potential to form adducts. The biological half-lives of hemoglobin adducts is about 4 months; however, DNA adducts may have longer half-lives. Protein adducts in serum were higher in exposed workers compared to controls (0.2–55 ppm vs. <0.01–0.5 ppm) (Lin et al., 2007a; Rappaport et al., 2005). Hemoglobin adducts of benzene oxide have been measured in dried blood spots of neonates and adults and suggested to be related to benzene exposure (Funk et al., 2008). While benzene adducts hold promise of being a valid biomarker of exposure, most laboratories do not have the analytical capability to measure them with the necessary sensitivity.

6.3.7   Acrylonitrile

Acrylonitrile (C_3H_3N, CAS No. 107-13-1) is widely used in the manufacture of synthetic fibers and rubber. The release of acrylonitrile into the air and water can occur during monomer and polymer production, transportation, and usage, which can potentially lead to acrylonitrile exposure (Cole et al., 2008). A major source for nonoccupational exposure to acrylonitrile is tobacco smoke (Leonard et al., 1999). In mainstream smoke, acrylonitrile yields range from 7.8 to 39.1 µg/cigarette (Humans, 2004), and the actual acrylonitrile concentration in ETS was estimated to be 0.49 µg/m^3 (Miller et al., 1998).

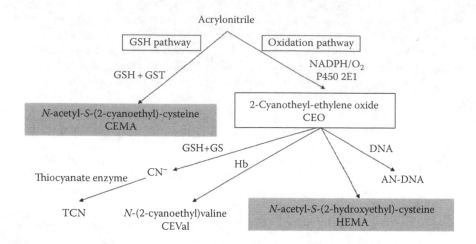

FIGURE 6.6 A simplified metabolic pathway of acrylonitrile. GSH, glutathione; GST, glutathione *S*-transferase; Hb, hemoglobin.

Acrylonitrile, although not directly carcinogenic/mutagenic, is a potentially carcinogenic organic compound through its metabolites. In carcinogenesis bioassays of rats, acrylonitrile caused tumors in the brain, forestomach, and Zymbal's gland (Maltoni et al., 1988). A number of epidemiological studies have been conducted to assess the possible carcinogenic activity of acrylonitrile in humans (Blair et al., 1998; Cole et al., 2008; Haber and Patterson, 2005; O'Berg, 1980; Starr et al., 2004). The IARC has classified the carcinogenicity of acrylonitrile as category 2B (IARC, 1999). Acrylonitrile can undergo metabolism through conjugation with GSH. This is a rapid nonenzymatic reaction with GSH, and the rate of the reaction can be enhanced by the presence of GSH transferase. In addition, acrylonitrile undergoes metabolism by oxidation to form cyanoethylene oxide (Guengerich et al., 1981), which can be further metabolized by conjugation with GSH, DNA, or hemoglobin or by hydrolysis (Figure 6.1) (Leonard et al., 1999). Two mercapturic acids have been identified as the final products of the reactions with GSH, 2-cyanoethylmercapturic acid (CEMA, major metabolite) (Langvardt et al., 1980) and 2-hydroxyethylmercapturic acid (HEMA, minor metabolite) (Figure 6.6) (VanBladeren et al., 1981b), in the urine of rats orally administered with acrylonitrile. Urinary metabolites have emerged as highly practical biomarkers for determining the uptake of specific carcinogens and toxicants from tobacco smoke (Carmella et al., 2007; Lee et al., 2007; Vatsavai et al., 2008; Waidyanatha et al., 2004).

6.3.7.1 CEMA and HEMA

Recently, urinary excretion of CEMA was found to be higher in smokers compared with nonsmokers and strongly correlated with the smoking dosage (Scherer et al., 2007a). Furthermore, Schettgen et al. (2009) found that a gradual increase in CEMA excretion was correlated to urinary cotinine. In 2011, Minet and coworkers observed that CEMA levels in smokers correlated significantly to the tar yield cigarettes, daily cigarette consumption, and urinary biomarkers of smoke exposure

(Minet et al., 2011). HEMA has also been used as a biomarker for acrylonitrile exposure from cigarettes (Carmella et al., 2009; Ding et al., 2009; Tardif et al., 1987). However, the excretion of HEMA might be influenced by the uptake of other industrial chemicals such as 1,2-dibromoethane, vinyl chloride, and ethylene oxide (Jones and Wells, 1981; Van Bladeren et al., 1981a,c). Therefore, HEMA is more suitable as a nonspecific marker for those hazardous chemicals (Calafat et al., 1999; De Rooij et al., 1998).

Methods published so far for the determination of urinary CEMA and HEMA mainly involve LC–MS/MS (Barr and Ashley, 1998; Schettgen et al., 2008, 2009) and gas chromatography (GC) (Jakubowski et al., 1987). However, most of these methods could determine only one mercapturic acid metabolite of acrylonitrile (Barr and Ashley, 1998; Schettgen et al., 2008). Moreover, some methods required either extensive manual sample preparation involving SPE with anion-exchange cartridges (Carmella et al., 2009; Scherer et al., 2010) or automated sample cleanup using a restricted access material (RAM) precolumn (Schettgen et al., 2009). Among the analytical methods described to detect low-level mercapturic acid metabolites, one of particular interest was developed by Schettgen et al. (2009). They employed an additional pump to load the sample onto a RAM-phase precolumn to purify the urine sample and enrich the analyte before transferring for LC–MS/MS. To the best of our knowledge, the correlation between the mainstream acrylonitrile content of Chinese Virginia cigarettes and urinary CEMA and HEMA excretion has not yet been reported in the literature.

We (Hou et al., 2012) had developed and validated an analytical method for the simultaneous determination of CEMA and HEMA in human urine using an easy, automated, column-switching LC–MS/MS instrument. This newly developed method was applied to study the dose–response relationship of smoking-related acrylonitrile exposure to CEMA and HEMA excretion (Figure 6.7). The levels of urinary CEMA and HEMA in smokers of different tar yield cigarettes under an ISO 3308 smoking condition (8, 10, and 13 mg/Chinese Virginia cigarettes) were compared with the tobacco-specific biomarker cotinine (Figures 6.7 and 6.8). Comparing the urinary CEMA levels in smokers reported by Minet et al. (2011) with those in subjects smoking 10 mg tar yield cigarettes, the average concentration of CEMA (645 and 221 nmol/24 h urine) was lower in this study, whereas the concentration of HEMA (12.6 and 10.7 nmol/24 h urine) was lower in the study by Calafat et al. (1999). This discrepancy might be explained by variations in the race and gender distributions of these different studies or, alternatively, by variations in the subjects' smoking behavior. It is important to note that the cigarettes used in our study were Chinese Virginia cigarettes, which are different from the blended cigarettes used in the other studies. To simulate the uptake of acrylonitrile by Chinese smokers, a rough dose–response model between urinary mean CEMA concentration and acrylonitrile yield in mainstream smoke was established ($y = 153 + 0.912x$, $R = 1.00$) based on the average urinary CEMA (213, 223, and 229 nmol/24 h) and cigarette smoke acrylonitrile (4.30, 5.10, and 5.50 lg/cigarette) from smokers with different tar cigarettes. On average, a smoker's uptake of acrylonitrile from Chinese Virginia cigarettes (15 cigarettes/day) was approximately 55.7 µg, assuming that the CEMA concentration reflects approximately 21% of

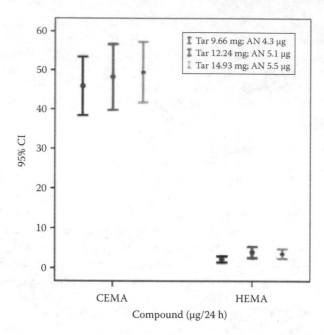

FIGURE 6.7 CMEA and HEMA in the 24 h urine samples from smokers after smoking different tar and acrylonitrile yield cigarettes.

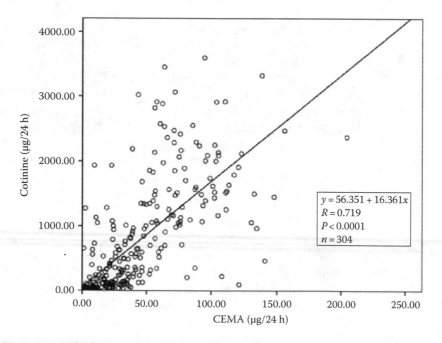

FIGURE 6.8 Correlation between urinary cotinine and CEMA in the study population.

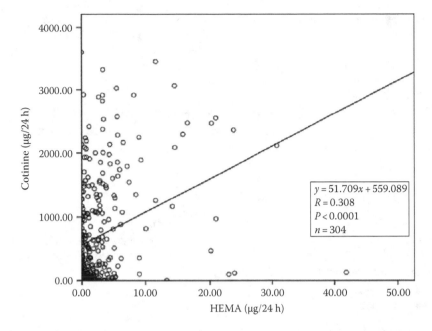

FIGURE 6.9   Correlation between urinary cotinine and HEMA in the study population.

the absorbed acrylonitrile dose. As a pilot study, the brand and daily consumption of cigarettes have been controlled in order to minimize differences between samples. This study validates acrylonitrile as a biomarker and provides a rough dose–response relationship between urinary CEMA and acrylonitrile from Chinese Virginia cigarette smoke. Based on this, a further study to get values closer to the actual uptake of smoke constituents with the real smoking behavior will be conducted using smokers who smoked their own cigarettes free of daily consumption requirements (Figure 6.9).

6.4   DISCUSSION

Long-term exposure to certain VOCs may increase the risk for cancer, birth defects, and neurocognitive impairment. Therefore, VOC exposure is an area of significant public health concern. Urinary VOC metabolites are useful biomarkers for assessing VOC exposure because of noninvasiveness of sampling and longer physiological half-lives of urinary metabolites compared with VOCs in blood and breath. A number of specific and sensitive assays for detection and quantification of urinary VOC metabolites have been published. Alwis et al. (2012) developed a method using reversed-phase UPLC–ESI–MS/MS to simultaneously quantify 28 urinary VOC metabolites as biomarkers of acrolein, acrylamide, acrylonitrile, benzene, 1-bromopropane, 1,3-butadiene, carbon disulfide, crotonaldehyde, cyanide, *N,N*-dimethylformamide, ethylbenzene, ethylene oxide, propylene oxide, styrene, tetrachloroethylene, toluene, trichloroethylene, vinyl chloride, and xylene exposure.

The research demonstrated that after adjustment for smoking intensity and duration, elevated urinary levels of the mercapturic acids MHBMA, HEMA, SPMA, and HPMA, biomarkers of the tobacco smoke gas phase constituents 1,3-butadiene, acrylonitrile, benzene, and acrolein, respectively, were associated with a statistically significant increased risk of developing lung cancer among Shanghai smokers. The positive association for lung cancer risk with HEMA, SPMA, and HPMA remained statistically significant after adjustment for urinary total NNAL and PheT (r-1,t-2,3,c-4-tetrahydroxy-1,2,3,4-tetrahydrophenanthrene), biomarkers of NNK and PAH, respectively (Yuan et al., 2012). However, the role of these biomarkers in the overall mutagenic and carcinogenic properties of VOCs is unknown.

6.5 CONCLUSION

MHBMA, HEMA, SPMA, and HPMA are useful biomarkers of the tobacco smoke gas phase constituents 1,3-butadiene, acrylonitrile, benzene, and acrolein, respectively. These can be used for further studies related to VOCs exposure and health effects such as cancer, birth defects, and neurocognitive impairment.

REFERENCES

ACGIH. Biological Exposure Indices (BEI). *American Conference of Governmental Industrial Hygienists (ACGIH)*, Cincinnati, OH, 2005, http://www.acgih.org/store/ProductDetail.cfm?id=652.

Alwis, K. U., Blount, B. C., Britt, A. S., Patel, D., Ashley, D. L. Simultaneous analysis of 28 urinary VOC metabolites using ultra high performance liquid chromatography coupled with electrospray ionization tandem mass spectrometry (UPLC-ESI/MSMS). *Analytica Chimica Acta.* 2012, 750, 152–160.

Aprea, C., Sciarra, G., Bozzi, N. et al. Reference values of urinary trans,trans-muconic acid: Italian multicentric study. *Archives of Environmental Contamination and Toxicology.* 2008, 55, 329–340.

Arif, A. A., Shah, S. M. Association between personal exposure to volatile organic compounds and asthma among US adult population. *International Archives of Occupational and Environmental Health.* 2008, 81, 503.

Barbieri, A., Violante, F. S., Sabatini, L. et al. Urinary biomarkers and low-level environmental benzene concentration: Assessing occupational and general exposure. *Chemosphere.* 2008, 74, 64–69.

Barr, D. B., Ashley, D. L. A rapid, sensitive method for the quantitation of N-acetyl-S-(2-hydroxyethyl)-L-cysteine in human urine using isotope-dilution HPLC-MS-MS. *Journal of Analytical Toxicology.* 1998, 22, 96–104.

Blair, A., Stewart, P. A., Zaebst, D. D. et al. Mortality of industrial workers exposed to acrylonitrile. *Scandinavian Journal of Work, Environment and Health.* 1998, 24, 25–41.

Boogaard, P. J., van Sittert, N. J., Megens, H. J. Urinary metabolites and haemoglobin adducts as biomarkers of exposure to 1,3-butadiene: A basis for 1,3-butadiene cancer risk assessment. *Chemico-Biological Interactions.* 2001, 135–136, 695–701.

Boysen, G., Georgieva, N. I., Upton, P. B. et al. Analysis of diepoxide-specific cyclic N-terminal globin adducts in mice and rats after inhalation exposure to 1,3-butadiene. *Cancer Research.* 2004, 64, 8517–8520.

Brunnemann, K. D., Kagan, M. R., Cox, J. E. et al. Analysis of 1,3-butadiene and other selected gas-phase components in cigarette mainstream and sidestream smoke by gas chromatography-mass selective detection. *Carcinogenesis.* 1990, 11, 1863–1868.

Calafat, A. M., Barr, D. B., Pirkle, J. L. et al. Reference range concentrations of N-acetyl-S-(2-hydroxyethyl)-L-cysteine, a common metabolite of several volatile organic compounds, in the urine of adults in the United States. *Journal of Exposure Analysis and Environmental Epidemiology*. 1999, 9, 336.

Cardinali, F. L., McCraw, J. M., Ashley, D. L. et al. Treatment of vacutainers for use in the analysis of volatile organic compounds in human blood at the low parts-per-trillion level. *Journal of Chromatographic Science*. 1995, 33, 557–560.

Carmella, S. G., Chen, M., Han, S. et al. Effects of smoking cessation on eight urinary tobacco carcinogen and toxicant biomarkers. *Chemical Research in Toxicology*. 2009, 22, 734–741.

Carmella, S. G., Chen, M., Zhang, Y. et al. Quantitation of acrolein-derived (3-hydroxypropyl) mercapturic acid in human urine by liquid chromatography-atmospheric pressure chemical ionization tandem mass spectrometry: Effects of cigarette smoking. *Chemical Research in Toxicology*. 2007, 20, 986–990.

Chen, L., Wang, M., Villalta, P. W. et al. Quantitation of an acetaldehyde adduct in human leukocyte DNA and the effect of smoking cessation. *Chemical Research in Toxicology*. 2007, 20, 108–113.

Cole, P., Mandel, J. S., Collins, J. J. Acrylonitrile and cancer: A review of the epidemiology. *Regulatory Toxicology and Pharmacology*. 2008, 52, 342–351.

Danzon, M. A., Van Leeuwen, R., Krzyzanowski, M. *Air Quality Guidelines for Europe*. 2000, World Health Organization, Regional Office for Europe, Copenhagen, Denmark.

Ding, Y. S., Blount, B. C., Valentin-Blasini, L. et al. Simultaneous determination of six mercapturic acid metabolites of volatile organic compounds in human urine. *Chemical Research in Toxicology*. 2009, 22, 1018–1025.

Esterbauer, H., Schaur, R. J., Zollner, H. Chemistry and biochemistry of 4-hydroxynonenal, malonaldehyde and related aldehydes. *Free Radical Biology and Medicine*. 1991, 11, 81–128.

Farmer, P. B., Kaur, B., Roach, J. et al. The use of S phenylmercapturic acid as a biomarker in molecular epidemiology studies of benzene. *Chemico-Biological Interactions*. 2005, 153, 97.

Faroon, O., Roney, N., Taylor, J. et al. Acrolein health effects. *Toxicology and Industrial Health*. 2008, 24, 447–490.

FCTC. *Conference of the Parties to the WHO Framework Convention on Tobacco Control: Third Session*. WHO Framework Convention on Tobacco Control, Bangkok, 2008, COP/3/6.

FCTC. Work in progress in relation to Articles 9 and 10 of the WHO Framework Convention on Tobacco Control. Report by WHO's Tobacco Free Initiative. 2010, COP/4/INF. DOC./2.

Feng, Z., Hu, W., Hu, Y. et al. Acrolein is a major cigarette-related lung cancer agent: Preferential binding at p53 mutational hotspots and inhibition of DNA repair. *Proceedings of the National Academy of Sciences*. 2006, 103, 15404–15409.

Fowles, J., Dybing, E. Application of toxicological risk assessment principles to the chemical constituents of cigarette smoke. *Tobacco Control*. 2003, 12, 424–430.

Funk, W. E., Waidyanatha, S., Chaing, S. H. et al. Hemoglobin adducts of benzene oxide in neonatal and adult dried blood spots. *Cancer Epidemiology, Biomarkers and Prevention*. 2008, 17, 1896–1901.

Ghittori, S., Maestri, L., Fiorentino, M. L. et al. Evaluation of occupational exposure to benzene by urinalysis. *International Archives of Occupational and Environmental Health*. 1995, 67, 195–200.

Gordon, S. M., Wallace, L. A., Brinkman, M. C. et al. Volatile organic compounds as breath biomarkers for active and passive smoking. *Environmental Health Perspectives*. 2002, 110, 689.

Guengerich, F. P., Geiger, L. E., Hogy, L. L. et al. In vitro metabolism of acrylonitrile to 2-cyanoethylene oxide, reaction with glutathione, and irreversible binding to proteins and nucleic acids. *Cancer Research.* 1981, 41, 4925–4933.

Haber, L., Patterson, J. Report of an independent peer review of an acrylonitrile risk assessment. *Human and Experimental Toxicology.* 2005, 24, 487–527.

Heck, Hd., Casanova, M. Isotope effects and their implications for the covalent binding of inhaled [^{3}H]-and [^{14}C] formaldehyde in the rat nasal mucosa. *Toxicology and Applied Pharmacology.* 1987, 89, 122–134.

Heck, Hd., Casanova-Schmitz, M., Dodd, P. B. et al. Formaldehyde (CH_2O) concentrations in the blood of humans and Fischer-344 rats exposed to CH_2O under controlled conditions. *The American Industrial Hygiene Association Journal.* 1985, 46, 1–3.

Hoet, P., De Smedt, E., Ferrari, M. et al. Evaluation of urinary biomarkers of exposure to benzene: Correlation with blood benzene and influence of confounding factors. *International Archives of Occupational and Environmental Health.* 2009, 82, 985–995.

Hoffmann, D., Djordjevic, M. V., Hoffmann, I. The changing cigarette. *Preventive Medicine.* 1997, 26, 427–434.

Hou, H., Xiong, W., Gao, N. et al. A column-switching LC-MS-MS method for quantitation of 2-cyanoethylmercapturic acid and 2-hydroxyethylmercapturic acid in Chinese smokers. *Analytical Biochemistry.* 2012, 430, 75–82.

IARC. *IARC Monographs on the Evaluation of Carcinogenic Risks to Humans. Some Industrial Chemicals and Dyestuffs.* 1982, World Health Organization International Agency for Research on Cancer, Lyon, France, p. 29.

IARC. *IARC Monographs on the Evaluation of the Carcinogenic Risk of Chemicals to Humans: 1,3-Butadiene.* 1992, World Health Organization International Agency for Research on Cancer, Lyon, France.

IARC. Acrolein. *IARC Monographs on the Evaluation of Carcinogenic Risks to Humans.* 1995, IARC, Lyon, France, Vol. 63, pp. 337–372.

IARC. *IARC Monographs on the Evaluation of Carcinogenic Risks to Humans.* 1999, World Health Organization International Agency for Research on Cancer, Lyon, France, Vol. 71, pp. 43–108.

IARC Working Group on the Evaluation of Carcinogenic Risks to Humans. *Tobacco Smoke and Involuntary Smoking.* 2004, IARC, Lyon, France.

IARC Working Group on the Evaluation of Carcinogenic Risks to Humans. *Formaldehyde, 2-Butoxyethanol and 1-Tert-Butoxypropan-2-ol.* 2006, World Health Organization/ IARC, Lyon, France.

International Agency for Research on Cancer. Allyl compounds, aldehydes, epoxides and peroxides. In: *IARC Monographs on the Evaluation of Carcinogenic Risks of Chemicals to Humans.* 1985, Vol. 36, pp. 101–132. IARC, Lyon, France.

International Agency for Research on Cancer. Re-evaluation of some organic chemicals, hydrazine and hydrogen peroxide. In: *IARC Monographs on the Evaluation of Carcinogenic Risks to Humans.* 1999, IARC, Lyon, France.

Jakubowski, M., Linhart, I., Pielas, G. et al. 2-Cyanoethylmercapturic acid (CEMA) in the urine as a possible indicator of exposure to acrylonitrile. *British Journal of Industrial Medicine.* 1987, 44, 834–840.

Johnson, E. S., Langård, S., Lin, Y.-S. A critique of benzene exposure in the general population. *Science of the Total Environment.* 2007, 374, 183.

Jones, A., Wells, G. The comparative metabolism of 2-bromoethanol and ethylene oxide in the rat. *Xenobiotica.* 1981, 11, 763–770.

Jösch, C., Klotz, L.-O., Sies, H. Identification of cytosolic leucyl aminopeptidase (EC 3.4.11.1) as the major cysteinylglycine-hydrolysing activity in rat liver. *Biological Chemistry.* 2003, 384, 213–218.

Kaminskas, L. M., Pyke, S. M., Burcham, P. C. Strong protein adduct trapping accompanies abolition of acrolein-mediated hepatotoxicity by hydralazine in mice. *The Journal of Pharmacology and Experimental Therapeutics.* 2004, 310, 1003–1010.

Kawai, Y., Furuhata, A., Toyokuni, S. et al. Formation of acrolein-derived 2′-deoxyadenosine adduct in an iron-induced carcinogenesis model. *The Journal of Biological Chemistry.* 2003, 278, 50346–50354.

Kaye, C. M. Biosynthesis of mercapturic acids from allyl alcohol, allyl esters and acrolein. *Biochemical Journal.* 1973, 134, 1093.

Kim, S., Vermeulen, R., Waidyanatha, S. et al. Using urinary biomarkers to elucidate dose-related patterns of human benzene metabolism. *Carcinogenesis.* 2006, 27, 772–781.

Koivisto, P., Kilpeläinen, I., Rasanen, I. et al. Butadiene diolepoxide-and diepoxybutane-derived DNA adducts at N7-guanine: A high occurrence of diolepoxide-derived adducts in mouse lung after 1,3-butadiene exposure. *Carcinogenesis.* 1999, 20, 1253–1259.

Korte, J. E., Hertz-Picciotto, I., Schulz, M. R. et al. The contribution of benzene to smoking-induced leukemia. *Environmental Health Perspectives.* 2000, 108, 333.

Kotapati, S., Matter, B. A., Grant, A. L. et al. Quantitative analysis of trihydroxybutyl mercapturic acid, a urinary metabolite of 1,3-butadiene, in humans. *Chemical Research in Toxicology.* 2011, 24, 1516–1526.

Langvardt, P., Putzig, C., Braun, W. et al. Identification of the major urinary metabolites of acrylonitrile in the rat. *Journal of Toxicology and Environmental Health, Part A Current Issues.* 1980, 6, 273–282.

Lee, H.-L., Wang, C., Lin, S. et al. Liquid chromatography/tandem mass spectrometric method for the simultaneous determination of tobacco-specific nitrosamine NNK and its five metabolites. *Talanta.* 2007, 73, 76–80.

Leonard, A., Gerber, G., Stecca, C. et al. Mutagenicity, carcinogenicity, and teratogenicity of acrylonitrile. *Mutation Research/Reviews in Mutation Research.* 1999, 436, 263–283.

Li, F., Liu, P., Wang, T. et al. Genotoxicity/mutagenicity of formaldehyde revealed by the *Arabidopsis thaliana* plants transgenic for homologous recombination substrates. *Mutation Research/Genetic Toxicology and Environmental Mutagenesis.* 2010, 699, 35–43.

Lieberman, M. W., Barrios, R., Carter, B. Z. et al. gamma-Glutamyl transpeptidase. What does the organization and expression of a multipromoter gene tell us about its functions? *American Journal of Pathology.* 1995, 147, 1175.

Lin, Y.-S., Egeghy, P., Rappaport, S. Relationships between levels of volatile organic compounds in air and blood from the general population. *Journal of Exposure Science and Environmental Epidemiology.* 2007b, 18, 421–429.

Lin, Y.-S., Vermeulen, R., Tsai, C. H. et al. Albumin adducts of electrophilic benzene metabolites in benzene-exposed and control workers. *Environmental Health Perspectives.* 2007a, 115, 28.

Linhart, I., Frantík, E., Vodičková, L. et al. Biotransformation of acrolein in rat: Excretion of mercapturic acids after inhalation and intraperitoneal injection. *Toxicology and Applied Pharmacology.* 1996, 136, 155–160.

List of MAK and BAT Values (2007) Commision for the Investigation of Health Hazards of Chemical Compounds in the Work Area, Report No. 43 DFG Deutsche Forschungsgemeinschaft, Wiley-VCH, Weinheim, 2007.

Lu, K., Boysen, G., Gao, L. et al. Formaldehyde-induced histone modifications in vitro. *Chemical Research in Toxicology.* 2008, 21, 1586–1593.

Lu, K., Craft, S., Nakamura, J. et al. Use of LC-MS/MS and stable isotopes to differentiate hydroxymethyl and methyl DNA adducts from formaldehyde and nitrosodimethylamine. *Chemical Research in Toxicology.* 2012, 25, 664–675.

Lu, K., Ye, W., Gold, A. et al. Formation of S-[1-(N^2-deoxyguanosinyl) methyl] glutathione between glutathione and DNA induced by formaldehyde. *Journal of the American Chemical Society.* 2009, 131, 3414–3415.

Maltoni, C., Ciliberti, A., Cotti, G. et al. Long-term carcinogenicity bioassays on acrylonitrile administered by inhalation and by ingestion to Sprague-Dawley rats. *Annals of the
New York Academy of Sciences*. 1988, 534, 179–202.

Matsushima, S., Ishguro, S., Sugawara, S. Composition studies on some varieties of tobacco
and their smoke. *Beitrage zur Tabakforschung International*. 1979, 10, 31–38.

McGhee, J. D., Von Hippel, P. H. Formaldehyde as a probe of DNA structure. II. Reaction with
endocyclic imino groups of DNA bases. *Biochemistry*. 1975, 14, 1297–1303.

Merk, O., Speit, G. Detection of crosslinks with the comet assay in relationship to genotoxicity
and cytotoxicity. *Environmental and Molecular Mutagenesis*. 1999, 33, 167–172.

Metz, B., Kersten, G. F., Hoogerhout, P. et al. Identification of formaldehyde-induced modifications in proteins reactions with model peptides. *The Journal of Biological Chemistry*.
2004, 279, 6235–6243.

Miller, S. L., Branoff, S., Nazaroff, W. W. Exposure to toxic air contaminants in environmental tobacco smoke: An assessment for California based on personal monitoring data.
Journal of Exposure Analysis and Environmental Epidemiology. 1998, 8, 287.

Minet, E., Cheung, F., Errington, G. et al. Urinary excretion of the acrylonitrile metabolite
2-cyanoethylmercapturic acid is correlated with a variety of biomarkers of tobacco
smoke exposure and consumption. *Biomarkers*. 2011, 16, 89–96.

Nath, R. G., Chung, F.-L. Detection of exocyclic 1, N^2-propanodeoxyguanosine adducts as
common DNA lesions in rodents and humans. *Proceedings of the National Academy of
Sciences*. 1994, 91, 7491–7495.

O'Berg, M. T. Epidemiologic study of workers exposed to acrylonitrile. *Journal of
Occupational and Environmental Medicine*. 1980, 22, 245–252.

Osterman-Golkar, S. M., Moss, O., James, A. et al. Epoxybutene–hemoglobin adducts in rats
and mice: Dose response for formation and persistence during and following long-term
low-level exposure to butadiene. *Toxicology and Applied Pharmacology*. 1998, 150,
166–173.

Pánková, K. IARC monographs on the evaluation of the carcinogen risk of chemical to
humans. *Biologia Plantarum*. 1986, 28, 354–354.

Parent, R. A., Paust, D. E., Schrimpf, M. K. et al. Metabolism and distribution of [2,3-^{14}C]
acrolein in Sprague-Dawley rats II. Identification of urinary and fecal metabolites.
Toxicological Sciences. 1998, 43, 110–120.

Patel, J., Wood, J. C., Leibman, K. C. The biotransformation of allyl alcohol and acrolein in
rat liver and lung preparations. *Drug Metabolism and Disposition*. 1980, 8, 305–308.

Pérez, H. L., Lähdetie, J., Landin, H. et al. Haemoglobin adducts of epoxybutanediol from
exposure to 1,3-butadiene or butadiene epoxides. *Chemico-Biological Interactions*.
1997, 105, 181.

Quievryn, G., Zhitkovich, A. Loss of DNA–protein crosslinks from formaldehyde-exposed
cells occurs through spontaneous hydrolysis and an active repair process linked to pro-
teosome function. *Carcinogenesis*. 2000, 21, 1573–1580.

Rappaport, S. M., Waidyanatha, S., Yeowell-O'Connell, K. et al. Protein adducts as biomarkers
of human benzene metabolism. *Chemico-Biological Interactions*. 2005, 153, 103–109.

Richardson, K. A., Peters, M., Wong, B. A. et al. Quantitative and qualitative differences in the
metabolism of ^{14}C-1,3-butadiene in rats and mice: Relevance to cancer susceptibility.
Toxicological Sciences. 1999, 49, 186–201.

Roethig, H. J., Munjal, S., Feng, S. et al. Population estimates for biomarkers of exposure to
cigarette smoke in adult US cigarette smokers. *Nicotine and Tobacco Research*. 2009,
11, 1216–1225.

Rustemeier, K., Stabbert, R., Haussmann, H. et al. Evaluation of the potential effects of ingre-
dients added to cigarettes. Part 2: Chemical composition of mainstream smoke. *Food
and Chemical Toxicology*. 2002, 40, 93–104.

Sanduja, R., Ansari, G., Boor, P. J. 3-Hydroxypropylmercapturic acid: A biologic marker of exposure to allylic and related compounds. *Journal of Applied Toxicology*. 1989, 9, 235–238.

Sanguinetti, G., Accorsi, A., Barbieri, A. et al. Failure of urinary trans,trans-muconic acid as a biomarker for indoor environmental benzene exposure at PPB levels. *Journal of Toxicology and Environmental Health Part A*. 2001, 63, 599–604.

Sapkota, A., Halden, R. U., Dominici, F. et al. Urinary biomarkers of 1,3-butadiene in environmental settings using liquid chromatography isotope dilution tandem mass spectrometry. *Chemico-Biological Interactions*. 2006, 160, 70–79.

Sarigiannis, D. A., Karakitsios, S. P., Gotti, A. et al. Exposure to major volatile organic compounds and carbonyls in European indoor environments and associated health risk. *Environment International*. 2011, 37, 743–765.

Scherer, G., Engl, J., Urban, M. et al. Relationship between machine-derived smoke yields and biomarkers in cigarette smokers in Germany. *Regulatory Toxicology and Pharmacology*. 2007a, 47, 171–183.

Scherer, G., Urban, M., Hagedorn, H. et al. Determination of two mercapturic acids related to crotonaldehyde in human urine: Influence of smoking. *Human and Experimental Toxicology*. 2007b, 26, 37–47.

Scherer, G., Urban, M., Hagedorn, H.-W. et al. Determination of methyl-, 2-hydroxyethyl- and 2-cyanoethylmercapturic acids as biomarkers of exposure to alkylating agents in cigarette smoke. *Journal of Chromatography B*. 2010, 878, 2520–2528.

Schettgen, T., Musiol, A., Alt, A. et al. A method for the quantification of biomarkers of exposure to acrylonitrile and 1,3-butadiene in human urine by column-switching liquid chromatography–tandem mass spectrometry. *Analytical and Bioanalytical Chemistry*. 2009, 393, 969–981.

Schettgen, T., Musiol, A., Kraus, T. Simultaneous determination of mercapturic acids derived from ethylene oxide (HEMA), propylene oxide (2-HPMA), acrolein (3-HPMA), acrylamide (AAMA) and N,N-dimethylformamide (AMCC) in human urine using liquid chromatography/tandem mass spectrometry. *Rapid Communications in Mass Spectrometry*. 2008, 22, 2629–2638.

Singh, M., Nam, D. T., Arseneault, M. et al. Role of by-products of lipid oxidation in Alzheimer's disease brain: A focus on acrolein. *Journal of Alzheimer's Disease*. 2010, 21, 741–756.

Speit, G., Neuss, S., Schütz, P. et al. The genotoxic potential of glutaraldehyde in mammalian cells in vitro in comparison with formaldehyde. *Mutation Research*. 2008, 649, 146–154.

Speit, G., Neuss, S., Schmid, O. The human lung cell line A549 does not develop adaptive protection against the DNA-damaging action of formaldehyde. *Environmental and Molecular Mutagenesis*. 2010, 51, 130–137.

Starr, T. B., Gause, C., Youk, A. O. et al. A risk assessment for occupational acrylonitrile exposure using epidemiology data. *Risk Analysis*. 2004, 24, 587–601.

Stedman, R. L. Chemical composition of tobacco and tobacco smoke. *Chemical Reviews*. 1968, 68, 153–207.

Stevens, J. F., Maier, C. S. Acrolein: Sources, metabolism, and biomolecular interactions relevant to human health and disease. *Molecular Nutrition and Food Research*. 2008, 52, 7–25.

Sumner, S. C. J., Selvaraj, L., Nauhaus, S. K., Fennell, T. R. Urinary metabolites from F344 rats and B6C3F1 mice coadministered acrylamide and acrylonitrile for 1 or 5 days. *Chemical Research in Toxicology*. 1997, 10, 1152–1160.

Swenberg, J. A., Boysen, G., Georgieva, N., Bird, M. G., Lewis, R. J. Future directions in butadiene risk assessment and the role of cross-species internal dosimetry. *Chemico-Biological Interactions*. 2007, 166, 78–83.

Swenberg, J. A., Christova-Gueorguieva, N., Upton, P. et al. 1,3-Butadiene: Cancer, mutations, and adducts. Part V: Hemoglobin adducts as biomarkers of 1,3-butadiene exposure and metabolism. *Research Report (Health Effects Institute)*. 2000, (92), 191–210.

Symanski, E., Stock, T. H., Tee, P. G. et al. Demographic, residential, and behavioral determinants of elevated exposures to benzene, toluene, ethylbenzene, and xylenes among the US population: Results from 1999–2000 NHANES. *Journal of Toxicology and Environmental Health, Part A*. 2009, 72, 903–912.

Tang, S., Frank, B. P., Lanni, T. et al. Unregulated emissions from a heavy-duty diesel engine with various fuels and emission control systems. *Environmental Science and Technology*. 2007, 41, 5037–5043.

Tardif, R., Talbot, D., Gerin, M. et al. Urinary excretion of mercapturic acids and thiocyanate in rats exposed to acrylonitrile: Influence of dose and route of administration. *Toxicology Letters*. 1987, 39, 255–261.

Uchida, K. Current status of acrolein as a lipid peroxidation product. *Trends in Cardiovascular Medicine*. 1999, 9, 109–113.

Van Bladeren, P., Breimer, D., van Huijgevoort, J. et al. The metabolic formation of *N*-acetyl-*S*-2-hydroxyethyl-L-cysteine from tetradeutero-1,2-dibromoethane. Relative importance of oxidation and glutathione conjugation in vivo. *Biochemical Pharmacology*. 1981a, 30, 2499.

Van Bladeren, P., Delbressine, L., Hoogeterp, J. et al. Formation of mercapturic acids from acrylonitrile, crotononitrile, and cinnamonitrile by direct conjugation and via an intermediate oxidation process. *Drug Metabolism and Disposition*. 1981b, 9, 246–249.

Van Bladeren, P., Hoogeterp, J., Breimer, D. et al. The influence of disulfiram and other inhibitors of oxidative metabolism on the formation of 2-hydroxyethyl-mercapturic acid from 1,2-dibromoethane by the rat. *Biochemical Pharmacology*. 1981c, 30, 2983–2987.

van Sittert, N. J., Megens, H. J., Watson, W. P., Boogaard, P. J. Biomarkers of exposure to 1,3-butadiene as a basis for cancer risk assessment. *Toxicological Sciences*. 2000, 56, 189–202.

Vasiliou, V., Pappa, A., Estey, T. Role of human aldehyde dehydrogenases in endobiotic and xenobiotic metabolism. *Drug Metabolism Reviews*. 2004, 36, 279–299.

Vasiliou, V., Pappa, A., Petersen, D. R. Role of aldehyde dehydrogenases in endogenous and xenobiotic metabolism. *Chemico-Biological Interactions*. 2000, 129, 1–19.

Vatsavai, K., Goicoechea, H. C., Campiglia, A. D. Direct quantification of monohydroxy-polycyclic aromatic hydrocarbons in synthetic urine samples via solid-phase extraction–room-temperature fluorescence excitation–emission matrix spectroscopy. *Analytical Biochemistry*. 2008, 376, 213–220.

Voulgaridou, G.-P., Anestopoulos, I., Franco, R. et al. DNA damage induced by endogenous aldehydes: Current state of knowledge. *Mutation Research/Fundamental and Molecular Mechanisms of Mutagenesis*. 2011, 711, 13–27.

Waidyanatha, S., Rothman, N., Li, G. et al. Rapid determination of six urinary benzene metabolites in occupationally exposed and unexposed subjects. *Analytical Biochemistry*. 2004, 327, 184–199.

Wang, M., Cheng, G., Balbo, S. et al. Clear differences in levels of a formaldehyde-DNA adduct in leukocytes of smokers and nonsmokers. *Cancer Research*. 2009, 69, 7170–7174.

Wang, M., McIntee, E. J., Cheng, G. et al. Identification of DNA adducts of acetaldehyde. *Chemical Research in Toxicology*. 2000, 13, 1149–1157.

WHO. Best practices in tobacco control-regulation of tobacco products Canada report. 2005, WHO Study Group on Tobacco Product Regulation, Geneva, Switzerland.

Wiwanitkit, V., Suwansaksri, J., Soogarun, S. Monitoring of urine trans,trans-muconic acid level among smokers and non-smokers. *Respiratory Medicine*. 2005, 99, 788–791.

Yin, R., Liu, S., Zhao, C. et al. An ammonium bicarbonate enhanced stable isotope dilution UHPLC-MS/MS method for sensitive and accurate quantification of acrolein-DNA adducts in human leukocytes. *Analytical Chemistry*. 2013, 85, 3190–3197.

Yuan, J. M., Gao, Y. T., Wang, R. et al. Urinary levels of volatile organic carcinogen and toxicant biomarkers in relation to lung cancer development in smokers. *Carcinogenesis*. 2012, 33, 804–809.

Zhang, S., Villalta, P. W., Wang, M. et al. Analysis of crotonaldehyde-and acetaldehyde-derived 1, N^2-propanodeoxyguanosine adducts in DNA from human tissues using liquid chromatography electrospray ionization tandem mass spectrometry. *Chemical Research in Toxicology*. 2006, 19, 1386–1392.

Zhong, W., Hee, S. Q. Quantitation of normal and formaldehyde-modified deoxynucleosides by high-performance liquid chromatography/UV detection. *Biomedical Chromatography*. 2004, 18, 462–469.

Zhong, W., Que Hee, S. S. Formaldehyde-induced DNA adducts as biomarkers of in vitro human nasal epithelial cell exposure to formaldehyde. *Mutation Research/Genetic Toxicology and Environmental Mutagenesis*. 2004, 563, 13–24.

7 Aromatic Amines

7.1 INTRODUCTION

Aromatic amines (AAs) are widespread occupational and environmental pollutants (Richter and Branner, 2002). Carcinogenic aromatic amines are yielded from mainstream smoke and side stream smoke during the smoking process (Hoffmann et al., 1969; Saha et al., 2009), such as 1-naphthylamine (1-NA), 2-naphthylamine (2-NA), 3-aminobiphenyl (3-ABP), and 4-aminobiphenyl (4-ABP). For the general population, the main exposure mediums are cigarette smoke and products synthesized from AAs. 4-ABP is a cigarette smoke AA classified as a class 1 International Agency for Research on Cancer (IARC) carcinogen and also present in the Food and Drug Administration (FDA) HPHCs list (Administration, 2012; Cancer, 2010). 4-ABP mainstream smoke yields ranged from 0.5 to 3.3 ng/cigarette in 48 commercial cigarettes smoked with three different regimes including ISO (Riedel et al., 2006). 2-NA has been classified as Group 1 (carcinogenic to humans) by IARC (Humans et al., 2004). The concentration of 2-NA in mainstream cigarette smoke was 1.53–13.8 ng/cigarette (Saha et al., 2009). It is well known that cigarette smoking and occupational exposure to 2-NA and 4-ABP are risk factors for a variety of diseases including bladder cancer (Letašiová et al., 2012).

7.2 4-ABP

7.2.1 Toxicological Evaluation

4-ABP is a carcinogenic exit component of cigarette smoke, and is widely used in the rubber industry. Previous researches indicated that oral administration of 4-ABP could induce formation of bladder carcinomas in rats and develop hepatocellular and bladder carcinomas in mice (Lyon, 1982). 4-ABP has been extensively studied to elucidate the underlying mechanism of bladder carcinogenesis (Beland and Kadlubar, 1990). A genotoxic mode of action for 4-ABP has been demonstrated both in vitro and in vivo that involves the induction of DNA adducts and mutation (Nauwelaers et al., 2011). 4-ABP requires metabolic activation to exert its genotoxic effects through the induction of reactive oxygen species (Makena and Chung, 2007). The biotransformation of 4-ABP consists of N-oxidation catalyzed primarily by the cytochrome P450 (CYP) (Adris and Chung, 2006; Adris et al., 2007) and, to a much lesser extent, N-methylation or peroxidation (Ketelslegers et al., 2009). The resulting hydroxyarylamine may undergo detoxification through N-acetylation or conjugation with acetate, sulfate, or glucuronate (Chou et al., 1995; Orzechowski et al., 1994). 4-ABP has been classified as Group 1 (carcinogenic to humans) by IARC (Humans et al., 2004).

7.2.2 Metabolic Pathway in the Body

N-hydroxylation of the parent amine is a key step in the activation of AAs into carcinogenic species. N-hydroxy-4-aminobiphenyl can react with a single cysteine residue (93b) of hemoglobin (Hb) to form a sulfonamide cross-link (Bryant et al., 1987). The adduct can be hydrolyzed from Hb in alkaline conditions, which releases 4-ABP (Bryant et al., 1987). The hydroxylamine can also be transported to the bladder in a stable N- and O-glucuronidated form. In the acidic environment of the bladder, the glucuronide moiety can be hydrolyzed leading to the formation of DNA reactive nitrenium ions (Bryant et al., 1987). The glucuronide can also be hydrolyzed in urine samples using glucuronidase allowing the measurement of urinary 4-ABP (Grimmer et al., 2000). The acetate and sulfate O-conjugates can readily interact with DNA or proteins, whereas the glucuronate O-conjugate can circulate in the body and reach the urinary tract, wherein it undergoes hydrolysis at the acidic pH of urine (Orzechowski et al., 1994; Poupko et al., 1979). The resultant electrophilic nitrenium cation can bind directly to the DNA of the uroepithelial cells and form covalent adducts, predominantly at the C8 position of guanine, N-(deoxyguanosine-8-yl)-4-ABP (Frederickson et al., 1992) (Figure 7.1).

7.2.3 Biomarker of 4-ABP

7.2.3.1 Urinary 4-ABP

4-ABP in urine cannot only reflect the ability of an individual's detoxification but also evaluate the influence of xenobiotic enzyme polymorphisms on health risk.

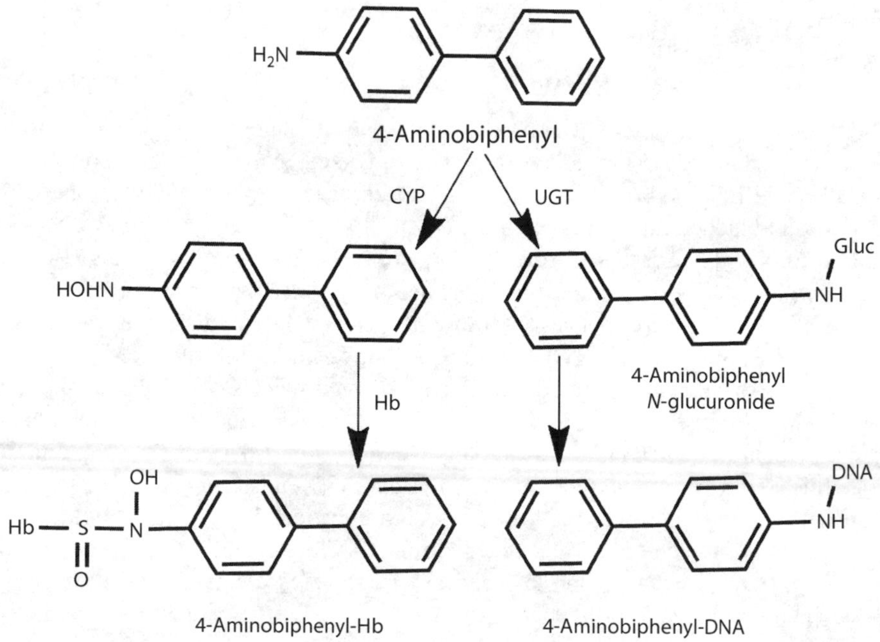

FIGURE 7.1 Metabolic pathway of 4-ABP.

Seyler and Bernert (2011) described a capillary gas chromatography/tandem mass spectrometry method for the measurement of urinary 4-ABP, the detection limit of the method was approximately 0.87 pg/mL. They also found that N-acetyl-4-ABP and 4-ABP glucuronide was present in urine samples from smokers. These metabolites have been identified in urine from animal, but have not been reported in human samples. However, this glucuronide was unstable and thus is unlikely to be used for biomonitoring. A validated method was used for the detection in samples obtained from smokers and nonsmokers. The 4-ABP concentration in smokers (geometric mean, 8.69 pg/mg creatinine) was significantly greater than that in nonsmokers (geometric mean, 1.64 pg/mg creatinine), $P < 0.001$. Jiang et al. (2014) developed a novel method using C_{18}-functional ultrafine magnetic silica nanoparticles (C_{18}-UMS NPs) as adsorbents to rapidly extract and enrich AAs (1-aminonaphthalene, 4-ABP, 4,4'-diaminodiphenylmethane, and 4-aminophenylthioether) from urine. The limits of detection (LODs) for 1-aminonaphthalene, 4-ABP, 4,4'-diaminodiphenylmethane, and 4-aminophenylthioether were 1.3, 0.88, 1.1, and 1.1 ng/mL, respectively. Yu et al. (2014) developed and validated a liquid chromatography–tandem mass spectrometry (LC–MS–MS) method for the determination of urinary 1-NA, 2-NA, 3-ABP, and 4-ABP in smokers and nonsmokers. A molecularly imprinted polymer solid-phase extraction (SPE) cartridge was applied to urine samples, and no derivatization reaction was involved. Lower LODs for four AAs were obtained and in the range of 1.5–5 ng/L. The method was applied to analyze urine samples of 40 smokers and 10 nonsmokers. They found that urinary 1-NA, 3-ABP, and 4-ABP excretion amounts showed significant differences ($P < 0.05$) between smokers and nonsmokers. Scherer et al. (2014) analyzed the relationship between urinary 4-ABP and 4-ABP-Hb, and a weak correlation between urinary and Hb biomarkers due to different accumulation and elimination rates was found. Time course analysis showed that a reduction in exposure was paralleled by a delayed reduction in Hb adducts.

7.2.3.2 Hemoglobin Adducts

Hb adduct levels of 4-ABP in smokers usually were 3–8-fold higher than that in nonsmokers (Bryant et al., 1987; Stillwell et al., 1987). An extreme situation existed for mothers at delivery: compared to nonsmoking women, smoking women from Louisville, KY, had 16-fold higher adduct contents (Pinorini-Godly and Myers, 1996), however, it was only 2.8-fold higher 4-ABP-Hb adduct concentrations in women from Homburg, Germany (Wickenpflug and Elmar, 1998). The relationship between smoking and 4-ABP adduct levels has an obvious dose response (Vineis et al., 1990; Wallin et al., 1995) and the adduct levels can be significantly decreased by controlled smoking cessation trials (Maclure et al., 1990). There are conflicting results about the influence of environmental tobacco smoke (ETS) on adduct levels. In New England, some studies indicated that passive smokers have a significant 1.3–1.6-fold higher 4-ABP of background levels than that found in adults (Maclure et al., 1989), children (Tang et al., 1999), and mothers at delivery (Hammond et al., 1993); however, there is no influence on pregnant women from Homburg, Germany (Wickenpflug and Elmar, 1998), and only a small number of insignificant increase was found in children from Upper Bavaria, Germany (Richter et al., 2001). In addition, the source of background levels of 4-ABP is still not clear. Possible sources of 4-ABP have been discussed. However, there is no convincing data to be presented. One possible source is the diet (Richter et al., 2000). Another source may be traffic exhaust containing 4-nitrobiphenyl (Bryant et al., 1987). Children who lived in a large city with high traffic density had 1.5-fold higher adduct levels than those who lived in a rural environment (Richter et al., 2001). The role of racial differences in influencing adducts is also very obvious (Mimi et al., 1994). Epileptic patients treated with phenobarbital were found to have lower levels of 4-ABP adducts than patients on other treatment. This effect was only apparent in smokers (Wallin et al., 1995). The mean 4-ABP-Hb adduct levels estimated from ETS exposure was 0.4–1.4 pg/g Hb (Schorp and Leyden, 2010), which account for 1%–4% of the median levels reported for nonsmokers.

7.2.3.3 DNA Adducts

Chemical carcinogenesis usually includes mutations and cell proliferation processes. Under the action of the cytochrome P-450 1A2 isozyme, aromatic amines are metabolized to form N-hydroxy derivatives in the liver (Murata et al., 2001). The oxidized forms of 4-ABP can bind with DNA to form DNA adducts. The N-hydroxy derivatives are metabolized to form N-glucuronides by the action of acetyltransferase or UDP-glucuronidase, and they are excreted into the blood and the urine (Kadlubar et al., 1977). These two derivatives can be converted to arylnitrenium ions, which then react with DNA to form DNA adducts (Ning and Xiaobai, 1997). In situations of Cu(II) and NADH, the N-hydroxy product of 4-ABP can induce to form 8-hydroxy-2-deoxyguanosine (8-OHdG), and it presents an obvious dose–effect relationship (Kadlubar et al., 1977). The 10 μM 4-ABP was incubated with cultured human hepatocytes. The levels of adducts ranged from 3.4 to 140 adducts per 10^7 DNA bases (Nauwelaers et al., 2011). The Shanghai Bladder Cancer Study enrolled 581 incident bladder cancer cases and 604 population controls, and they found ORs (95% CI) of bladder cancer for third and fourth versus first/second quartiles of 4-ABP Hb adducts were 1.30 (0.76–2.22) and 2.29 (1.23–4.24) among lifelong nonsmokers, respectively ($P_{trend}=0.009$). The two associations were independent of each other, which indicated that Hb adduct of 4-ABP was significantly and independently associated with increased bladder cancer risk among lifelong nonsmokers in Shanghai, China (Tao et al., 2013).

7.3 2-NA

7.3.1 Toxicological Evaluation

2-NA (Figure 7.2) is a well-known carcinogen AA found in amounts of 1.0–20 ng/cigarette in cigarette smoke (Patrianakos and Hoffmann, 1979).

The LD_{50} of 2-NA was 727 mg/kg in rat. Epidemiological studies have indicated that there was a strong relationship between occupational exposure to 2-NA and the occurrence of bladder cancer (Group, 1973). Oral administration of 2-NA has been reported to produce bladder carcinomas in the dog and monkey and at high dosage

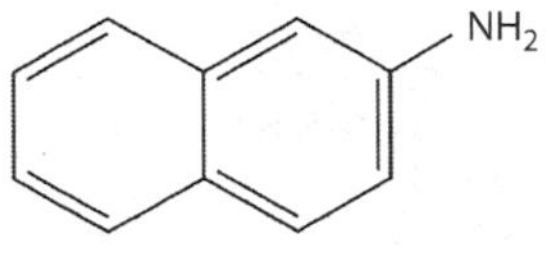

2-Naphthylamine

FIGURE 7.2 The structure of 2-NA.

levels in the hamster (Group, 1973). During the process of its toxic interaction with DNA, 2-NA can penetrate into the stack base pairs of DNA, where its intrinsic fluorescence can be quenched by DNA via a static pathway (Lin et al., 2013; Wang and Guengerich, 2013). 2-NA has been classified as a Group 1 carcinogen by the International Agency for Research on Cancer (IARC) that is carcinogenic to humans (Group, 1973). The commercial production and usage of 2-NA have been prohibited, whereas smokers who smoked every cigarette are exposed to 1–22 ng of 2-NA (IARC, 1985). There is increasing evidence that the excess of bladder cancer in smokers is attributable to AAs (Vineis and Pirastu, 1997). Smoking is a risk factor for bladder cancer, in addition to lung cancer (Bartsch et al., 1993; Swoboda and Friedl, 1992).

7.3.2 Metabolic Pathway in the Body

2-NA is metabolically activated through N-oxidation by the NADPH-dependent monooxygenase system of the liver, that is, the CYP monooxygenase, or by a prostaglandin H synthase–catalyzed ring oxygenation (Boyd and Eling, 1987; Hammons et al., 1989). 2-Nitroso-1-naphthol (NO-naphthol) may be produced as a metabolite of 2-NA. NO-Naphthol is formed by conversion of the N-hydroxy-2-naphthylamine (N-OH-NA) into nitroso compounds, followed by hydroxylation (Boyd and Eling, 1987; Manson, 1974).

7.3.3 Biomarkers of 2-NA

Urinary 2-NA in male and female rats showed a mean baseline concentration below 10 ng/L, equivalent to a daily urinary excretion of less than 0.1 ng 2-NA (Weiss et al., 2013). Urinary excretion levels of 2-NA between smokers (47.40 ± 50.68 ng/24 h) and nonsmokers (10.18 ± 7.25 ng/24 h) showed statistically significant differences ($P < 0.05$) (Yu et al., 2014). Urinary 2-NA and 4-ABP could be used as biomarker to reflect the exposure to 2-NA in cigarette smoke.

7.4 CONCLUSION

Previous research indicated that urinary 4-ABP and 2-NA excretion amounts showed significant differences between smokers and nonsmokers. Urinary 4-ABP and 2-NA are useful biomarkers to assess cigarette 4-ABP and 2-NA exposure.

REFERENCES

Adris, P., Chung, K.-T. Metabolic activation of bladder procarcinogens, 2-aminofluorene, 4-aminobiphenyl, and benzidine by *Pseudomonas aeruginosa* and other human endogenous bacteria. *Toxicology In Vitro.* 2006, 20, 367–374.

Adris, P., Lopez-Estraño, C., Chung, K.-T. The metabolic activation of 2-aminofluorine, 4-aminobiphenyl, and benzidine by cytochrome P-450-107S1 of *Pseudomonas aeruginosa. Toxicology In Vitro.* 2007, 21, 1663–1671.

Bartsch, H., Malaveille, C., Friesen, M. et al. Black (air-cured) and blond (flue-cured) tobacco cancer risk IV: Molecular dosimetry studies implicate aromatic amines as bladder carcinogens. *European Journal of Cancer.* 1993, 29, 1199–1207.

Beland, F., Kadlubar, F. Metabolic activation and DNA adducts of aromatic amines and nitroaromatic hydrocarbons. In: Cooper, C. S. et al. (Eds.), *Chemical Carcinogenesis and Mutagenesis I.* 1990, Springer, Berlin, Germany.

Boyd, J. A., Eling, T. E. Prostaglandin H synthase-catalyzed metabolism and DNA binding of 2-naphthylamine. *Cancer Research.* 1987, 47, 4007–4014.

Bryant, M. S., Skipper, P. L., Tannenbaum, S. R. et al. Hemoglobin adducts of 4-aminobiphenyl in smokers and nonsmokers. *Cancer Research.* 1987, 47, 602–608.

Chou, H.-C., Lang, N., Kadlubar, F. Metabolic activation of the N-hydroxy derivative of the carcinogen 4-aminobiphenyl by human tissue sulfotransferases. *Carcinogenesis.* 1995, 16, 413–417.

Food and Drug Administration. Guidance for Industry and FDA Staff: "Harmful and Potentially Harmful Constituents" in Tobacco Products as Used in Section 904(e) of the Federal Food, Drug, and Cosmetic Act. 2012. http://www.fda.gov/downloads/TobaccoProducts/GuidanceComplianceRegulatoryInformation/UCM241352.pdf.

Frederickson, S. M., Hatcher, J. F., Reznikoff, C. A. et al. Acetyl transferase-mediated metabolic activation of *N*-hydroxy-4-aminobiphenyl by human uroepithelial cells. *Carcinogenesis.* 1992, 13, 955–961.

Grimmer, G., Dettbarn, G., Seidel, A. et al. Detection of carcinogenic aromatic amines in the urine of non-smokers. *Science of the Total Environment.* 2000, 247, 81–90.

Hammond, S. K., Coghlin, J., Gann, P. H. et al. Relationship between environmental tobacco smoke exposure and carcinogen-hemoglobin adduct levels in nonsmokers. *Journal of the National Cancer Institute.* 1993, 85, 474–478.

Hammons, G. J., Alworth, W. L., Hopkins, N. E. et al. 2-Ethynylnaphthalene as a mechanism-based inactivator of the cytochrome P-450 catalyzed N-oxidation of 2-naphthylamine. *Chemical Research in Toxicology.* 1989, 2, 367–374.

Hoffmann, D., Masuda, Y., Wynder, E. L. α-Naphthylamine and β-naphthylamine in cigarette smoke. *Nature.* 1969, 221, 255–256.

IARC Working Group. Some aromatic amines, hydrazine and related substances, *N*-nitroso compounds and miscellaneous alkylating agents. *Lyon.* 1973, 4, 97–111.

IARC Working Group on the Evaluation of Carcinogenic Risks to Humans. Tobacco smoke and involuntary smoking. 2004, IARC, Lyon, France. This publication represents the views and expert opinions of an IARC Working Group on the Evaluation of Carcinogenic Risks to Humans, Lyon, France, June 11–18, 2002.

IARC. Tobacco habits other than smoking; betel-quid and areca nut chewing; and some related nitrosamines. Appendix 2. *IARC Monographs on the Evaluation of the Carcinogenic Risk of Chemicals to Man.* 1985, IARC, Lyon, France, Vol. 38, pp. 389–394.

International Agency for Research on Cancer. Some aromatic amines, organic dyes, and related exposures. *IARC Monographs on the Evaluation of Carcinogenic Risks to Humans.* 2010, Vol. 99, pp. 1–706, IARC, Lyon, France. http://monographs.iarc.fr/ENG/Monographs/vol99/mono99.pdf.

Jiang, C., Sun, Y., Yu, X. et al. Application of C_{18}-functional magnetic nanoparticles for extraction of aromatic amines from human urine. *Journal of Chromatography B*. 2014, 947, 49–56.

Kadlubar, F. F., Miller, J. A., Miller, E. C. Hepatic microsomal N-glucuronidation and nucleic acid binding of N-hydroxyarylamines in relation to urinary bladder carcinogenesis. *Cancer Research*. 1977, 37, 805–814.

Ketelslegers, H. B., Godschalk, R. W., Eskens, B. J. et al. Potential role of cytochrome P450-1B1 in the metabolic activation of 4-aminobiphenyl in humans. *Molecular Carcinogenesis*. 2009, 48, 685–691.

Letašiová, S., Medve'ová, A., Šovčíková, A. et al. Bladder cancer, a review of the environmental risk factors. *Environmental Health*. 2012, 11, S11.

Lin, J., Liu, Y., Liu, L. et al. Studies on the toxic interaction mechanism between 2-naphthylamine and herring sperm DNA. *Journal of Biochemical and Molecular Toxicology*. 2013, 27, 279–285.

Lyon, F. Some aromatic amines, anthraquinones and nitroso compounds, and inorganic fluorides used in drinking-water and dental preparations, *IARC Monographs on the Evaluation of Carcinogenic Risks to Humans*. 1982, Vol. 27, p. 341, IARC, Lyon, France.

Maclure, M., Bryant, M. S., Skipper, P. L. et al. Decline of the hemoglobin adduct of 4-aminobiphenyl during withdrawal from smoking. *Cancer Research*. 1990, 50, 181–184.

Maclure, M., Katz, R., Bryant, M. S. et al. Elevated blood levels of carcinogens in passive smokers. *American Journal of Public Health*. 1989, 79, 1381–1384.

Makena, P. S., Chung, K. T. Evidence that 4-aminobiphenyl, benzidine, and benzidine congeners produce genotoxicity through reactive oxygen species. *Environmental and Molecular Mutagenesis*. 2007, 48, 404–413.

Manson, D. Oxidation of N-naphthylhydroxylamines to nitrosonaphthols by air. *Journal of the Chemical Society, Perkin Transactions 1*. 1974, 2, 192–194.

Mimi, C. Y., Skipper, P. L., Taghizadeh, K. et al. Acetylator phenotype, aminobiphenyl-hemoglobin adduct levels, and bladder cancer risk in white, black, and Asian men in Los Angeles, California. *Journal of the National Cancer Institute*. 1994, 86, 712–716.

Murata, M., Tamura, A., Tada, M. et al. Mechanism of oxidative DNA damage induced by carcinogenic 4-aminobiphenyl. *Free Radical Biology and Medicine*. 2001, 30, 765–773.

Nauwelaers, G., Bessette, E. E., Gu, D. et al. DNA adduct formation of 4-aminobiphenyl and heterocyclic aromatic amines in human hepatocytes. *Chemical Research in Toxicology*. 2011, 24, 913–925.

Ning, S., Xiaobai, X. Reductive metabolism of 4-nitrobiphenyl by rat liver fraction. *Carcinogenesis*. 1997, 18, 1233–1240.

Orzechowski, A., Schrenk, D., Bock-Hennig, B. S. et al. Glucuronidation of carcinogenic arylamines and their N-hydroxy derivatives by rat and human phenol UDP-glucuronosyltransferases of the UGT1 gene complex. *Carcinogenesis*. 1994, 15, 1549–1553.

Patrianakos, C., Hoffmann, D. Chemical studies on tobacco smoke LXIV. On the analysis of aromatic amines in cigarette smoke. *Journal of Analytical Toxicology*. 1979, 3, 150–154.

Pinorini-Godly, M. T., Myers, S. R. HPLC and GC/MS determination of 4-aminobiphenyl haemoglobin adducts in fetuses exposed to the tobacco smoke carcinogen in utero. *Toxicology*. 1996, 107, 209–217.

Poupko, J. M., Hearn, W. L., Radomski, J. L. N-Glucuronidation of N-hydroxy aromatic amines: A mechanism for their transport and bladder-specific carcinogenicity. *Toxicology and Applied Pharmacology*. 1979, 50, 479–484.

Richter, E., Branner, B. Biomonitoring of exposure to aromatic amines: Haemoglobin adducts in humans. *Journal of Chromatography B*. 2002, 778, 49–62.

Richter, E., Rösler, S., Becker, A. Effect of diet on haemoglobin adducts from 4-aminobiphenyl in rats. *Archives of Toxicology*. 2000, 74, 203–206.

Richter, E., Rösler, S., Scherer, G. et al. Haemoglobin adducts from aromatic amines in children in relation to area of residence and exposure to environmental tobacco smoke. *International Archives of Occupational and Environmental Health*. 2001, 74, 421–428.

Riedel, K., Scherer, G., Engl, J. et al. Determination of three carcinogenic aromatic amines in urine of smokers and nonsmokers. *Journal of Analytical Toxicology*. 2006, 30, 187–195.

Saha, S., Mistri, R., Ray, B. Rapid and sensitive method for simultaneous determination of six carcinogenic aromatic amines in mainstream cigarette smoke by liquid chromatography/electrospray ionization tandem mass spectrometry. *Journal of Chromatography A*. 2009, 1216, 3059–3063.

Scherer, G., Newland, K., Papadopoulou, E. et al. A correlation study applied to biomarkers of internal and effective dose for acrylonitrile and 4-aminobiphenyl in smokers. *Biomarkers*. 2014, 19, 1–11.

Schorp, M. K., Leyden, D. E. Biomonitoring of smoke constituents: Exposure to 4-aminobiphenyl and 4-aminobiphenyl hemoglobin adduct levels in nonsmokers and smokers. *Inhalation Toxicology*. 2010, 22, 725–737.

Seyler, T. H., Bernert, J. T. Analysis of 4-aminobiphenyl in smoker's and nonsmoker's urine by tandem mass spectrometry. *Biomarkers*. 2011, 16, 212–221.

Stillwell, W., Bryant, M. S., Wishnok, J. S. GC/MS analysis of biologically important aromatic amines. Application to human dosimetry. *Biomedical and Environmental Mass Spectrometry*. 1987, 14, 221–227.

Swoboda, H., Friedl, H.-P. Tobacco-related cancer in relation to prevalence of drinking and smoking in Eastern Austria. *Journal of Cancer Research and Clinical Oncology*. 1992, 118, 621–625.

Tang, D., Warburton, D., Tannenbaum, S. R. et al. Molecular and genetic damage from environmental tobacco smoke in young children. *Cancer Epidemiology Biomarkers and Prevention*. 1999, 8, 427–431.

Tao, L., Day, B. W., Hu, B. et al. Elevated 4-aminobiphenyl and 2, 6-dimethylaniline hemoglobin adducts and increased risk of bladder cancer among lifelong nonsmokers—The Shanghai Bladder Cancer Study. *Cancer Epidemiology Biomarkers and Prevention*. 2013, 22, 937–945.

Vineis, P., Caporaso, N., Tannenbaum, S. R. et al. Acetylation phenotype, carcinogen-hemoglobin adducts, and cigarette smoking. *Cancer Research*. 1990, 50, 3002–3004.

Vineis, P., Pirastu, R. Aromatic amines and cancer. *Cancer Causes and Control*. 1997, 8, 346–355.

Wallin, H., Skipper, P. L., Tannenbaum, S. R. et al. Altered aromatic amine metabolism in epileptic patients treated with phenobarbital. *Cancer Epidemiology Biomarkers and Prevention*. 1995, 4, 771–773.

Wang, K., Guengerich, F. P. Reduction of aromatic and heterocyclic aromatic N-hydroxylamines by human cytochrome P450 2S1. *Chemical Research in Toxicology*. 2013, 26, 993–1004.

Weiss, T., Bolt, H. M., Schlüter, G. et al. Metabolic dephenylation of the rubber antioxidant N-phenyl-2-naphthylamine to carcinogenic 2-naphthylamine in rats. *Archives of Toxicology*. 2013, 87, 1265–1272.

Wickenpflug, B. B. C. K. W., Elmar, G. S. W.-D. H. Haemoglobin adducts from aromatic amines and tobacco specific nitrosamines in pregnant smoking and non smoking women. *Biomarkers*. 1998, 3, 35–47.

Yu, J., Wang, S., Zhao, G. et al. Determination of urinary aromatic amines in smokers and nonsmokers using a MIPs-SPE coupled with LC–MS/MS method. *Journal of Chromatography B*. 2014, 958, 130–135.

8 Biomarkers of Catechol and Hydroquinone

8.1 INTRODUCTION

Catechol, also known as pyrocatechol or 1,2-dihydroxybenzene, is an organic compound with the molecular formula $C_6H_4(OH)_2$ (Figure 8.1). In humans, catechols can occur as metabolites in the degradation of benzene or estrogens (Bolton et al., 1998; Porteous et al., 1949) or as endogenous compounds, such as neurotransmitter and their precursors (adrenaline, noradrenaline, dopamine, and L-3,4-dihydroxyphenylalaline [L-DOPA]) (Schweigert et al., 2001). Additionally, catechols can be taken up in the form of tobacco smoke (as catechol, catechol semiquinones, and polymerized catechols) (Pryor et al., 1998; Stone et al., 1995). The toxicity of catechols for microorganisms has been demonstrated in the past years (Boyd et al., 1997; Capasso et al., 1995; Fritz et al., 1991; Hellmér et al., 1992) and has been suggested to be the reason for the difficulties in cultivating microorganisms on benzene, toluene, or chlorobenzene (Fritz et al., 1991). Several studies additionally indicated the toxicity of catechols for water flea, zebra fish, trout, rabbit, cat, rat, mouse, and human cell lines (Garton and Williams, 1948; Hattula et al., 1981; Pellack-Walker et al., 1985; Rahouti et al., 1999; Schweigert et al., 2001; Van Den Heuvel et al., 1999). Indeed, in the environment toxic concentrations of catechols have been found (Capasso et al., 1995). Despite the facts that catechols are ubiquitous and their toxicity has been observed in a variety of organisms, the modes of action causing the toxicity are hardly understood. Catechols can form stable complexes with various di- and trivalent metal ions, the complexes with trivalent ions being the most stable. Catechols can also undergo redox reactions, cycling between catechols, semiquinone radicals, and orthobenzoquinone (Schweigert et al., 2001).

Hydroquinone, also benzene-1,4-diol or quinol, is an aromatic organic compound that is a type of phenol, having the chemical formula $C_6H_4(OH)_2$ (Figure 8.1). Hydroquinone is autoxidized by two successive one-electron oxidations, producing an extremely reactive semiquinone intermediate, which is the most reactive and most toxic intermediate of the quinone species (Enguita and Leitao, 2013). Dihydroxybenzene and quinones are recognized to induce oxidative stress as well as to nonspecifically bind both DNA and protein (North et al., 2011). Hydroquinone can form complexes with various di- and trivalent metal ions, such as copper and iron. In the case of copper, the complex formed increased H_2O_2 production by hydroquinone and enhances its autoxidation to benzoquinone (Sarkar et al., 2009).

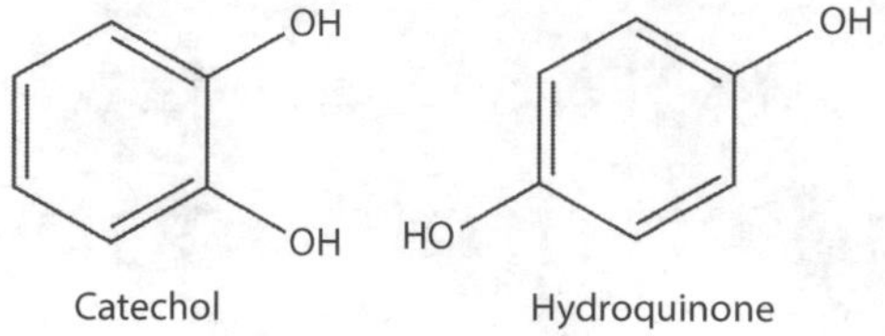

FIGURE 8.1   Chemical structure of catechol and hydroquinone.

8.2   TOXICOLOGICAL EVALUATION

Catechol and hydroquinone are present in the weakly acidic (phenolic) fraction of cigarette smoke, which has both cocarcinogenic and tumor-promoting activity (Leanderson and Tagesson, 1990). Catechol is strongly cocarcinogenic on mouse skin when applied together with benzo[a]pyrene (B[a]P) (Hecht et al., 1981; Vanduuren and Goldschmidt, 1976), and both hydroquinone and catechol are genotoxic (Robertson et al., 1991; Weisburger, 1992) and induce sister chromatid exchanges in human lymphocytes (Morimoto et al., 1983) and enzyme-altered foci in rat liver (Stenius et al., 1989).

Several studies have shown the toxicity of catechol on animal and human cell lines (Garton and Williams, 1948; Lai and Yu, 1997). Catechol can be oxidized to generate reactive oxygen species (ROS), semiquinone radicals, and quinines, which lead to oxidative stress and cell death (Bagchi, 1997; Benndorf et al., 2001). The semiquinone radicals and quinines can combine covalently with essential proteins and DNA to cause tissue damage (Chouchane et al., 2006). At micromolar concentration, catechol induces a time-dependent release of iron from ferritin in vitro and causes lipid peroxidation in rat brain homogenates (Agrawal et al., 2001). Furthermore, catechol, regulating the oxidation of protein kinase C, can influence the invasive capacity and metastatic spread of lung carcinoma cells (Gopalakrishna et al., 1994) and can also be genotoxic (Fabiani et al., 2001). Another study demonstrated that catechol-treated human Müller cells (MIO-M1) have decreased cell viability and mitochondrial function, increased caspase-3/7 activity, higher production of ROS/RNS, and decreased level of ATP compared to the control cultures (Mansoor et al., 2010).

At levels below 10 μmol/L, catechol and hydroquinone can suppress the mitogenic reaction of rat spleen cells to phytohemagglutinin (Pfeifer and Irons, 1981, 1982). It is universally accepted that catechol and hydroquinone could be further oxidized to quinines, so that they are in a position to alter most lymphoid responses (Bodell et al., 1993; Irons, 1985; Greenlee et al., 1981). However, one study found that hydroquinone could block IL-2-dependent proliferation of human T lymphoblasts (HTLs) keeping the ability of HTLs to produce IL-2 by mitogenic stimulation (Li et al., 1996). Furthermore, hydroquinone did not diminish intracellular glutathione levels, and its effect on DNA synthesis could be reversed by washing, suggesting that its effects are not mediated by the thiol-reactive p-BQ. These observations suggest that hydroquinone may interfere with either an IL-2-mediated event or a rate-limiting step in DNA synthesis (Li et al., 1997). IL-2 can induce the expression of transferrin receptors (TfR, CD71). TfR can help with the uptake of extracellular iron. The suppression of TfR expression might explicate the antiproliferative effects of hydroquinone, because TfR expression is essential for the induction of S-phase of the cell cycle (Neckers and Cossman, 1983; Seiser et al., 1993).

Exposure of HTLs in vitro to 50 µmol/L hydroquinone or 50 µmol/L catechol decreased IL-2-dependent DNA synthesis and cell proliferation by >90% with no effect on cell viability. The addition of catechol or hydroquinone to proliferate HTL could block ^3H-TdR uptake by more than 90% within 2 h, while having no significant effect on ^3H-UR uptake, indicating that these two compounds could suppress a rate-limiting step in DNA synthesis. Yet, the effects of catechol and hydroquinone may involve different mechanisms. Ferric chloride ($FeCl_3$) could reverse the inhibition effect of catechol without hydroquinone, in line with the well-known ability of catechol to chelate iron (Li et al., 1997). Hydroquinone could induce a reduction in TfR (CD71) expression, comparable to the concentration discovered in IL-2-starved cells. Hydroquinone could inhibit DNA synthesis in cultures of primary and transformed fibroblasts, mink lung epithelial cells, and transformed Jurkat T lymphocytes, suggesting that the antiproliferative effect of hydroquinone was not restricted to the proliferation mediated by IL-2 (Li et al., 1997). Nevertheless, compared to the primary human fibroblasts (IC50, 45 µmol/L) or the transformed Jurkat T cell line (IC50, 37 µmol/L), DNA synthesis by primary lymphocytes was more sensitive to hydroquinone (IC50, 6 µmol/L), indicating that normal lymphocytes appear to be especially sensitive to hydroquinone. In addition, the effects of catechol and hydroquinone on DNA synthesis might be partially reversed by a combination of guanosine deoxyribose and adenosine deoxyribose, indicating that ribonucleotide reductase might be inhibited by these two compounds (Li et al., 1997).

8.3   BIODEGRADATION AND METABOLISM

Catechols may serve as antioxidants, preventing lipid peroxidation, and may also act as prooxidant damaging macromolecules including DNA and proteins. Catechols may destroy membrane functioning through the redox cycling activity (Schweigert et al., 2001). Catechol itself does not cause oxidative DNA damage in vitro. In the case of molecular oxygen, DNA strand breaks can be found by combining with heavy metals. (e.g. Fe^{3+}, Cu^{2+}) (Li and Trush, 1994; Schweigert et al., 1999). DNA strand breaks are caused by a redox reaction of Cu(II) and catechol to yield Cu(I) and the semiquinone radical and a subsequent copper-catalyzed reduction of molecular oxygen, where superoxide and hydrogen peroxide are formed. A DNA–copper–oxo complex (DNA–Cu(I)–OOH) finally causes the DNA strand breaks by splitting of hydroxyl radicals in the vicinity of the DNA (Schweigert et al., 2000).

The DNA damage induced by catechol and hydroquinone, including a fraction of aqueous cigarette tar (ACT), has also been studied for the DNA-damaging activity in in vitro assays. This fraction can produce superoxide, hydrogen peroxide, and hydroxyl radicals and cause oxidative DNA damage (Pryor et al., 1998). This ACT also contains the tar radical consisting of polymerized catechol, suggesting that catechol is involved in the production of the ROS. The tar radicals bind to DNA and DNA adducts are formed (Stone et al., 1995). The DNA-damaging activity of

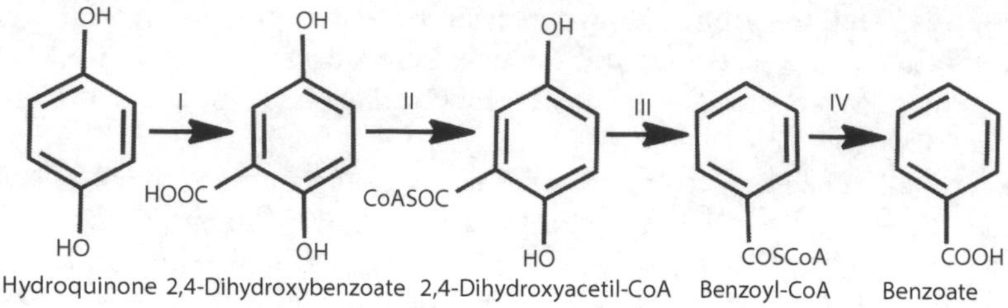

FIGURE 8.2 Anaerobic pathway for the metabolization of hydroquinone. I, hydroquinone carboxylase; II, hydroquinone acetyl CoA transferase; III, benzoyl-CoA oxidoreductase; IV, benzoyl-CoA hydrolase. (From Enguita, F.J. and Leitao, A.L., *BioMed. Res. Int.*, 2013, 542168, 2013.)

catechol estrogens (CEs) has been intensively studied. During the redox cycling of CEs, ROS and semiquinones are formed in the presence of Cu(II), and an enhanced rate of oxidative DNA damage can be determined (Mobley et al., 1999; Seacat et al., 1997). Through direct covalent binding to the DNA, quinines and/or semiquinones formed from catechol estrogens may also damage DNA (Akanni and Abul-Hajj, 1997; Dwivedy et al., 1992). This catechol moiety is responsible for the formation of CE–DNA adducts (Akanni and Abul-Hajj, 1997; Bolton et al., 1998).

Hydroquinone can be degraded by two different pathways depending on the oxygen availability. However, the anaerobic metabolization of hydroquinone is a less frequent process in nature, mainly restricted to a specific group of bacteria. It involves the conversion of hydroquinone to benzoate with an intermediate carboxylation and activation of the products by their linkage to acetyl CoA (Figure 8.2)

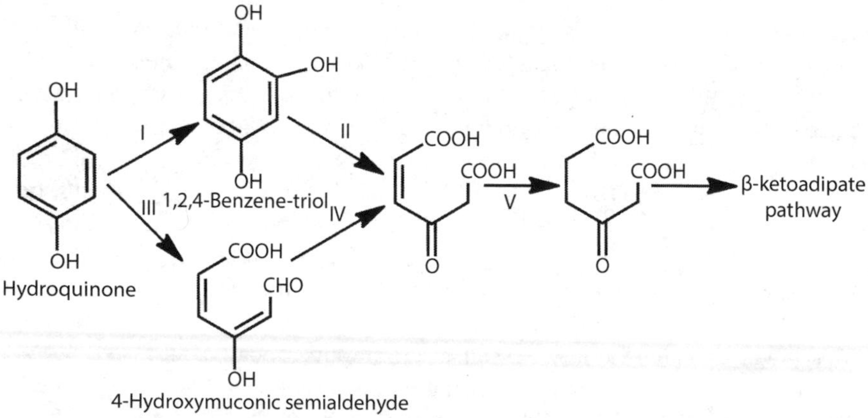

FIGURE 8.3 Two different branched pathways for the biodegradation of hydroquinone under aerobic conditions. I, hydroquinone hydroxylase; II, 1,2,4-trihydroxybenzene 1,2-dioxygenase; III, hydroquinone dioxygenase; IV, 4-hydroxymuconic semialdehyde dehydrogenase; V, beta-ketoadipate oxidoreductase. (From Enguita, F.J. and Leitao, A.L., *BioMed. Res. Int.*, 2013, 542168, 2013.)

(Enguita and Leitao, 2013). Cells can either employ benzoate as an anabolic fundamental brick or introduce the CoA-activated metabolites in the beta-oxidative catabolic pathway. In aerobic conditions, hydroquinone is channeled to the beta-ketoadipate pathway through two different metabolic branches (Figure 8.3) (Enguita and Leitao, 2013). The first pathway involves the initial hydroxylation of hydroquinone to 1,2,4-trihydroxybenzene followed by a ring-fission reaction catalyzed by a 1,2-dioxygenase (Anderson and Dagley, 1980; Takenaka et al., 2003; van Berkel et al., 1994). The second pathway of hydroquinone degradation is less common in nature. In this pathway, hydroquinone ring is directly cleaved by a specific hydroquinone 1,2-dioxygenase and the generated semialdehyde oxidized to maleylacetate (Darby et al., 1987; Spain and Gibson, 1991). The first aerobic branch has been characterized in bacteria and fungi; meanwhile, the second is exclusive of prokaryotic organisms (Darby et al., 1987; Spain and Gibson, 1991).

8.4 BIOMARKERS OF HYDROQUINONE AND CATECHOL

In one study, malondialdehyde (MDA) excretion in urine as an index for toxicological effects of hydroquinone was evaluated. The results indicated that the MDA assay was a selective and accurate marker for toxicological effects induced by hydroquinone (Ekström et al., 1988).

Hydroquinone formed one single detectable deoxyguanosine DNA adduct, which was a minor product of the reaction of DNA with p-benzoquinone (Gaskell et al., 2005a,b). After in vitro reaction of deoxyguanosine with hydroquinone, deoxyguanosine–benzoquinone adducts have been found and partly characterized (Snyder and Hedli, 1996; Snyder et al., 1987).

However, DNA adducts of hydroquinone have not been detected in humans. The covalent binding index of hydroquinone to DNA may be too low to form measurable hydroquinone DNA adduct levels by applying the currently available analytical methods. To date, DNA adducts of hydroquinone cannot be used as biomarkers mainly due to the lack of sensitive and specific analytical methods to measure such adducts (Enguita and Leitao, 2013).

Analytical methodologies for hydroquinone and catechol are specific and sensitive enough to evaluate general population exposure to these chemicals (Kerzic et al., 2010; Lee et al., 1993; Wittig et al., 2001). Overall, hydroquinone and catechol in blood and urine are the most relevant and valid biomarkers for interpreting hydroquinone and catechol exposure.

Hydroquinone and catechol in blood and urine indicate the exposure of hydroquinone and catechol. However, because of their short half-life, the concentrations in these biological matrices reflect only recent exposure.

REFERENCES

Agrawal, R., Sharma, P. K., Rao, G. S. Release of iron from ferritin by metabolites of benzene and superoxide radical generating agents. *Toxicology*. 2001, 168, 223–230.
Akanni, A., Abul-Hajj, Y. J. Estrogen-nucleic acid adducts: Reaction of 3,4-estrone-o-quinone radical anion with deoxyribonucleosides. *Chemical Research in Toxicology*. 1997, 10, 760–766.

Anderson, J. J., Dagley, S. Catabolism of aromatic acids in *Trichosporon cutaneum*. *Journal of Bacteriology*. 1980, 141, 534–543.

Benndorf, D., Loffhagen, N., Babel, W. Protein synthesis patterns in *Acinetobacter calcoaceticus* induced by phenol and catechol show specificities of responses to chemostress. *FEMS Microbiology Letters*. 2001, 200, 247–252.

Bodell, W. J., Levay, G., Pongracz, K. Investigation of benzene-DNA adducts and their detection in human bone marrow. *Environmental Health Perspectives*. 1993, 99, 241–244.

Bolton, J. L., Pisha, E., Zhang, F. et al. Role of quinoids in estrogen carcinogenesis. *Chemical Research in Toxicology*. 1998, 11, 1113–1127.

Boyd, E. M., Meharg, A. A., Wright, J. et al. Assessment of toxicological interactions of benzene and its primary degradation products (catechol and phenol) using a lux-modified bacterial bioassay. *Environmental Toxicology and Chemistry*. 1997, 16, 849–856.

Capasso, R., Evidente, A., Schivo, L. et al. Antibacterial polyphenols from olive oil mill waste waters. *Journal of Applied Bacteriology*. 1995, 79, 393–398.

Chouchane, S., Wooten, J. B., Tewes, F. J. et al. Involvement of semiquinone radicals in the in vitro cytotoxicity of cigarette mainstream smoke. *Chemical Research in Toxicology*. 2006, 19, 1602–1610.

Darby, J. M., Taylor, D. G., Hopper, D. J. Hydroquinone as the ring-fission substrate in the catabolism of 4-ethylphenol and 4-hydroxyacetophenone by *Pseudomonas putida* JD1. *Journal of General Microbiology*. 1987, 133, 2137–2146.

Dwivedy, I., Devanesan, P., Cremonesi, P. et al. Synthesis and characterization of estrogen 2,3- and 3,4-quinones. Comparison of DNA adducts formed by the quinones versus horseradish peroxidase-activated catechol estrogens. *Chemical Research in Toxicology*. 1992, 5, 828–833.

Ekström, T., Warholm, M., Kronevi, T. et al. Recovery of malondialdehyde in urine as a 2,4-dinitrophenylhydrazine derivative after exposure to chloroform or hydroquinone. *Chemico-Biological Interactions*. 1988, 67, 25–31.

Enguita, F. J., Leitao, A. L. Hydroquinone: Environmental pollution, toxicity, and microbial answers. *BioMed Research International*. 2013, 2013, 542168.

Fabiani, R., De Bartolomeo, A., Rosignoli, P. et al. Influence of culture conditions on the DNA-damaging effect of benzene and its metabolites in human peripheral blood mononuclear cells. *Environmental and Molecular Mutagen*. 2001, 37, 1–16.

Fritz, H., Reineke, W., Schmidt, E. Toxicity of chlorobenzene on *Pseudomonas* sp. strain RHO1, a chlorobenzene-degrading strain. *Biodegradation*. 1991, 2, 165–170.

Garton, G. A., Williams, R. T. Studies in detoxication. 17. The fate of catechol in the rabbit and the characterization of catechol monoglucuronide. *Biochemical Journal*. 1948, 43, 206–211.

Gaskell, M., McLuckie, K. I. E., Farmer, P. B. Comparison of the repair of DNA damage induced by the benzene metabolites hydroquinone and *p*-benzoquinone: A role for hydroquinone in benzene genotoxicity. *Carcinogenesis*. 2005a, 26, 673–680.

Gaskell, M., McLuckie, K. I. E., Farmer, P. B. Genotoxicity of the benzene metabolites para-benzoquinone and hydroquinone. *Chemico-Biological Interactions*. 2005b, 153–154, 267–270.

Gopalakrishna, R., Chen, Z. H., Gundimeda, U. Tobacco smoke tumor promoters, catechol and hydroquinone, induce oxidative regulation of protein kinase C and influence invasion and metastasis of lung carcinoma cells. *Proceedings of the National Academy of Sciences USA*. 1994, 91, 12233–12237.

Greenlee, W. F., Sun, J. D., Bus, J. S. A proposed mechanism of benzene toxicity: Formation of reactive intermediates from polyphenol metabolites. *Toxicology and Applied Pharmacology*. 1981, 59, 187–195.

Hattula, M. L., Wasenius, V. M., Reunanen, H. et al. Acute toxicity of some chlorinated phenols, catechols and cresols to trout. *Bulletin of Environmental Contamination and Toxicology*. 1981, 26, 295–298.

Hecht, S. S., Carmella, S., Mori, H. et al. A study of tobacco carcinogenesis. 20. Role of catechol as a major cocarcinogen in the weakly acidic fraction of smoke condensate. *Journal of the National Cancer Institute.* 1981, 66, 163–169.

Hellmér, L., Bolcsfodi, G. An evaluation of the *E. coli* K-12 uvrB/recA DNA repair host-mediated assay. I. In vitro sensitivity of the bacteria to 61 compounds. *Mutation Research.* 1992, 272, 145–160.

Irons, R. D. Quinones as toxic metabolites of benzene. *Journal of Toxicology and Environmental Health.* 1985, 16, 673–678.

Kerzic, P. J., Liu, W. S., Pan, M. T. et al. Analysis of hydroquinone and catechol in peripheral blood of benzene-exposed workers. *Chemico-Biological Interactions.* 2010, 184, 182–188.

Lai, C. T., Yu, P. H. Dopamine- and L-beta-3,4-dihydroxyphenylalanine hydrochloride (L-Dopa)-induced cytotoxicity towards catecholaminergic neuroblastoma SH-SY5Y cells. Effects of oxidative stress and antioxidative factors. *Biochemical Pharmacology.* 1997, 53, 363–372.

Leanderson, P., Tagesson, C. Cigarette smoke-induced DNA-damage: Role of hydroquinone and catechol in the formation of the oxidative DNA-adduct, 8-hydroxydeoxyguanosine. *Chemico-Biological Interactions.* 1990, 75, 71–81.

Lee, B. L., Ong, H. Y., Shi, C. Y. et al. Simultaneous determination of hydroquinone, catechol and phenol in urine using high-performance liquid chromatography with fluorimetric detection. *Journal of Chromatography B: Biomedical Sciences and Applications.* 1993, 619, 259–266.

Li, Q., Aubrey, M. T., Christian, T. et al. Differential inhibition of DNA synthesis in human T cells by the cigarette tar components hydroquinone and catechol. *Fundamental and Applied Toxicology.* 1997, 38, 158–165.

Li, Q., Geiselhart, L., Mittler, J. N. et al. Inhibition of human T lymphoblast proliferation by hydroquinone. *Toxicology and Applied Pharmacology.* 1996, 139, 317–323.

Li, Y., Trush, M. A. Reactive oxygen-dependent DNA damage resulting from the oxidation of phenolic compounds by a copper-redox cycle mechanism. *Cancer Research.* 1994, 54, 1895s–1898s.

Mansoor, S., Gupta, N., Luczy-Bachman, G. et al. Protective effects of memantine and epicatechin on catechol-induced toxicity on Müller cells in vitro. *Toxicology.* 2010, 271, 107–114.

Mobley, J. A., Bhat, A. S., Brueggemeier, R. W. Measurement of oxidative DNA damage by catechol estrogens and analogues in vitro. *Chemical Research in Toxicology.* 1999, 12, 270–277.

Morimoto, K., Wolff, S., Koizumi, A. Induction of sister-chromatid exchanges in human-lymphocytes by microsomal activation of benzene metabolites. *Mutation Research.* 1983, 119, 355–360.

Neckers, L. M., Cossman, J. Transferrin receptor induction in mitogen-stimulated human T lymphocytes is required for DNA synthesis and cell division and is regulated by interleukin 2. *Proceedings of the National Academy of Sciences USA.* 1983, 80, 3494–3498.

North, M., Tandon, V. J., Thomas, R. et al. Genome-wide functional profiling reveals genes required for tolerance to benzene metabolites in yeast. *PLoS One.* 2011, 6, e24205.

Pellack-Walker, P., Walker, J. K., Evans, H. H. et al. Relationship between the oxidation potential of benzene metabolites and their inhibitory effect on DNA synthesis in L5178YS cells. *Molecular Pharmacology.* 1985, 28, 560–566.

Pfeifer, R. W., Irons, R. D. Inhibition of lectin-stimulated lymphocyte agglutination and mitogenesis by hydroquinone: Reactivity with intracellular sulfhydryl groups. *Experimental and Molecular Pathology.* 1981, 35, 189–198.

Pfeifer, R. W., Irons, R. D. Effect of benzene metabolites on phytohemagglutinin-stimulated lymphopoiesis in rat bone marrow. *Journal of Reticuloendothelial Society.* 1982, 31, 155–170.

Porteous, J. W., Williams, R. T. Studies in detoxication. 20. The metabolism of benzene. II. The isolation of phenol, catechol, quinol and hydroxyquinol from the ethereal sulphate fraction of the urine of rabbits receiving benzene orally. *Biochemical Journal*. 1949, 44, 56–61.

Pryor, W. A., Stone, K., Zang, L. Y. et al. Fractionation of aqueous cigarette tar extracts: Fractions that contain the tar radical cause DNA damage. *Chemical Research in Toxicology*. 1998, 11, 441–448.

Rahouti, M., Steiman, R., Seigle-Murandi, F. et al. Growth of 1044 strains and species of fungi on 7 phenolic lignin model compounds. *Chemosphere*. 1999, 38, 2549–2559.

Robertson, M. L., Eastmond, D. A., Smith, M. T. 2 Benzene metabolites, catechol and hydroquinone, produce a synergistic induction of micronuclei and toxicity in cultured human-lymphocytes. *Mutation Research*. 1991, 249, 201–209.

Sarkar, C., Mitra, P. K., Saha, S. et al. Effect of copper-hydroquinone complex on oxidative stress-related parameters in human erythrocytes (in vitro). *Toxicology Mechanisms and Methods*. 2009, 19, 86–93.

Schweigert, N., Acero, J. L., von Gunten, U. et al. DNA degradation by the mixture of copper and catechol is caused by DNA-copper-hydroperoxo complexes, probably DNA-Cu(I) OOH. *Environmental and Molecular Mutagenesis*. 2000, 36, 5–12.

Schweigert, N., Belkin, S., Leong-Morgenthaler, P. et al. Combinations of chlorocatechols and heavy metals cause DNA degradation in vitro but must not result in increased mutation rates in vivo. *Environmental and Molecular Mutagenesis*. 1999, 33, 202–210.

Schweigert, N., Zehnder, A. J. B., Eggen, R. I. L. Chemical properties of catechols and their molecular modes of toxic action in cells, from microorganisms to mammals. *Environmental Microbiology*. 2001, 3, 81–91.

Seacat, A. M., Kuppusamy, P., Zweier, J. L. et al. ESR identification of free radicals formed from the oxidation of catechol estrogens by Cu^{2+}. *Archives of Biochemistry and Biophysics*. 1997, 347, 45–52.

Seiser, C., Teixeira, S., Kühn, L. C. Interleukin-2-dependent transcriptional and post-transcriptional regulation of transferrin receptor mRNA. *Journal of Biological Chemistry*. 1993, 268, 13074–13080.

Snyder, R., Hedli, C. C. An overview of benzene metabolism. *Environmental Health Perspectives*. 1996, 104, 1165–1171.

Snyder, R., Jowa, L., Witz, G. et al. Formation of reactive metabolites from benzene. *Archives of Toxicology*. 1987, 60, 61–64.

Spain, J. C., Gibson, D. T. Pathway for biodegradation of *p*-nitrophenol in a *Moraxella* sp. *Applied and Environmental Microbiology*. 1991, 57, 812–819.

Stenius, U., Warholm, M., Rannug, A. et al. The role of GSH depletion and toxicity in hydroquinone-induced development of enzyme-altered foci. *Carcinogenesis*. 1989, 10, 593–599.

Stohs, S. J., Bagchi, D., Bagchi, M. Toxicity of trace elements in tobacco smoke. *Inhalation Toxicology*. 1997, 9, 867–890.

Stone, K., Bermudez, E., Zang, L. Y. et al. The ESR properties, DNA nicking, and DNA association of aged solutions of catechol versus aqueous extracts of tar from cigarette smoke. *Archives of Biochemistry and Biophysics*. 1995, 319, 196–203.

Takenaka, S., Okugawa, S., Kadowaki, M. et al. The metabolic pathway of 4-aminophenol in *Burkholderia* sp. strain AK-5 differs from that of aniline and aniline with C-4 substituents. 2003.

van Berkel, W. J., Eppink, M. H., Middelhoven, W. J. et al. Catabolism of 4-hydroxybenzoate in *Candida parapsilosis* proceeds through initial oxidative decarboxylation by a FAD-dependent 4-hydroxybenzoate 1-hydroxylase. *FEMS Microbiology Letters*. 1994, 121, 207–215.

Van Den Heuvel, R. L., Leppens, H., Schoeters, G. E. Lead and catechol hematotoxicity in vitro using human and murine hematopoietic progenitor cells. *Cell Biology and Toxicology*. 1999, 15, 101–110.

Vanduuren, B. L., Goldschmidt, B. M. Cocarcinogenic and tumor-promoting agents in tobacco carcinogenesis. *Journal of the National Cancer Institute*. 1976, 56, 1237–1242.

Weisburger, J. H. Mutagenic, carcinogenic, and chemopreventive effects of phenols and catechols—The underlying mechanisms. *ACS Symposium Series*. 1992, 507, 35–47.

Wittig, J., Wittemer, S., Veit, M. Validated method for the determination of hydroquinone in human urine by high-performance liquid chromatography–coulometric-array detection. *Journal of Chromatography B: Biomedical Sciences and Applications*. 2001, 761, 125–132.

9 Biomarkers of Hydrogen Cyanide, Carbon Monoxide, and Nitrogen Oxides

9.1 INTRODUCTION

Tobacco is the only product legally available in the market even if it is harmful to humans (World Health Organization, 2008). During smoking, a complex mixture of compounds is inhaled into the respiratory system affecting different organs. In 1959, some 400 compounds were known to be present in tobacco leaves and tobacco smoke; today, the figure has risen to more than 4000 (Stedman, 1968). The topic on tobacco and health was discussed and focused on by the public for a long time. There are 44 harmful constituents belonging to 13 different classes of chemical compounds revealed by Hoffmann, hydrocyanic acid, carbon monoxide and nitrogen oxides included; on the other hand, these three constituents are regarded as key controlled target compounds by the FCTC. The biomarkers of these three hazardous constituents reveal evidences that prove themselves and their metabolites present in human bodies.

9.2 HYDROGEN CYANIDE

The cyanide anion consists of a carbon atom triply bonded to a nitrogen atom with a net negative charge (Erik et al., 2001). Hydrogen cyanide (HCN) and simple cyanide salts such as NaCN and KCN are among the most toxic (Kroto et al., 1985). In tobacco smoke, cyanide occurs in considerable amounts, the concentration of which differs from one sort of tobacco to the next (Newsome et al., 1965; Stedman, 1968; Wynder and Hoffmann, 1967). In the mainstream smoke of cigarettes without filter, the value of HCN was found 150–300 µg (Scherer, 2006), which were in the range of the data reported (Newsome et al., 1965). In cigarettes with cellulose acetate filter, only an eligible retention by the filter was observed. Suitable absorbent filters can reduce the original cyanide content by 20%–80%. The cyanide content in the particle phase amounts to approximately 100 µg CN^-/cig, as well as in Blend cigarettes as in Maryland cigarettes. The cyanide values in the gaseous phase are slightly higher. For Blend cigarettes, they amount to about 150 µg CN^-/cig, for Maryland cigarettes about 120 µg CN^-/cig.

9.2.1 Toxicological Evaluation

As the archetypal poison, it is probably perceived that the lethal dose of cyanide is in the order of only 1 mg/kg (Scherer, 2006). The recognition of its toxicity can be traced back to antiquity. HCN was isolated from Prussian blue in 1782. In 1786, Scheele, the chemist who was the first to isolate the material, fully demonstrated its toxicity by accidentally breaking a vial and dying as a result. Cyanide poisoning, which is clinically dramatic, has lost much of the prominence it formerly had in forensic toxicology as a homicidal or suicidal etiology, of which there are very few instances today (Michael et al., 2004). However, the current massive use of cyanide salts for industrial purposes, and the release of HCN through burning plastics and other nitrogen containing materials (Yves et al., 1985), results in frequent cases of accidental poisoning.

Inhalation of HCN compounds in the chemical and galvanic industries as well as during gold extraction is another possibility for humans to take up considerable amounts of cyanide. The poisoning that means the liberation of HCN from organic thiocyanates is of importance. It is accomplished in insects and mammals by the enzyme glutathione S-transferase, mainly from lower aliphatic homologues. This fact explains the high toxicity of some lower alkyl derivatives of SCN. Also, liberation of cyanide from succinonitrile has been described recently (Wakefield et al., 2010).

Some features of acute exposure to HCN at below fatal concentrations may include headache, nausea, dizziness, confusion, muscle weakness, loss of coordination, hyperventilation, cardiac arrhythmia, bradycardia, rapid loss of consciousness, and coma. The concentration of HCN, which is fatal to humans following inhalation, is dependent upon the duration of exposure. It has been widely reported that: (1) a concentration of 130 ppm for 30 min is likely to be fatal, (2) a concentration of 180 ppm is likely to be fatal after just 10 min, (3) a concentration of 270 ppm is immediately fatal. A cyanide concentration of greater than 1 µg/mL in blood samples taken postmortem from fire fatalities is considered to suggest significant toxicity of HCN.3 µg/mL is considered to be a lethal level of cyanide. However, the measurement of blood cyanide levels is problematic. The analysis should be cautious with respect to factors including the time between sample removal and analysis and the storage method (Wakefield et al., 2010).

The toxicological diagnosis of cyanide poisoning usually has little impact on clinical therapies because the analysis for cyanide ion is usually performed with some delay; however, it is always important with a view to the diagnostic confirmation of clinical and forensic poisonings. The greatest hurdles posed by cyanide ion analysis result from the typically low concentrations involved in poisoning cases and the intrinsic instability of cyanide ion, which is relatively easily volatilized. These two hindrances have raised the need to develop a straightforward and expeditious method for the quantitative determination of cyanide ion in a small volume of sample and with minimal manipulation (to avoid losses) (Michael et al., 2004).

9.2.2 Metabolic Pathway in the Body

Following exposure and systemic uptake of HCN, it undergoes dissociation in the blood, to form the cyanide ion. The cyanide ion is easy to distribute within the body.

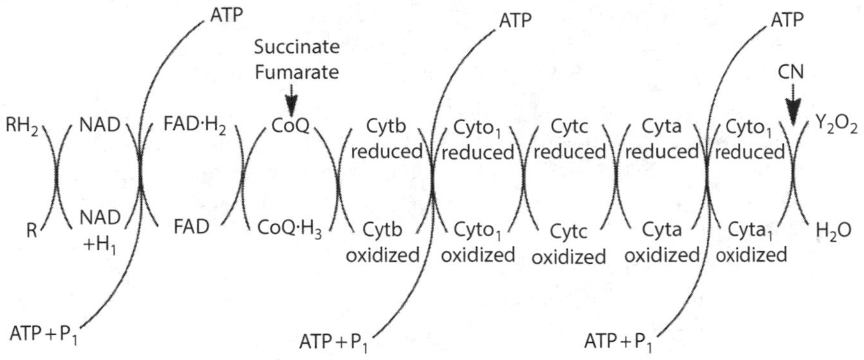

FIGURE 9.1   Outline of respiratory chain oxidation. Pi, inorganic phosphate; FAD, flavo-protein; CoQ, coenzyme Q; R1, proton donor; Cyt, cytochrome.

It is responsible for the toxicity of HCN by reducing the cellular utilization of oxygen (cellular respiration). Cytochrome oxidase is a principal enzyme involved in the utilization of oxygen in most cells throughout the body. The cyanide ion binds to cytochrome oxidase and inhibits it by forming a cytochrome oxidase cyanide complex. The inhibition of cytochrome oxidase results in a rapid onset of cytotoxic hypoxia and loss of cellular function. The cardiac and cerebral tissues are particularly susceptible to the effects of cyanide on cellular respiration. The most common cause of death from HCN intoxication is due to depression of the respiratory system resulting from the cytotoxic hypoxia effect of the cyanide ion on the central nervous system, but effects on the cardiovascular system may also be a cause of death (Wakefield et al., 2010).

Cyanide exerts its toxicity by the inhibition of cytochrome oxidase causing a cytotoxic hypoxia. It is toxic to a number of enzyme systems. Mechanisms include combination with essential metal ions, formation of cyanohydrins with carbonyl compounds, and the sequestration of sulfur as thiocyanate. However, the main target enzyme is cytochrome C oxidase, the terminal oxidase of the respiratory chain and involves interaction with the ferric ion of cytochrome (Figure 9.1). The net effect is the prevention of oxygen uptake at the intracellular tissue level. Cyanide intoxication features a bright red arterialization of venous blood as oxygen is not absorbed on passage through tissue (John et al., 2013), but this effect appears to be theoretical only and has not been recorded in fact (Cummings, 2004).

9.2.3   BIOMARKERS

Thiocyanate (SCN): SCN is a metabolite of HCN. As one of the most extensively used indicators of tobacco smoke, it can be monitored in various body mediums (such as in urine, saliva, and blood). It is a component of tobacco smoke, but usually can be found in nonsmokers (Gerhard et al., 2006), resulting from environmental and endogenous sources.

Cyanide is extensively present in the environment. For example, in the metal industry (John et al., 2013), HCN can be found in the working atmosphere.

Some intestinal tract bacteria also can produce HCN. The level of SCN may vary with the seasons resulting from variations in the intake of foodstuffs that contain small amounts of SCN. The half-life of SCN in plasma or saliva is 10–14 days.

Nitric oxide (NO): There have been relatively little studies on the possible health effects arising from exposure to NO. As an airway irritant it is recognized to be significantly less active than NO_2. NO is irritating to the eyes and upper respiratory tract. Deep inhalation can result in the delayed onset of pulmonary edema occurring a few hours postexposure and may be aggravated by physical exertion. It is reported that healthy human volunteers exposed to NO at concentrations above approximately 20 ppm (24.6 mg/m^3) have demonstrated a significant increase (10%) in total airway resistance (Wakefield et al., 2010).

9.3   CARBON MONOXIDE

CO is a colorless, odorless, tasteless, and nonirritant gas in nature. It is the most common asphyxiant gas that is poisonous to humans. It is formed during both smoldering and flaming combustion of all organic materials. It can be produced naturally or by human activity (Dimitri et al., 2010). A common source of CO for the general population is tobacco smoke. CO inhalation through the lungs results in the formation of carboxy hemoglobin (COHb). It reduces the oxygen-carrying capacity of the blood resulting in hypoxia (James et al., 2000).

The CO concentration in tobacco smoke is approximately 4.5% (45,000 ppm) (James et al., 2000). Regular cigarette smoking may produce COHb levels ranging from 5% (1 pack per day) to 9% (2–3 packs per day), whereas heavy cigar consumption can produce COHb levels up to 20%. In healthy individuals, concentrations of COHb at levels of 20%–30% can lead to headaches, dizziness, and shortness of breath, whereas levels of 30%–50% can lead to confusion and unconsciousness, and levels in excess of 50% are usually life threatening. Cigarette smokers often have low tolerable levels of COHb (up to 10%) (Dimitri et al., 2010).

9.3.1   TOXICOLOGICAL EVALUATION

The affinity of hemoglobin for CO is around 200–250 times greater than the affinity for O_2 (Gerhard et al., 2006), which is widely regarded as the main factor for the toxicity of CO. CO inhalation through the lungs results in the formation of COHb, which reduces the oxygen-carrying capacity of the blood resulting in hypoxia (James et al., 2000). The concentration of COHb in the blood will increase in most cases in individuals exposed to a combustion atmosphere. The concentration of COHb is dependent upon the duration of exposure and the concentration of CO in the environment (Wakefield et al., 2010).

Most fatalities following inhalation of CO have reported the concentration of COHb in the blood to be greater than 50%. Following acute exposure to CO, a postmortem COHb concentration of approximately 70% can be associated with CO poisoning alone. A COHb concentration in the range of 30%–70% is likely to be associated with cause of death being due to a combination of both CO poisoning and other factors, such as the presence of additional toxic combustion products. In a fatality in which there is a COHb concentration of less than 30%, the main cause of death is likely to be due to effects other than CO poisoning (Wakefield et al., 2010).

Acute health effects resulting from CO-induced hypoxia at concentrations below that causing lethality can include neurological effects such as headache, dizziness, confusion and disorientation, loss of coordination, memory loss, fainting, cerebral edema, and coma. In a fire environment, the neurological effects of CO exposure may hinder the ability to perform tasks, recognize danger and escape from a hazardous situation. Neurological symptoms following severe acute toxicity may appear 2–40 days postexposure, including lethargy, irritability, lack of concentration, and possible severe effects including dementia and psychosis. These are not all related to CO-induced hypoxia (Wakefield et al., 2010).

Inhalation of CO is also likely to give rise to metabolic acidosis. The heart is particularly sensitive to the effects of CO. An acute exposure may give rise to cardiovascular effects, reduced myocardial function, hypotension, vasodilation, cyanosis, cardiac arrhythmias, shock, circulatory failure, and cardiac arrest included (Wakefield et al., 2010).

CO can be measured in both expired alveolar air and blood (Stewart, 1975). Further research is needed to determine the toxicological importance of CO alone and in combination with the other components of tobacco smoke. Researchers have demonstrated high correlations among CO, self-reported smoking, and urinary cotinine. Associations between CO and urinary cotinine have ranged from $r = 76$ to $r = 79$; correlation coefficients between CO and self-report have ranged from $r = 65$ to $r = 70$ ($P < 0.001$ for all) (Secker-Walker et al., 1997). Exhaled CO has been successfully used to corroborate self-report data, with concordance approaching 100% (Becoña and Vázquez, 1998). In many studies currently cited to justify ambient air standards for CO, neither the smoking habits of the subjects nor their exposure to passive smoking has been taken into account (James et al., 2000).

9.3.2 METABOLIC PATHWAY IN THE BODY

The actual quantity of CO entering the lung depends on the form of tobacco, the pattern of smoking and the depth of inhalation. Very little CO is absorbed in the mouth and upper airways; therefore, most of the CO available for binding to hemoglobin in blood must reach the distal respiratory tract to raise the level of COHb present in blood (James et al., 2000). The formation of COHb also leads to a left shift in the dissociation of oxygen from hemoglobin (Gerhard et al., 2006), which further increase the likelihood of hypoxia. Hypoxia following exposure to CO results from the conversion of hemoglobin to COHb due to competition between O_2 and CO for the heme binding sites. The affinity of hemoglobin for CO is around 200–250 times greater than the affinity for O_2 (Wakefield et al., 2010).

It is reported that CO readily crosses the placenta and binds to fetal hemoglobin with a higher affinity than that for maternal hemoglobin. CO is also cleared from fetal blood much slower than that from maternal blood, resulting in 10%–15% increase in COHb formation in the fetus relative to the mother (Wakefield et al., 2010).

9.3.3 CO in Exhaled Air

CO is the most common asphyxiant product and is formed during both smoldering and flaming combustion containing organic materials. The production of CO in the combustion environment is dependent upon the availability of oxygen, with an increase in CO formation following from a decrease in O_2. Therefore, the production of CO is greater in cases of ventilation-controlled combustion than that with well-ventilated combustion. So the CO concentrations detected are greater in the case of a rotary smoking machine than that of a linear smoking machine. At the point of flashover, the production of CO increases significantly due to the combustion becoming ventilation-controlled and the rapid increase in the mass burning rate (Wakefield et al., 2010).

CO inhalation through the lungs results in the formation of COHb. In addition, as the result of higher baseline COHb levels, smokers actually may be exhaling more CO into the air than they are inhaling from the ambient environment. Smokers may even show an adaptive response to the elevated COHb levels, as evidenced by increased red cell volumes and reduced plasma volumes. As a consequence, it is not clear if incremental increases in COHb caused by environmental exposure actually would be additive to the chronically elevated COHb levels caused by tobacco smoke (James et al., 2000).

Environmental sources of CO can result in CO levels indistinguishable from those produced by direct cigarette use, thereby confounding the measurement (James et al., 2000). Sources of inhaled CO readily absorbed into the bloodstream include exhaust from internal combustion engines, industrial emissions, and SHS (Becoña and Vázquez, 1998; Velicer et al., 1992). Another disadvantage of CO measurement is the relatively short half-life of 4–5 h approximately. Notably, a person wishing to foil the measure can abstain from smoking for several hours prior to CO testing, in which case the measured CO level could be that of a nonsmoker because of this short half-life.

In the general population, false-negative rates of CO measurements have ranged from 2% to 16% (Velicer et al., 1992). Another disadvantage of using CO as a biomarker is that sensitivity decreases with infrequent and irregular smoking patterns, causing those who are light or atypical smokers to appear indistinguishable from nonsmokers (Jarvis et al., 1987; Lando et al., 1991; Vogt et al., 1977).

9.3.4 COHb in Blood

CO in the blood originates to a great extent from tobacco smoke. The catabolism of heme, proteins, and combustion products such as passive smoking and automobile exhaust, as well as exposure at different workplaces, is a less significant source. During cigarette smoke inhalation, CO is rapidly absorbed into the bloodstream, binds to oxygen binding sites of hemoglobin molecules in red blood cells and forms COHb. The average half-life of COHb is reported to be 3–5 h, depending on the respiration rate (James et al., 2000).

A cigarette smoker may be exposed to 400–500 ppm CO. It takes 6 min to smoke a typical cigarette, producing an average baseline COHb of 4%, with a typical range of 3%–8%. Heavy smokers may achieve COHb levels as high as 15%. In comparison, nonsmokers average about 1% COHb in their blood (James et al., 2000).

Exposure to tobacco smoke not only increases COHb concentrations in smokers, but also affects nonsmokers. For example, acute exposure (1–2 h) to smoke-polluted environments has been reported to cause an incremental increase in nonsmokers' COHb of about 1%. Available data strongly suggests that acute and chronic CO exposure attributed to tobacco smoke can affect the cardiopulmonary system, but the potential interaction of CO with other products of tobacco smoke confounds the results (James et al., 2000).

9.4   NITROGEN OXIDES

Nitrogen oxides are defined as the class of chemical compounds that consist of nitrogen and oxygen, including nitric oxide (NO), nitrogen dioxide (NO_2), and nitrous oxide (N_2O). But only NO and NO_2 are usually referred to as nitrogen oxides in general, for it is most likely the two compounds are common atmospheric pollutants. Because of their pharmacological importance, the presence of nitrogen oxides in tobacco smoke has aroused public concern. The combustion process of nitrogenous compounds emits nitrogen oxides, mostly in the form of NO, while NO_2 is almost completely absent in the mainstream of cigarette smoke (Wakefield et al., 2010). According to a paper from Vilcins and Lepharth (1975), the rapid decrease of NO in fresh smoke is accompanied by an increase of NO_2 due to the oxidation of NO during the aging period.

NO is found to be approximately five times less toxic than NO_2 (Wakefield et al., 2010) and six times less active as a mammalian ciliostatic agent. It is irritating to the eyes and upper respiratory tract. Deep inhalation can result in the delayed onset of pulmonary edema occurring a few hours postexposure and may be aggravated by physical exertion. NO_2 is even more toxic by inhalation. Symptoms of poisoning (lung edema) tend to appear several hours after inhalation of a low but potentially fatal dose. Also, low concentrations (4 ppm) will anesthetize the nose, thus creating a potential for overexposure.

9.4.1   TOXICOLOGICAL EVALUATION

9.4.1.1   Nitric Oxide

NO plays important physiological roles ranging from blood pressure modulation to neurotransmission, but an excess can be toxic. It is present in cigarette smoke at up to 500 ppm and probably represents one of the greatest exogenous sources of NO to which humans are exposed. It has been known that NO reacts quickly with superoxide radical to give peroxynitrite and with organic peroxyl radicals to give alkyl peroxynitrites, both of which are cytotoxic species (Eiserich et al., 1994). Otherwise, NO has oxidative effects in the lung. It can decrease neutrophil accumulation and surfactant function. Peroxynitrite and NO_2 generated from NO may also induce genotoxic alterations and/or tissue injury. Even outside of the lung, inhaled NO may alter vascular tone and platelet function, and methemoglobinemia can adversely affect tissue oxygenation.

9.4.1.2 Nitrogen Dioxide

Although NO undergoes oxidation to NO_2 by the well-known termolecular reaction $2NO + O_2 \rightarrow 2NO_2$, but the rate of conversion is found to be relatively slow at the concentration existing in cigarette smoke. Nearly 500 s would be required in undiluted smoke, and more than 5000 s would be required in the more diluted system in human lungs, for the oxidation of half NO to NO_2 (the rate constant for NO_2 formation has been shown to be $(1.2 \pm 0.1) \times 10^{11}$ ppm^{-2} s^{-1}) (Sokol et al., 1999).

NO_2 is known to be directly toxic to the respiratory tract, and the Occupational Safety and Health Administration limits human peak exposure to 5 ppm. The acute pulmonary toxicity of NO_2, thought to be several times greater than nitric oxide, has been studied more extensively. Increased airway reactivity has been reported in humans at exposures as low as 1.5 ppm NO_2 (Frampton et al., 1991). Other toxic effects observed following inhalation of NO_2 ($\leq$5 ppm) include altered surfactant chemistry and metabolism (Müller et al., 1994), epithelial hyperplasia of the terminal bronchioles, increased cellularity of the alveoli in rats (Evans et al., 1972), and diffuse inflammation (Müller et al., 1994). At higher doses, the major toxicological effect of NO_2 is pulmonary edema. The epidemiological studies provide some evidence that long-term exposure to NO_2 at concentrations of 40–100 µg/m³ may decrease lung function and increase the risk of respiratory symptoms (World Health Organization, 2003).

9.4.2 METABOLIC PATHWAY IN THE BODY

9.4.2.1 Nitric Oxide

The amounts of NO inhaled by smokers differ depending on the types of cigarettes, the pattern of smoking, and the depth of inhalation. The lower solubility of NO may result in little amounts absorbed in the upper airways and greater amounts reaching the pulmonary region, where it then diffuses to blood and reacts with hemoglobin (Yoshida and Kasama, 1987). The strong affinity of NO to hemoglobin has been studied in relation to air pollution. In vitro data indicate that the affinity of NO is several thousand times higher than that for CO. The increasing affinity from O_2 to CO and NO is mainly due to a decrease of the dissociation velocity constant, which varies about 2×10^5 times from O_2 to NO. Furthermore, NO constitute is the most rapidly binding ligand to hemoglobin so far discovered. The velocity constant of a combination of NO and hemoglobin is about 280 times faster than that for CO and also about 5 times higher than that for oxygen (Meyer and Piiper, 1989). It has also been shown that NO produces nitrosylhemoglobin (NOHb) in the absence of O_2 and methemoglobin (MetHb) in the presence of O_2. In a study by Wennmalm et al. (1992), plasma or whole venous or arterialized blood from healthy human donors was incubated with NO, and the resulting formation of MetHb, NOHb, and plasma nitrite (NO_2^-) and nitrate (NO_3^-) were measured. The results were as follows: (1) In plasma, NO was converted to NO_2^- and NO_3^- in a ratio of 5:1. (2) In arterial blood (oxygen saturation 94%–99%), NO was almost quantitatively converted to MetHb and NO_3^-. No NO_2^- was detected and NOHb formation was low. (3) In venous blood (oxygen saturation 36%–85%), more NOHb and less NO_3^- were formed, in comparison with the corresponding formation in arterialized blood.

9.4.2.2   Nitrogen Dioxide

NO_2 is mainly formed in the aging period. Once deposited, NO_2 dissolves in lung fluids, and various chemical reactions occur and give rise to products found in the blood and other body fluids. Using estimation of radiolabeled NO_3^- and NO_2^- levels in the blood and urine of rabbits following exposure to NO_2, Svorcova and Kaut (1971) suggested that inhaled NO_2 enters the blood stream. When studying isolated perfused rat lungs, Postlethwait and Bidani (1989) found that 70% of absorbed $^{15}NO_2^-$ appeared in the perfusate and 30% appeared in the lung tissue and that NO_2^- accounted for the ^{15}N-nitrogen. Oda et al. (1981) noted a dose-dependent increase in both NO_2^- and NO_3^- levels in the blood of mice during exposures to NO_2. The blood levels of NO_2^- and NO_3^- declined rapidly after the exposures ended, with decay half-times of a few minutes for NO_2^- and about 1 h for NO_3^-. The shorter time for the former was ascribed to its rapid oxidation reaction with hemoglobin, producing NO_3^- and MetHb. They also exposed mice to an NO_2 concentration of 75,200 $\mu g/m^{-3}$ and determined NOHb and MetHb. In the result, only 0.2% NOHb was found, and MetHb did not increase. A linear relationship between the concentration of NO_2 and NOHb could be observed; MetHb did not increase at any concentration of NO_2 (Oda et al., 1981).

9.4.3   BIOMARKERS

9.4.3.1   8-Nitroguanine

There are evidences that nitrogen oxides gases can cause DNA damages in mammalian cells. Hsieh et al. (2001) have found that nitrogen oxides gases can induce 8-nitroguanine ($8\text{-}NO_2\text{-}G$) formation in human lung fibroblast cells. A parallel correlation between the formation of $8\text{-}NO_2\text{-}G$ and the number of tobacco cigarette smoke was observed (Hsieh et al., 2002). The results showed a dose-dependent increase in $8\text{-}NO_2\text{-}G$ in smokers' blood and lungs of rats after exposure to cigarette smoke, suggesting that smoke-caused DNA damage is due to the formation of $8\text{-}NO_2\text{-}G$ by nitrogen oxides attacking DNA. Therefore, $8\text{-}NO_2\text{-}G$ could act as a specific marker for DNA damage induced by gaseous nitrogen oxides.

9.4.3.2   Methemoglobin

Inhaled NO can combine with hemoglobin to form nitrosylhemoglobin (NOHb), which is rapidly oxidized to methemoglobin (MetHb), and the binding and formation of MetHb is concentration and time dependent (Ripple et al., 1989). In adults with pulmonary hypertension, peak MetHb levels of 9.6% and 14% were reported after 10–18 h of 80 ppm NO inhalation. Inhalation of NO (<45 ppm) resulted in a toxic level of 67% MetHb in a patient with hydrochlorothiazide-induced pulmonary edema.

9.4.3.3   Urinary NO_2^- and NO_3^-

Inhaled NO_2 can react with oxidizable tissue to form NO_2^-. NO_2^- is then further oxidized in the blood by oxyhemoglobin to form NO_3^-, which is excreted in the urine. The ^{15}NO inhalation experiments on rats (Yoshida et al., 1983) demonstrated that metabolites of NO are NO_3^- and NO_2^-; the 15N recovery in urine after 48 h was 55%, 75% of which appeared as NO_3^- and 24% as urea.

9.4.3.4   Formation of *N*-Nitroso Compounds

Humans are exposed to NO_2^- by the inhalation of nitrogen oxides in urban air, tobacco smoking and ingestion. Once inhaled, the stomach could provide the acidic environment required for the formation of *N*-nitrosamines from NO_2^- and secondary amines. Ingested NO_3^- can be converted to NO_2^- by bacterial reduction in the saliva, the gastrointestinal tract, and the urinary bladder. Iqbal et al. (1980) were the first to demonstrate a linear time-dependent and concentration-dependent relationship between the amounts of *N*-nitroso morpholine (NMOR) and NO_2. In the research, NMOR was found in whole-mouse homogenates after the mice were gavaged with 2 mg of morpholine and exposed to NO_2. Garland et al. (1986) found there was a positive relationship between the atmospheric NO_2 levels and the urinary excretion of *N*-nitrosomethylamine in normal human study objects.

9.5   DISCUSSION

The most important hazardous combustion products are CO, HCN, nitrogen oxides and the low availability of oxygen as these may be hazardous in the long term (Wakefield et al., 2010). Tobacco present in this area is less likely to pose an immediate danger to health, but should be considered as hazardous. The asphyxiant gas CO is present in the effluent at much lower concentrations and is therefore likely to be less of a hazard to health, unless individuals are directly in contact with the effluent. The major immediate hazard to public health in a room is expected to be the exposure to irritants generated in the effluent. The adverse effects resulting from exposure to these irritants are likely to be resolved following removal from the exposure.

If a person has preexisting respiratory diseases, such as asthma or chronic obstructive pulmonary disease, they are most at risk from exposure to mainstream smoke and environmental tobacco smoke. The existing respiratory condition increases the susceptibility of the individual to the adverse effects of exposure to asphyxiant gases such as CO. Therefore, it is likely that acute exposure to smoke containing mixtures of asphyxiant and irritant gases is likely to exacerbate these conditions.

For the special group, pregnant women are particularly at risk following exposure to smoke, as unborn infants are also particularly susceptible to COHb. Following exposure to CO, the fetal circulation would be expected to have a greater concentration of COHb than the maternal circulation due to differences in the uptake and elimination of CO. Compared to the mother, this increased level of COHb may cause a potentially serious hypoxia to the fetus at COHb levels that are less harmful to the mother. Similarly, newborn infants and children may also be at increased risk of adverse effects such as hypoxia and respiratory irritation following exposure to hazardous environmental tobacco smoke. Infants are more susceptible to toxicity because the hazardous gas is likely to result in toxicity to infants and children at lower concentrations than those required to cause similar effects in adults. Elderly individuals exposed to CO, HCN, and nitrogen oxides would also be at greater risk of potentially life-threatening effects due to conditions associated with age, such as reduced lung function.

Shortly, any reduction in the utilization of oxygen increases the risk of the individual to hypoxia resulting from exposure to asphyxiant gases such as CO, HCN and low oxygen concentration, thereby there is a positive relationship between the potential for adverse effects and the duration of exposure. Rather than nonsmokers, smokers are likely to be at greater risk to toxicity following inhalation of combustion products. It is observed the baseline level of COHb is likely to be greater than that for nonsmokers. Therefore, the inhalation of combustion products at concentrations that do not cause significant toxicity to nonsmokers may cause considerable toxicity to typical smokers.

9.6   CONCLUSION

The prediction on the toxicity of cigarettes is still a complex area. There is a potential for the generation of a huge number of pyrolysis products depending on the conditions of burning. Although each type of cigarette has individual characteristics, it will ultimately need to be considered on case-by-case commonalities, particularly with regard to the most important components relating to toxicity. The generalization may assist in rapidly identifying which hazardous combustion products are of most concern to public health.

If tobacco combustion is incomplete, due to low temperature, lack of ventilation, and absence of flaming, tobacco would be expected to form the quantities of hazardous combustion products. Asphyxiant gases including CO and HCN and irritant gases such as nitrogen oxides are most likely to be generated by cigarette smoke, but are less likely to pose a major hazard to public health and individuals due to dispersion and dilution.

REFERENCES

Alarie, Y. The toxicity of smoke from polymeric materials during thermal decomposition. *Annual Review of Pharmacology and Toxicology*. 1985, 25, 325–347.

Becoña, E., Vázquez, F. L. Self-reported smoking and measurement of expired air carbon monoxide in a clinical treatment. *Psychological Reports*. 1998, 83, 316–318.

Cummings, T. The treatment of cyanide poisoning. *Occupational Medicine*. 2004, 54, 82–85.

Dimitri, G., Jochen, B., Katherine, W. et al. Carbon monoxide concentrations in the 2009 Victorian Bushfire disaster victims. *Forensic Science International*. 2010, 205, 69–72.

Egekeze, J. O., Oehme, F. W. Cyanides and their toxicity: A literature review. 2013, 2, 104–114.

Eiserich, J. P., Vossen, V., O'Neill, C. A. et al. Molecular mechanisms of damage by excess nitrogen oxides: Nitration of tyrosine by gas-phase cigarette smoke. *FEBS Letters*. 1994, 353, 53–56.

Evans, M. J., Stephens, R. J., Cabral L. J. et al. Cell renewal in the lungs of rats exposed to low levels of NO_2. *Archives of Environmental Health*. 1972, 24, 180–188.

Frampton, M. W., Morrow, P. E., Cox, C. et al. Effects of nitrogen dioxide exposure on pulmonary function and airway reactivity in normal humans. *American Review of Respiratory Disease*. 1991, 143, 522–527.

Garland, W. A., Kuenzig, W., Rubio, F. et al. Urinary excretion of nitrosodimethylamine and nitrosoproline in humans: Interindividual and intraindividual differences and the effect of administered ascorbic acid and α-tocopherol. *Cancer Research*. 1986, 46, 5392–5400.

Hsieh, Y.-S., Chen, B.-C., Shiow, S.-J. et al. Formation of 8-nitroguanine in tobacco cigarette smokers and in tobacco smoke-exposed Wistar rats. *Chemico-Biological Interactions*. 2002, 140, 67–80.

Hsieh, Y. S., Wang, H. C., Tseng, T. H. et al. Gaseous nitric oxide-induced 8-nitroguanine formation in human lung fibroblast cells and cell-free DNA. *Toxicology and Applied Pharmacology.* 2001, 172, 210–216.

Jarvis, M. J., Tunstall-Pedoe, H., Feyerabend, C. et al. Comparison of tests used to distinguish smokers from nonsmokers. *American Journal of Public Health.* 1987, 77, 1435–1438.

Kroto, H. W., Heath, J. R., O'Brien, S. C., Curl, R. F., Smalley, R. E. C60: Buckminsterfullerene. *Nature.* 1985, 318, 162–163.

Lando, H. A., McGovern, P. G., Kelder, S. H. et al. Use of carbon monoxide breath validation in assessing exposure to cigarette smoke in a worksite population. *Health Psychology.* 1991, 10, 296.

Meyer, M., Piiper, J. Nitric oxide (NO), a new test gas for study of alveolar-capillary diffusion. *European Respiratory Journal.* 1989, 2, 494–496.

Müller, B., Schäfer, H., Barth, P. et al. Lung surfactant components in bronchoalveolar lavage after inhalation of NO_2 as markers of altered surfactant metabolism. *Lung.* 1994, 172, 61–70.

Newsome, J., Norman, V., Keith, C. Vapor phase analysis of tobacco smoke. *Tobacco Science.* 1965, 9, 102–110.

Oda, H., Tsubone, H., Suzuki, A. et al. Alterations of nitrite and nitrate concentrations in the blood of mice exposed to nitrogen dioxide. *Environmental Research.* 1981, 25, 294–301.

Postlethwait, E. M., Bidani, A. Pulmonary disposition of inhaled NO_2-nitrogen in isolated rat lungs. *Toxicology and Applied Pharmacology.* 1989, 98, 303–312.

Raub, J. A., Mathieu-Nolf, M., Hampson, N. B. et al. *Toxicology.* 2000, 145, 1–14.

Ripple, G., Mundie, T., Stavert, D. M. et al. Kinetics of methemoglobin formation and elimination as a function of inhaled nitric oxide concentration and minute ventilation. *Toxicologist.* 1989, 9, 754.

Scherer, G. Carboxyhemoglobin and thiocyanate as biomarkers of exposure to carbon monoxide and hydrogen cyanide in tobacco smoke. *Experimental and Toxicologic Pathology.* 2006, 58, 101–124.

Secker-Walker, R. H., Vacek, P. M., Flynn, B. S. et al. Exhaled carbon monoxide and urinary cotinine as measures of smoking in pregnancy. *Addictive Behaviors.* 1997, 22, 671–684.

Sokol, G. M., Van Meurs, K. P., Wright, L. L. et al. Nitrogen dioxide formation during inhaled nitric oxide therapy. *Clinical Chemistry.* 1999, 45, 382–387.

Stedman, R. L. Chemical composition of tobacco and tobacco smoke. *Chemical Reviews.* 1968, 68, 153–207.

Stewart, R. D. The effect of carbon monoxide on humans. *Annual Review of Pharmacology.* 1975, 15, 409–423.

Svorcova, S. and Kaut, V. Arterio-venous differences in the nitrite and nitrate ion concentrations in rabbits after inhalation of nitrogen oxide. *Ceskoslovenska Hygiena.* 1971, 16, 71–76.

Thostenson, E. T., Ren, Z., Chou, T.-W. Advances in the science and technology of carbon nanotubes and their composites: A review. *Composites Science and Technology.* 2001, 61, 1899–1912.

Velicer, W. F., Prochaska, J. O., Rossi, J. S. et al. Assessing outcome in smoking cessation studies. *Psychological Bulletin.* 1992, 111, 23.

Vogt, T. M., Selvin, S., Widdowson, G. et al. Expired air carbon monoxide and serum thiocyanate as objective measures of cigarette exposure. *American Journal of Public Health.* 1977, 67, 545–549.

Wakefield, J. C. et al. A toxicological review of the products of combustion. Health Protection Agency. 2010.

Wennmalm, A., Benthin, A., Petersson, A. S. Dependence of the metabolism of nitric oxide (NO) in healthy human whole blood on the oxygenation of its red cell haemoglobin. *British Journal of Pharmacology.* 1992, 106, 507–508.

World Health Organization. Health aspects of air pollution with particulate matter, ozone and nitrogen dioxide. Report on a WHO Working Group. January 13–15, 2003, WHO, Bonn, Germany.

World Health Organization. WHO report on the global tobacco epidemic, 2008: The MPOWER package. 2008. World Health Organization, Geneva, Switzerland.

Wynder, E. L., Hoffmann, D. *Tobacco and Tobacco Smoke: Studies in Experimental Carcinogenesis*. 1967, Academic Press, New York.

Yeoh, M. J., M. B., B.S., D.A. (UK), Braitberg, G. Carbon monoxide and cyanide poisoning in fire related deaths in Victoria, Australia. *Journal of Toxicology*. 2004, 42, 855–863.

Yoshida, K., Kasama, K. Biotransformation of nitric oxide. *Environmental Health Perspectives*. 1987, 73, 201–206.

Yoshida, K., Kasama, K., Kitabatake, M. et al. Biotransformation of nitric oxide, nitrite and nitrate. *International Archives of Occupational and Environmental Health*. 1983, 52, 103–115.

10 Biomarkers of Heavy Metal Exposure

10.1 INTRODUCTION

The main threats to human health from heavy metals are associated with exposure to lead, cadmium, mercury, and arsenic (arsenic is a metalloid, but is usually classified as a heavy metal). These metals have been extensively studied and their effects on human health regularly are reviewed by international bodies such as the World Health Organization (WHO) and International Agency for Research on Cancer (IARC). There is no clear definition of what a heavy metal is; density is in most cases taken to be the defining factor. Heavy metals are thus commonly defined as those having a specific density of more than 5 g/cm^3.

Cadmium exposure is mainly from cigarette smoking and diet. In smokers, cigarette smoking is a major source of cadmium exposure. In nonsmokers, food is the most important source of cadmium exposure. Recent data indicate that adverse health effects of cadmium exposure may occur at lower exposure levels than previously anticipated, primarily in the form of kidney damage but possibly also bone effects and fractures.

Mercury exposure is primarily via food, especially, fish consumption being a major source of methyl mercury and dental amalgam exposure. The general population does not face a significant health risk from methyl mercury, although certain groups with high fish consumption may attain blood levels associated with a low risk of neurological damage to adults.

Lead exposure for the general population is from air and food in roughly equal proportions. During the last century, lead emissions to ambient air have caused considerable pollution, mainly due to lead emissions from petrol. Although lead in petrol has dramatically decreased over the last decades, thereby reducing environmental exposure, phasing out any remaining uses of lead additives in motor fuels should be encouraged. In smokers, cigarette smoking is another source of lead exposure. Our research indicates that lead in a smoker's urine was significantly higher than a nonsmoker.

Arsenic exposure is mainly via intake of food and drinking water, food being the most important source in most populations. Long-term exposure to arsenic in drinking water is mainly related to increased risks of skin cancer, but also some other cancers, as well as other skin lesions such as hyperkeratosis and pigmentation changes. In smokers, cigarette smoking is a source of arsenic exposure. Although a clear exposure–response model of cigarette smoke and arsenic in smokers has not been observed, smokers' urine arsenic levels were slightly higher than nonsmokers.

Heavy metals have been used in many different areas for thousands of years. Lead has been used for at least 5000 years in early applications including building materials, pigments for glazing ceramics, and pipes for transporting water. In ancient Rome, lead acetate was used to sweeten old wine, and some Romans might have consumed as much as a gram of lead a day. Mercury was allegedly used by the Romans as a salve to alleviate teething pain in infants and was later (from the 1300s to the late 1800s) employed as a remedy for syphilis. Although adverse health effects of heavy metals have been known for a long time, exposure to heavy metals continues and is even increasing in some areas. For example, mercury is still used in gold mining in many parts of Latin America. Arsenic is still common in wood preservatives, and tetraethyl lead remains a common additive to petrol, although tetraethyl lead in petrol has decreased dramatically in developed countries. Since the middle of the nineteenth century, heavy metals consumption rapidly increased, with concomitant emissions to the environment. At the end of the twentieth century, however, emissions of heavy metals started to decrease in developed countries: in the United Kingdom, emissions of heavy metals fell by over 50% between 1990 and 2000. But in developing countries, many problems caused by heavy metal emissions are getting more attention. In pace with the course of industry in China speeding up, the problem of heavy metal pollution becomes severe day by day. In China, emissions of heavy metals to the environment occur via a wide range of processes and pathways, including air (e.g., during combustion, extraction, and processing), surface waters (via runoff and releases from storage and transport), and soil (and hence into groundwaters and crops). Atmospheric emissions tend to be of greatest concern in terms of human health, because of both the quantities involved and the widespread dispersion and potential for exposure that often ensues. People exposed to potentially harmful physical, chemical, and biological agents in cigarette smoke, food, and water. However, exposure does not result only from the presence of a harmful agent in the environment. The key word in the definition of exposure is *contact* (Berglund et al., 2001). There must be contact between the potential harmful agent and the human body, such as the mouth, the skin, or the nose. Exposure is often defined as a function of concentration and time: "occurs when there is contact at a boundary between a human and the environment with a contaminant of a specific concentration for enough time." For exposure to happen, coexistence of heavy metals and people has to occur.

10.2   METABOLISM PATHWAY

Heavy metals, such as cadmium (Cd), arsenic (As), mercury (Hg), and lead (Pb), may be absorbed from the soil by tobacco plants, then distributed into the smoke from the cigarette with the process of burning, and finally enter into the smoker's body with the smoke that is inhaled by a smoker during a puff (mainstream smoke). Heavy metals are nephrotoxic and carcinogenic. Chromium is considered to be carcinogenic by the WHO IARC Group 1, while Pb and Hg are of special concern due to their neurotoxicity. As and Se can add burden to kidney metabolism.

For heavy metal ions undergoing metabolism in body, the pathway is shown in Figure 10.1. Ions are excreted from the body as metal prototype through excrement and exhalation from the lung. The absorption rate of any element from inhaled air varies from 7% to 10%; the rest of the heavy metal ions are excreted from the body

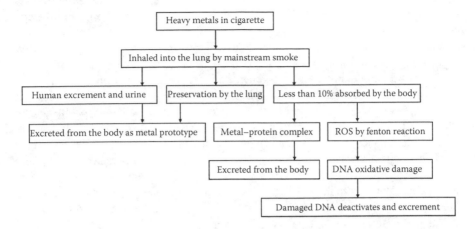

FIGURE 10.1 The pathway of heavy metals.

(i.e., chromium, which does not accumulate, is metabolized and removed from the body much faster than others). The ions are also combined with protein or DNA to form metal–protein or metal–DNA complex. For example, Pb may react with melatonin, serotonin, and tryptophan in the body, or Cr could combine with DNA to from Cr–DNA adducts. Damaged DNA activation is initiated by metal ion–catalyzed formation of reactive oxygen species (ROS) through the Fenton reaction. This metabolic could produce oxyhydrogen free radicals, which are carcinogenic and may cause oxidative damage to DNA.

10.3 CADMIUM

10.3.1 Occurrence, Exposure, and Dose

Cadmium occurs naturally in ores together with zinc, lead, and copper. Cadmium is also present as a pollutant in phosphate fertilizers, and crops such as tobacco take up cadmium from fertilizers. Notwithstanding reductions in Europe, however, cadmium production, consumption, and emissions to the environment worldwide have increased dramatically in China. Natural and anthropogenic sources of cadmium, including industrial emissions and the application of fertilizer and sewage sludge to farmland, may lead to contamination of soils and to increased cadmium uptake by crops, vegetables, and tobacco, grown for human consumption. The uptake process of soil cadmium by plants is enhanced at low pH value (Jarup, 2003). Cigarette smoking is a major source of cadmium exposure. Because cadmium excretion is slow, cadmium accumulation in the body can be significant. Cadmium concentration in blood reflects recent exposure; urinary cadmium concentration more closely reflects total body burden. However, when renal damage from cadmium exposure occurs, the excretion rate increases sharply, and urinary cadmium levels no longer reflect body burden. Cadmium (or metabolite) concentrations in blood and urine have been used as biomarkers of exposure. Biological monitoring of cadmium in the general population has shown that cigarette smoking may cause significant increases in blood cadmium (B-Cd) levels, the concentrations in smokers being on average four to five times higher than those in nonsmokers (Hossn et al., 2001). Despite evidence of exposure from environmental tobacco smoke, however, this is probably contributing little to the total cadmium body burden. Crops are the most important source of cadmium exposure in the general nonsmoking population in most countries. Cadmium is present in most foodstuffs, but concentrations vary greatly, and individual intake also varies considerably due to differences in dietary habits (Jarup, 2003). B-Cd generally reflects current exposure, but partly also lifetime body burden. The cadmium concentration in urine (U-Cd) is mainly influenced by the body burden, U-Cd being proportional to the kidney concentration. Smokers and nonsmokers living in contaminated areas have higher urinary cadmium concentrations, smokers having about twice as high concentrations as nonsmokers.

Cadmium levels in cigarette mainstream smoke are getting attention; they are at a rate of around 0.15 µg/cigarette. Therefore, cadmium in tobacco smoke does not immediately significantly contribute to the total dose of cadmium absorbed by the body, except in heavy smokers. Exposure of the general population to cadmium should be kept as low as possible. The EU has proposed that urinary cadmium concentration should be below 0.66 µg/g of creatinine, which reflects the recent findings on the adverse effects of low-level cadmium exposure. In a priority list of hazardous substances established by the U.S. Comprehensive Environmental Response, Compensation, and Liability Act, cadmium ranks seventh. Cigarette smoke is the largest source of nonoccupational exposure, because cadmium tends to accumulate in tobacco leaves and between 40% and 60% of inhaled cadmium can enter blood circulation (Akesson et al., 2005; Moriguchi et al., 2004).

10.3.2 Physical Effects

Inhalation of cigarette smoke containing cadmium particles can be life threatening. Although acute pulmonary effects or deaths are uncommon, sporadic cases still occur. Exposure to fumes containing cadmium particulate may cause chronic kidney damage. The sign of renal damage is usually a tubular dysfunction, evidenced by an increased excretion of low-molecular-weight proteins or enzymes. It has been suggested that tubular damage is reversible, but there is overwhelming evidence that cadmium-induced tubular damage is indeed irreversible. The WHO considers that a urinary excretion of 10 nmol/mmol creatinine (corresponding to circa 200 mg Cd/kg kidney cortex) would constitute a *critical limit* below which kidneys would not sustain damage. However, WHO calculated that circa 10% of individuals with this kidney concentration would be affected by tubular damage. Several reports have since shown that kidney damage and bone effects are likely to occur at lower kidney cadmium levels (WHO, 1992). European studies have shown signs of cadmium-induced kidney damage in the general population at urinary cadmium levels around 2–3 µg Cd/g creatinine (Buchet et al., 1990; Ferraro et al., 2010; Jarup et al., 2000; Noonan et al., 2002). The initial tubular damage may progress to more severe kidney damage, and already in 1950 it was reported that some cadmium-exposed workers had developed decreased glomerular filtration rate (GFR) (Friberg, 1950). This has been confirmed in later studies of occupationally exposed workers (Bernard et al., 1992).

10.3.3 CADMIUM IN OXIDATIVE STRESS

Cadmium can increase oxidative stress and lipid peroxidation (LPO) in cultured mice hepatocytes. Cadmium-induced production of reactive nitrogen species (RNS) and ROS through interaction with critical subcellular sites such as mitochondria, peroxisomes, and microsomes resulted in the generation of free radicals and LPO in subcellular membranous structures. Production of RNS and ROS has been reported in a variety of cell culture systems, as well as in intact animals via routes of exposure. Oxidative stress was found in the liver of mice exposed to a single oral Cd dose (20 mg/kg b. w. in the form of $CdCl_2$) through increased LPO level, expressed as malondialdehyde after 6, 12, and 24 h. Since Cd has no redox activity, it may enhance RNS and ROS production by suppressing free-radical scavengers such as catalase (CAT), GSH peroxidase (GSH-Px), and superoxide dismutase (SOD). In this way, cadmium could induce the formation of reactive species including ROS and RNS. Many studies confirm that the formation of free radicals such as superoxide ion, hydrogen peroxide, and hydroxyl radicals involves depletion of GSH and changes in the activity of antioxidant enzymes. An acute oral Cd dose (20 mg Cd/kg b. w.) exposure significantly decreased the GSH content in mice liver and increased GSH in the kidney, but did not cause significant GSH changes in the testis. These results show that the effect of Cd on GSH tissue levels varies with animal species, exposure dose, metabolism route, and duration of exposure. In general, acute exposure to metals decreases GSH levels due to the formation of metal–GSH complexes and/or consumption by the GSH-Px under oxidative stress induced by metals (Figure 10.2).

Cadmium impairs enzyme activity of antioxidative defense system (SOD, CAT, GSH-Px, GSH-S-transferase, and GSH reductase) and of the nonenzymatic component GSH and GSSG. Cadmium also elevates the levels of Fenton metals

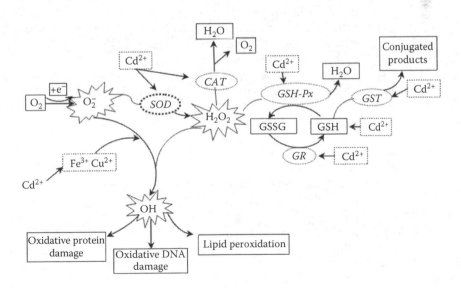

FIGURE 10.2 Pathways of Cd-induced generation of ROS.

(Cu^{2+}, Co^{2+}, and Fe^{3+}), which can break down hydrogen peroxide, H_2O_2, to form a reactive hydroxyl radical, $OH·$.

10.4 ARSENIC

10.4.1 OCCURRENCE, EXPOSURE, AND DOSE

Arsenic is widely distributed in the earth's surface in small amounts, occurring in soil, rock, and water. In several countries, inorganic arsenic is present in groundwater used for drinking (e.g., China, Bangladesh, and Chile), whereas organic arsenic (such as arsenobetaine) is mainly found in fish, which thus may give rise to human exposure. There are many arsenic complexes in nature, over 200 crystalline or mineral forms, such as realgar (As_4S_4) and arsenopyrite (FeAsS); the smelting of nonferrous metals and elemental metalloids and the production of energy from fossil fuel are the two major industrial processes that lead to arsenic contamination of water, soil, and air, smelting activities being the largest source of atmospheric arsenic pollution. Water concentrations are usually <10 µg/L, although higher concentrations may occur near anthropogenic sources, but waste disposal and pesticide application can result in much higher concentrations (WHO, 2001). General population exposure to arsenic is mainly via intake of food and drinking water. Food is the most important source, but in some areas, arsenic in drinking water is a significant source of exposure to inorganic arsenic. Contaminated soils such as mine tailings are also a potential source of arsenic exposure. Absorption of arsenic in inhaled tobacco smoke is highly dependent on the solubility and the size of particles. Soluble arsenic compounds are easily absorbed from the gastrointestinal tract. However, inorganic arsenic is extensively methylated in humans and the metabolites are excreted in the urine. Arsenic (or metabolite) concentrations in blood, hair, nails, and urine have been used as biomarkers of exposure. Arsenic in hair and nails can be useful indicators of past arsenic exposure, if care is taken to avoid external arsenic contamination of the samples. Speciated metabolites in urine expressed as either inorganic arsenic or the sum of metabolites (inorganic arsenic, monomethylarsonic acid [MMA], and dimethylarsinic acid [DMA]) are generally the best estimate of recent arsenic dose (Chilvers et al., 1987). However, consumption of certain seafood may confound estimation of inorganic arsenic exposure and should thus be avoided before urine sampling.

10.4.2 PHYSICAL EFFECTS

Arsenic can combine with such noncarbon chemicals as sulfur and oxygen to form arsenides, arsenites, and arsenates (oxidation states of –3, +3 and +5), referred to as inorganic arsenic compounds, which are acutely toxic and intake of large quantities could lead to gastrointestinal symptoms, severe disturbances of the cardiovascular and central nervous systems, and eventually death. In survivors, bone marrow depression, hemolysis, hepatomegaly, melanosis, polyneuropathy, and encephalopathy may be observed. Inorganic arsenic exposure may induce peripheral vascular disease (black foot disease reported in Taiwan). Human ingestion of arsenic via drinking water shows a high risk of mortality from lung, bladder, and kidney cancer, the risk increasing with increasing exposure. There is also an increased risk of skin cancer and other skin lesions, such as hyperkeratosis and pigmentation changes. Studies on occupational exposure to arsenic by inhalation, such as smelter workers, pesticide manufacturers, and miners in many different countries, consistently demonstrate an excess of lung cancer (CDC, 2009). Although these groups are exposed to other harmful chemicals in addition to arsenic, there is no other common factor that could explain the findings. The lung cancer risk increases with increasing arsenic exposure and confounding by cigarette smoking does not explain the findings. The latest WHO research shows that arsenic exposure via drinking water is causally related to cancer in the lungs, kidney, bladder, and skin, the last of which is preceded by directly observable precancerous lesions and arsenic in mainstream smoke is not clear. The estimation of past exposures are important when assessing the exposure–response relationships, but it would seem that drinking water arsenic concentrations of approximately 100 µg/L have led to cancer at these sites and that precursors of skin cancer have been associated with levels of 50–100 µg/L and cigarette smoking arsenic concentrations are less clear. The relationships between arsenic exposure and other health effects are less clear. There is relatively strong evidence for hypertension and cardiovascular disease. There is strong evidence for cardiovascular disease and hypertension, but the evidence is only suggestive for diabetes and weak for cerebrovascular disease, long-term neurological effects, and cancer at sites other than lung, bladder, kidney, and skin (Pappas, 2011).

10.4.3 METABOLISM PATHWAY

These concerns can be addressed, in part, by focusing on the role that metabolism plays in arsenic toxicity and carcinogenesis. The biotransformation of arsenic in mammals is based on several generally accepted steps (Figure 10.3). For methylation, arsenate (As^{5+}) species must be reduced to arsenite (As^{3+}) at first, a process that occurs through reactions involving GSH. Then arsenite is sequentially methylated, first to form MMA and subsequently to form DMA. Oxidative addition of methyl groups to arsenic occurs by, as yet, only partially characterized methyltransferase enzymes, with S-adenosyl methionine as the methyl-donating cofactor (Aposhian et al., 1997; Thompson et al., 1993). Inorganic arsenic has been considered a detoxification mechanism, since arsenic-methylated compounds are less acutely toxic and less reactive with tissue macromolecules and are excreted faster in urine than compounds of inorganic arsenic (NAS, 1989). In that some evidence suggests that arsenic metabolites may contribute to the carcinogenicity or toxicity of the metalloid (Yamamoto et al., 1995).

10.5 MERCURY

10.5.1 OCCURRENCE, EXPOSURE, AND DOSE

The mercury compound cinnabar (HgS) was used in prehistoric cave paintings for red colors, and metallic mercury was known in ancient Greece where it (as well as white lead) was used as a cosmetic to lighten the skin. In medicine, apart from the previously mentioned use of mercury as a cure for syphilis, mercury compounds have also been used as diuretics (calomel [Hg_2Cl_2]), and mercury amalgam is still

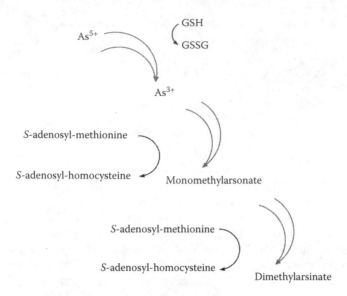

FIGURE 10.3   The pathway of arsenic.

used for filling teeth in many countries (WHO, 1991). Metallic mercury is used in thermometers, barometers, and instruments for measuring blood pressure. A major use of mercury is in the chlor-alkali industry, in the electrochemical process of manufacturing chlorine, where mercury is used as an electrode. The largest occupational group exposed to mercury is dental care staff. Air concentrations in some dental surgeries reached 20 μg/m³, but since then, levels have generally fallen to about one-tenth of those concentrations. Inorganic mercury is converted to organic compounds, such as methyl mercury, which is very stable and accumulates in the food chain. Methyl mercury was commonly used for control of fungi on seed grain. The general population is primarily exposed to mercury via food, fish being a major source of methyl mercury exposure, and dental amalgam. Several experimental studies have shown that mercury vapor is released from amalgam fillings and that the release rate may increase with chewing. Mercury in urine is primarily related to (relatively recent) exposure to inorganic compounds, whereas blood mercury may be used to identify exposure to methyl mercury. A number of studies have correlated the number of dental amalgam fillings or amalgam surfaces with the mercury content in tissues from human autopsy, as well as in samples of blood, urine, and plasma. Mercury in hair may be used to estimate long-term exposure, but potential contamination may make interpretation difficult (Sallsten et al., 1996).

10.5.2   PHYSICAL EFFECTS

10.5.2.1   Inorganic Mercury

Acute mercury vapor exposure may lead to lung damage. Chronic poisoning is characterized by neurological and psychological symptoms, such as tremor, changes in personality, restlessness, anxiety, sleep disturbance, and depression, and these symptoms are reversible after cessation. There is no central nervous involvement related to inorganic mercury exposure, because of the protection by the blood–brain barrier. The kidney may also be damaged from mercury exposure, which is reversible after exposure has stopped. The proteinuria is relatively low in occupational exposure. Metallic mercury is an allergen, which may cause contact eczema, and mercury from amalgam filings may give rise to oral lichen. Mercury in amalgam may cause a variety of symptoms. This so-called amalgam disease is, however, controversial, and although some authors claim proof of symptom relief after removal of dental amalgam fillings, there is no scientific evidence of this.

10.5.2.2 Organic Mercury

Methyl mercury poisoning has a latency of 1 month or longer after acute exposure, and the main symptoms relate to nervous system damage. The earliest symptoms are paresthesias and numbness in the hands and feet. Later on, coordination difficulties and concentric constriction of the visual field as well as auditory symptoms may develop. High doses may lead to death, usually 2–4 weeks after onset of symptoms (Apostoli et al., 2002; Becker et al., 2003; McKelvey et al., 2011). The Minamata catastrophe in Japan was caused by methyl mercury poisoning from fish contaminated by mercury discharges to the surrounding sea. In Iraq, more than 10,000 people were poisoned by eating bread baked from mercury-polluted grain, and several thousand people died. Without high seafood consumption, the general population does not face significant health risks from methyl mercury exposure. A high quantity of fish in the diet may also increase the risk of cardiovascular disease. In a recent study, the association of mercury levels in toenail clippings and docosahexaenoic acid levels in adipose tissue with the risk of a first myocardial infarction in humans was evaluated. Mercury levels were 15% higher than those in controls (95% CI, 5%–25%), and the adjusted odds ratio for myocardial infarction associated with the highest compared with the lowest quintile of mercury was 2.16 (95% CI, 1.09–4.29; P for trend = 0.006) (Zubero et al., 2011). The association between mercury levels in toenails and the risk of coronary heart disease among male health professionals with no previous history of cardiovascular disease. But the factors for coronary heart disease had been eliminated, mercury levels were not significantly associated with the risk of coronary heart disease (Leroux-Roels et al., 2007). These intriguing contradictory findings need to be followed up by more studies of other similarly exposed populations.

10.6   LEAD

10.6.1   Occurrence, Exposure, and Dose

The general population is exposed to lead from air and food in roughly equal proportions. During the last century, lead emissions to ambient air have further polluted our environment, with over 50% of lead emissions originating from petrol. In China in the last decade, lead emissions have decreased rapidly due to the use of unleaded petrol. Subsequently, blood lead levels in the general population have decreased.

Occupational lead exposure in the glass industry may have low or moderate levels, and high levels of lead exposure may be in lead mines and smelteries. Lead in tobacco can reach smoker via mainstream smoke. Up to 50% of inhaled inorganic lead may be absorbed in the lungs. Adults take up 10%–15% of lead in food, whereas children may absorb up to 50% via the gastrointestinal tract. The blood level of lead is bound to erythrocytes, and elimination is slow principally via urine. If lead accumulates in the skeleton, it is only slowly released from this body compartment. The biological half-life of lead in blood is about 1 month and in the skeleton 20–30 years (WHO, 1995). Blood lead was used as a biomarker of lead exposure. In adults, inorganic lead does not penetrate the blood–brain barrier, whereas this barrier is less developed in children. The permeable blood–brain barrier and the high levels of gastrointestinal uptake make children especially susceptible to lead exposure that can result in brain damage. Organic lead compounds penetrate body and cell membranes (Mortada et al., 2001). Tetramethyl lead and tetraethyl lead penetrate the skin easily. Especially, organic lead compounds may also cross the blood–brain barrier, and thus adults may be exposed to lead encephalopathy related to acute poisoning.

10.6.2 Physical Effects

The symptoms of acute lead poisoning are headache, irritability, abdominal pain, and various symptoms related to the nervous system. The symptoms of lead encephalopathy are characterized by sleeplessness and restlessness. Children may be affected by behavioral disturbances and learning and concentration difficulties. In severe cases of lead encephalopathy, the affected person may suffer from acute psychosis, confusion, and reduced consciousness. Lead exposure for a long time may result in memory deterioration, prolonged reaction time, and reduced ability to understand. Blood lead levels under 3 μmol/L may show signs of peripheral nerve symptoms with reduced nerve conduction velocity and reduced dermal sensibility. If the neuropathy is severe, the lesion may be permanent. The classical picture includes a dark blue lead sulfide line at the gingival margin. In less serious cases, the most obvious sign of lead poisoning is disturbance of hemoglobin synthesis, and long-term lead exposure may lead to anemia. Recent research has shown that long-term low-level lead exposure in children may also lead to diminished intellectual capacity (Elliott et al., 1999). Figure 10.3 shows a meta-analysis of four prospective studies using mean blood lead level over a number of years. The combined evidence suggests a weighted mean decrease in IQ of 2 points for a 0.48 μmol/L (10 μg/dL) increase in blood lead level (95% confidence interval from 0.3 points to 3.6 points) (Lidsky and Schneider, 2003; Steenland and Boffetta, 2000). Acute exposure to lead is known to cause proximal renal tubular damage. Long-term lead exposure may also give rise to kidney damage, and in a recent study of Egyptian policemen, urinary excretion of NAG was positively correlated with duration of exposure to lead from automobile exhaust, blood lead, and nail lead. Despite intensive efforts to define the relationship between body burden of lead and blood pressure or other effects on the cardiovascular system, no causal relationship has been demonstrated in humans. Using routinely collected data on mortality, hospital episode statistics data, and statutory returns to the Health and

Safety Executive (RIDDOR), one death and hospital cases were identified. The authors found that mortality and hospital admission ascribed to lead poisoning in England were rare, but that cases continue to occur and that some seem to be associated with considerable morbidity. Blood lead levels in children below 10 µg/dL have so far been considered acceptable, but recent data indicate that there may be toxicological effects of lead at lower levels of exposure than previously anticipated (Castano et al., 2012; Richter et al., 2009). There is also evidence that certain genetic and environmental factors can increase the detrimental effects of lead on neural development, thereby rendering certain children more vulnerable to lead neurotoxicity. IARC classified lead as a "possible human carcinogen" based on sufficient animal data and insufficient human data. Since then, a few studies have been published, the overall evidence for lead as a carcinogen being only weak; the most likely candidates are lung cancer, stomach cancer, and gliomas.

10.7   DISCUSSION

Harmful metals enter the body via food, drinking water, and the air we breathe, or by skin contact. They could accumulate in the liver, kidneys, bones, pancreas, and the central nervous system where they decrease health without being noticed and diagnosed. Heavy metals can and do cause cancer without ever being implicated in the diagnosis. Heavy metals cause sodium retention leading to high blood pressure. Heavy metals can and do cause heart disease and mental retardation. Everyone is contaminated with heavy metals, some seriously, without ever knowing it. Nonessential heavy metal ions cause aging in addition to serious diseases and death. People who are otherwise very healthy will have increased aging caused by heavy metal ions cross-linking between normal molecules in the body. These ions are sometimes referred to as *free radicals*. The cross-linking has been identified in diseases such as hardening of the arteries, skin ailments, carpal tunnel syndrome, degeneration of organs, and nerve damage. When heavy metal poisoning is suspected, it is important to begin the treatment as soon as possible to minimize long-term damage to the patient's nervous system and digestive tract. Heavy metal poisoning is considered a medical emergency, and the patient should be taken to a hospital emergency room.

10.8   CONCLUSION

Heavy metal ions could induce free radicals by Fenton reaction-induced tissue damage and carcinogenesis. Cadmium exposure could cause renal tubular damage and bone fractures even at lower exposure levels. Therefore, the levels of heavy metal exposure in the general population should be reduced in order to minimize the risk of adverse health effects. The general population does not face a significant health risk from methyl mercury by tobacco, but certain groups with high fish consumption may attain blood levels associated with a low risk of neurological damage to adults. Children are particularly vulnerable to lead exposure. Blood levels in children should be reduced below the levels so far considered acceptable, recent data indicating that there may be neurotoxic effects of lead at lower levels of exposure than previously anticipated. Although lead in petrol has dramatically declined over the last decades, thereby reducing environmental exposure, there is a need to reduce any remaining content of lead products in cigarettes. The use of lead-based fertilizer in growing tobacco should also be abandoned. Long-term exposure to arsenic in drinking water is mainly related to increased risks of skin cancer, but also some other cancers and other skin lesions such as hyperkeratosis and pigmentation changes. Environmental exposure to arsenic, primarily by inhaled cigarette smoke, is causally associated with lung cancer, though clear exposure–response relationships and high risks have not been observed.

REFERENCES

Akesson, A., Lundh, T., Vahter, M., Bellerup, P., Lidfeldt, J., Nerbrand, C., Samsioe, G., Stromberg, U., Skerfving, S. Tubular and glomerular kidney effects in Swedish women with low environmental cadmium exposure. *Environmental Health Perspectives.* 2005, 113, 1627–1631.

Aposhian, H. V. Enzymatic methylation of arsenic species and other new approaches to arsenic toxicity, *Pharmacology and Toxicology.* 1997, 37, 397–419.

Apostoli, P., Cortesi, I., Mangili, A., Elia, G., Drago, I., Gagliardi, T., Soleo, L. et al. Assessment of reference values for mercury in urine: The results of an Italian polycentric study. *Science of the Total Environment.* 2002, 289, 13–24.

Becker, K., Schulz, C., Kaus, S., Seiwert, M., Seifert, B. German Environmental Survey 1998 (GerES III): Environmental pollutants in the urine of the German population. *International Journal of Hygiene and Environmental Health.* 2003, 206, 15–24.

Berglund, M., Elinder, C. G., Järup, L. *Humans Exposure Assessment: An Introduction.* 2001, World Health Organization, Geneva, Switzerland.

Bernard, A., Roels, H., Buchet, J. P., Cardenas, A., Lauwerys, R. Cadmium and health: The Belgian experience. *IARC Scientific Publications.* 1992, 118, 15–33.

Buchet, J. P., Lauwerys, R., Roels, H., Bernard, A., Bruaux, P., Claeys, F., Ducoffre, G. et al. Renal effects of cadmium body burden of the general population. *Lancet.* 1990, 336, 699–702.

Castano, A., Sanchez-Rodriguez, J., Canas, A., Esteban, M., Navarro, C., Rodriguez-Garcia, A., Arribas, M., Diaz, G., Rodriguez-Garcia, J. A. Mercury, lead and cadmium levels in the urine of 170 Spanish adults: A pilot human biomonitoring study. *International Journal of Hygiene and Environmental Health.* 2012, 215, 191–195.

Centers for Disease Control and Prevention (CDC). Fourth national report on human exposure to environmental chemicals. 2009, Department of Health and Human Services, Centers for Disease Control and Prevention, Atlanta, GA.

Chilvers, D. C., Peterson P. J. Global cycling of arsenic. In: Hutchinson, T. C., Meema, K. M. (eds.), *Lead, Mercury, Cadmium and Arsenic in the Environment.* Chichester, U.K.: John Wiley & Sons, 1987, 279–303.

Elliott, P., Arnold, R., Barltrop, D., Thornton, I., House, I. M., Henry, J. A. Clinical lead poisoning in England: An analysis of routine sources of data. *Occupational and Environmental Medicine.* 1999, 56, 820–824.

Ferraro, P. M., Costanzi, S., Naticchia, A., Sturniolo, A., Gambaro, G. Low level exposure to cadmium increases the risk of chronic kidney disease: Analysis of the NHANES 1999–2006. *BMC Public Health.* 2010, 304, 1186–1471.

Friberg, L. Health hazards in the manufacture of alkaline accumulators with special reference to chronic cadmium poisoning. *Acta Medica Scandinavica.* 1950, 240, 1–124.

Hossn, E., Mokhtar, G., El-Awady, M., Ali, I., Morsy, M., Dawood, A. Environmental exposure of the pediatric age groups in Cairo City and its suburbs to cadmium pollution. *Science of the Total Environment.* 2001, 273, 135–146.

Hutchinson, T. C., Meema, K. M. Lead, mercury, cadmium and arsenic in the environment. *Energy Citations Database.* 1988, 17, 279–303.

Jarup, L., Berglund, M., Elinder, C. G., Nordberg, G., Vahter, M. Health effects of cadmium exposure a review of the literature and a risk estimate. *Scandinavian Journal of Work, Environment and Health.* 1998, 24, 1–51.

Järup, L. Hazards of heavy metal contamination, *British Medical Bulletin.* 2003, 68, 167–182.

Jarup, L., Hellstrom, L., Alfven, T., Carlsson, M. D., Grubb, A., Persson, B. Low level exposure to cadmium and early kidney damage: The OSCAR study. *Occupational and Environmental Medicine.* 2000, 57, 668–672.

Leroux-Roels, I., Borkowski, A., Vanwolleghem, T. Antigen sparing and cross-reactive immunity with an adjuvanted rH5N1 prototype pandemic influenza vaccine: A randomised controlled trial, *The Lancet,* 2007, 370(9587), 580–589.

Lidsky, T. I., Schneider, J. S. Lead neurotoxicity in children: Basic mechanisms and clinical correlates. *Brain.* 2003, 126, 5–19.

McKelvey, W., Jeffery, N., Clark, N., Kass, D., Parsons, P. J. Population-based inorganic mercury biomonitoring and the identification of skin care products as a source of exposure in New York City. *Environmental Health Perspectives.* 2011, 119, 203–209.

Moriguchi, J., Ezaki, T., Tsukahara, T., Furuhi, K., Fukui, Y., Okamoto, S., Ukai, H., Saikurai, H., Ikeda, M. α1-Microglobulin as a promising marker of cadmium-induced tubular dysfunction, possibly better than β2-microglobulin. *Toxicology Letter.* 2004, 148, 11–20.

Mortada, W. I., Sobh, M. A., El-Defrawy, M. M., Farahat, S. E. Study of lead exposure from automobile exhaust as a risk for nephrotoxicity among traffic policemen. *American Journal of Nephrology.* 2001, 21, 274–279.

National Academy of Sciences (NAS). *Recommended Dietary Allowances,* 10th edn. 1989, National Academy Press, Washington, DC.

Noonan, C. W., Sarasua, S. M., Campagna, D., Kathman, S. J., Lybarger, J. A., Mueller, P. W. Effects of exposure to low levels of environmental cadmium on renal biomarkers. *Environmental and Health Perspectives.* 2002, 110, 151–155.

Pappas, R. S. Toxic elements in tobacco and in cigarette smoke: Inflammation and sensitization. *Metallomics.* 2011, 11, 1181–1198.

Richter, P. A., Bishop, E. E., Wang, J. T. Tobacco smoke exposure and levels of urinary metals in the U.S. youth and adult population: The National Health and Nutrition Examination Survey (NHANES) 1999–2004. *International Journal of Hygiene and Environmental Health.* 2009, 6, 1930–1964.

Roels, H., Lauwerys, R., Konings, J., Buchet, J. P., Bernard, A., Green, S., Bradley. D., Morgan, W., Chettle, D. Renal function and hyperfiltration capacity in lead smelter workers with high bone lead. *Occupational and Environmental Medicine.* 1994, 51, 505–512.

Sallsten, G., Thoren, J., Barregard, L., Schutz, A., Skarping, G. Long-term use of nicotine chewing gum and mercury exposure from dental amalgam fillings. *Journal of Dental Research.* 1996, 75, 594–598.

Steenland, K., Boffetta, P. Lead and cancer in humans: Where are we now? *American Journal of Indian Medicine.* 2000, 38, 295–299.

Thompson, D. J. A chemical hypothesis for arsenic methylation in mammals, *Chemico-Biological Interactions,* 1993, 88(2–3), 89–114.

WHO. Lead. *Environmental Health Criteria,* vol. 165. Geneva: World Health Organization, 1995.

World Health Organization. Inorganic mercury. *Environmental Health Criteria.* 1991, Geneva, Switzerland, p. 118.

World Health Organization. Cadmium. *Environmental Health Criteria.* 1992, Geneva, Switzerland, p. 134.

World Health Organization. Arsenic and arsenic compounds. *Environmental Health Criteria,* 2001, WHO, Geneva, Switzerland, p. 224.

Yamamoto, S., Yoshitsugu, Ki, Tsutomu, M. Cancer induction by an organic arsenic with five carcinogens. *Cancer Research*. 1995, 55, 1271–1285.

Zubero, M. B., Aurrekoetxea, J. J., Ibarluzeac, B. B., et al. Evolution of PCDD/Fs and dioxin-like PCBs in the general adult population living close to a MSW incinerator. *Science of the Total Environment*. 2011, 410, 241–247.

Zubero, M. B., Aurrekoetxea, J. J., Ibarluzea, J. M., Arenaza, M. J., Rodríguez, C., Sáenz, J. R. Heavy metal levels (Pb, Cd, Cr and Hg) in the adult general population near an urban solid waste incinerator. *Science of the Total Environment*. 2010, 408, 4468–4474.

11 Application of Smoke Exposure Biomarkers in Epidemiologic Research

11.1 SMOKE EXPOSURE BIOMARKER AND HEALTH EFFECTS

Biomarkers of tobacco smoke exposure are related to lung cancer and cardiovascular diseases (Hecht, 2003), in which specific carcinogens are responsible for lung cancer and their contribution to risk is less clear. It makes them attractive surrogate end points to evaluate the effectiveness of smoking reduction interventions at improving health. For example, the polycyclic aromatic hydrocarbons (PAHs) and tobacco-specific carcinogen 4-(methylnitrosamino)-1-(3-pyridyl)-1-butanone (NNK) present in tobacco smoke are believed to be the major causative chemicals for lung cancer in smokers (Hecht, 1999). A PAH metabolite 1-hydroxypyrene (1-HOP) and NNK metabolite 4-(methylnitrosamino)-1-(3-pyridyl)-1-butanol (NNAL), its glucuronide (NNAL-Gluc) are biomarkers of smoking-related PAHs and TSNAs exposure (Hecht et al., 2004a,b). Such biomarkers could ultimately become part of a predictive model for identifying those smokers most likely to get lung cancer and cardiovascular diseases. Furthermore, doing so could improve understanding of lung cancer etiology in smokers and provide a basis for rational approaches to prevention and even therapy. This model, which has evaded researchers to date, could serve not only to identify long-term smokers needing more vigorous intervention or surveillance but also perhaps to identify among smokers newly acquiring the habit those with demonstrably increased susceptibility, in hopes that the increased risk may help motivate them to give up smoking.

From the complexity of the exposure agent (tobacco smoke has several thousand constituents) and the diversity of the biological end points (various smoking-related diseases such as cancer, cardiovascular diseases, and chronic obstructive pulmonary disease), it is sure that no single biomarker can adequately reflect tobacco smoke exposure and the risk associated with the use of tobacco products. A panel of biomarkers, including biomarkers of internal exposure, biologically effective dose, and potential harm, would be most useful for the assessment of harm reduction. This panel should be studied in the context of genetic susceptibilities for cancer risk and smoking behavior. Also, biomarkers that are specific for individual carcinogens and those that consider tobacco smoke as a complex mixture are complementary, and both should be used. These markers should be used in laboratories with appropriate quality control and quality assurance procedures (Shields, 2002).

11.2 APPLICATION OF SMOKE EXPOSURE BIOMARKERS IN EPIDEMIOLOGIC RESEARCH

Assessment of biomarkers is an appropriate way to estimate exposure to cigarette smoke and smokeless tobacco (SLT) constituents in tobacco consumers. Naufal et al. (2011) used the U.S. National Health and Nutrition Examination Survey (NHANES, 1999–2008) to evaluate biomarkers of PAHs, NNK, volatile organic compounds (VOC), halogenated aromatic hydrocarbons (HAHs), acrylamide, and metals. Results show that biomarker levels in SLT consumers were significantly lower than in smokers (excluding NNK and some HAHs) and were not significantly different compared with nonconsumers (excluding NNK and some PAHs). Strasser's study (Strasser et al., 2011) found that faster nicotine metabolizers (third and fourth quartiles vs. first quartile) based on the ratio of the metabolites cotinine and 3′-hydroxycotinine exhibited significantly greater total puff volume and total NNAL; the total puff volume by daily cigarette consumption interaction was a significant predictor of total NNAL level. These results provide useful information for science-based risk assessment and regulation of tobacco products.

The carcinogenic activity of NNK and its metabolites in the lung has been proved, and there are a few studies that explore these chemicals that induce cancer risk. A nested case–control study of 210 male lung cancer cases and 630 matched controls aged 40–75 years participating in the Health Professionals Follow-up Study was conducted (Al-Delaimy and Willett, 2011). Toenail samples collected in 1987 were analyzed for nicotine levels, and incident lung cancer cases were diagnosed between 1988 and 2000. The toenail nicotine biomarker was found to be a strong predictor of lung cancer independent of smoking history, suggesting that the adverse effects of cigarette smoke may be underestimated in studies based on smoking history only.

Hecht et al. (Church et al., 2009) conducted a case–control study nested in the Prostate, Lung, Colorectal, and Ovarian Cancer Screening Trial. They randomly selected 100 lung cancer cases and 100 controls who smoked at baseline and analyzed their baseline serum for total NNAL, cotinine, and r-1,t-2,3,c-4-tetrahydroxy-1,2,3,4-tetrahydrophenanthrene (PheT), a biomarker of PAH exposure and metabolic activation to examine the association of the biomarkers with all lung cancers and for histologic subtypes. Their results indicated that individual associations of age, smoking duration, and total NNAL with lung cancer risk were statistically significant. After adjustment, total NNAL was the only biomarker significantly associated with risk (odds ratio, 1.57 per unit SD increase; 95% confidence interval, 1.08–2.28). A similar statistically significant result was obtained for adenocarcinoma risk, but not for nonadenocarcinoma. It provided insight into the etiology of smoking-related lung cancer and reinforced targeting NNK for cancer prevention. Chung et al. (2011) conducted a case–control study to assess whether the tobacco-specific nitrosamine NNK metabolites, including total NNAL, free NNAL, and NNAL-Gluc, were associated with an increased risk of urothelial carcinoma (UC). There were 127 pairs of UC cases and matched healthy participants recruited for a hospital-based case–control study. Participants completed questionnaires of medical and social information, including smoking history, and provided 50 mL urine samples. Urine samples were analyzed for free NNAL and NNAL-Gluc using the liquid chromatography–tandem mass spectrometry method. Nonparametric analysis and multivariate logistic regression were applied to compare the differences in NNK-related metabolites between UC cases and controls and to estimate the UC risk associated with certain risk factors. Overall, controls with higher cumulative cigarette smoking exposure had higher total NNAL, free NNAL, and NNAL-Gluc. In addition, a decreased NNAL-Gluc/free NNAL ratio corresponded to a significantly increased UC risk. The association between the NNAL-Gluc/free NNAL ratio and UC risk was significant in a dose–response manner. Furthermore, cumulative cigarette smoking exposure was found to interact significantly with low NNAL-Gluc/free NNAL ratio to affect UC risk in this study. This is the first study to conclude that the metabolic products of total NNAL, free NNAL, and NNAL-Gluc might be measured as biomarkers of cigarette smoking exposure. Furthermore, the NNAL-Gluc/free NNAL ratio was a better biomarker to evaluate UC risk than total NNAL.

PAHs are believed to be among the principal causative agents for lung cancer in smokers, but no epidemiologic studies have evaluated the relationship of PAH uptake and metabolism to lung cancer. Hecht et al. (Yuan et al., 2011) quantified prediagnostic urinary levels of PheT, a validated biomarker of PAH uptake and metabolism, as well as NNAL and its glucuronides (total NNAL), and cotinine and its glucuronides (total cotinine), validated biomarkers of uptake of NNK, and nicotine, respectively, in relation to lung cancer risk among current smokers in a nested case–control study within a cohort of 18,244 Chinese men in Shanghai, China. Their study confirmed that urinary total NNAL and total cotinine are independently related to lung cancer risk. By using a prospective study design comparing biomarker levels in cancer cases and controls, all of whom were smokers, the results demonstrate that several of these biomarkers—total cotinine, total NNAL, PheT, and total N-nitrosonornicotine (NNN)—are biomarkers of cancer risk (Hecht et al., 2013). Therefore, these biomarkers have the potential to become part of a cancer risk prediction algorithm for smokers.

REFERENCES

Al-Delaimy, W. K., Willett, W. C. Toenail nicotine level as a novel biomarker for lung cancer risk. *American Journal of Epidemiology*. 2011, 173, 822–828.

Chung, C.-J., Lee, H.-L., Yang, H.-Y., Lin, P., Pu, Y.-S., Shiue, H.-S., Su, C.-T., Hsueh, Y.-M. Low ratio of 4-(methylnitrosamino)-1-(3-pyridyl)-1-butanol-glucuronides (NNAL-Gluc)/free NNAL increases urothelial carcinoma risk. *Science of the Total Environment*. 2011, 409, 1638–1642.

Church, T. R., Anderson, K. E., Caporaso, N. E., Geisser, M. S., Le, C. T., Zhang, Y., Benoit, A. R., Carmella, S. G., Hecht, S. S. A prospectively measured serum biomarker for a tobacco-specific carcinogen and lung cancer in smokers. *Cancer Epidemiology Biomarkers and Prevention*. 2009, 18, 260–266.

Hecht, S. S. Tobacco smoke carcinogens and lung cancer. *Journal of the National Cancer Institute*. 1999, 91, 1194–1210.

Hecht, S. S. Tobacco carcinogens, their biomarkers and tobacco-induced cancer. *Nature Reviews Cancer*. 2003, 3, 733–744.

Hecht, S. S., Carmella, S. G., Le, K.-A., Murphy, S. E., Li, Y. S., Le, C., Jensen, J., Hatsukami, D. K. Effects of reduced cigarette smoking on levels of 1-hydroxypyrene in urine. *Cancer Epidemiology Biomarkers & Prevention.* 2004a, 13, 834–842.

Hecht, S. S., Murphy, S. E., Carmella, S. G., Zimmerman, C. L., Losey, L., Kramarczuk, I., Roe, M. R., Puumala, S. S., Li, Y. S., Le, C. Effects of reduced cigarette smoking on the uptake of a tobacco-specific lung carcinogen. *Journal of the National Cancer Institute.* 2004b, 96, 107–115.

Hecht, S. S., Murphy, S. E., Stepanov, I., Nelson, H. H., Yuan, J.-M. Tobacco smoke biomarkers and cancer risk among male smokers in the Shanghai Cohort Study. *Cancer Letter.* 2013, 334, 34–38.

Naufal, Z. S., Marano, K. M., Kathman, S. J., Wilson, C. L. Differential exposure biomarker levels among cigarette smokers and smokeless tobacco consumers in the National Health and Nutrition Examination Survey 1999–2008. *Biomarkers.* 2011, 16, 222–235.

Shields, P. G. Tobacco smoking, harm reduction, and biomarkers. *Journal of the National Cancer Institute.* 2002, 94, 1435–1444.

Strasser, A. A., Benowitz, N. L., Pinto, A. G., Tang, K. Z., Hecht, S. S., Carmella, S. G., Tyndale, R. F., Lerman, C. E. Nicotine metabolite ratio predicts smoking topography and carcinogen biomarker level. *Cancer Epidemiology Biomarkers & Prevention.* 2011, 20, 234–238.

Yuan, J.-M., Gao, Y.-T., Murphy, S. E., Carmella, S. G., Wang, R., Zhong, Y., Moy, K. A., Davis, A. B., Tao, L., Chen, M. Urinary levels of cigarette smoke constituent metabolites are prospectively associated with lung cancer development in smokers. *Cancer Research.* 2011, 71, 6749–6757.

12 Suggestions for the Use of Specific Biomarkers of Exposure to Cigarette Smoke for Regulation Purposes

12.1 BACKGROUND

Tobacco use status and the intensity of cigarette smoke exposure have traditionally been estimated using self-report of whether an individual uses tobacco and the amount and frequency of use that the individual reports. These methods have been used in epidemiological studies as valid markers of increased disease risks.

Biomarkers of cigarette smoke exposure can also be used to assess tobacco use status and the intensity of cigarette smoke exposure. Biomarker-based validation of smoke status is different from self-reported status for a modest fraction of former users and smaller fractions of current and nonusers (FCTC, 2007). In addition, while a clear relationship between daily cigarette consumption and levels of smoke constituents or their metabolites measured in bodily fluids, there is substantial variation in the amount of smoke constituent present for any specific number of cigarettes smoked per day (Benowitz et al., 1983; Jarvis et al., 2001; Joseph et al., 2005). The discrepancy in smoking status and the individual variability of constituent/metabolite levels among smokers of similar numbers of cigarettes smoked per day call into question the accuracy of self-reported smoking status and number of cigarettes smoked per day as measures of smoke exposure for individual smokers. While less fully documented, similar concerns also exist for self-reported status and intensity of use in relation to other forms of tobacco (FCTC, 2007).

Measurement of tobacco or tobacco smoke–related biomarkers in bodily fluids, particularly those constituents specific to tobacco (e.g., TSNAs and nicotine), has been used to improve the accuracy of self-reported tobacco use status in intervention and epidemiological studies. This method has also been used to quantify tobacco exposure intensity in experimental and other experimental settings. Cotinine, a metabolite of nicotine, used for verification of tobacco use status is now standard in settings where the accuracy of individual-level smoking status is critical, such as in studies of smoking cessation therapies or in determining the eligibility of applicants for lower life insurance rates as nonsmokers. However, to date, only a few epidemiological studies of disease outcomes have been conducted to demonstrate that quantitative estimates of smoke exposure using biomarkers improve the accuracy of disease risk prediction compared with self-reported data (Baltar et al., 2011; Boffetta et al., 2006; Whincup et al., 2004).

The current standard for assessing smoking status and the amount of smoke exposure in the general population remains self-report, in part due to the expense and difficulty of collecting biological samples for large populations and in part due to the unresolved research questions as to how to use these data to estimate population exposure.

The reality that more accurate definition of tobacco use status is possible and the prospect that more accurate quantification of exposure is attainable support the use of biomarkers of tobacco exposure in those settings where the question being asked necessitates greater accuracy of use status or quantitative exposure than can be achieved by self-reported data.

An additional concern for scientists has been the long durations of exposure required to demonstrate the effects of tobacco exposure on many types of disease risks using traditional epidemiological approaches with disease manifestation as the outcome. Evaluating the risks resulting from changes in tobacco product designs using epidemiological approaches could take decades of exposure, making this approach of very limited value for regulatory assessment of the risks or claims for newer tobacco products and designs. Measures of cellular or organ changes consistent with tobacco-related injury and disease clearly precede disease manifestation, and they offer the potential for more rapid demonstration of differences in risk resulting from changes in the design or use of tobacco products (Stratton et al., 2001). A number of these cellular and organ changes have been suggested as potential biomarkers of tobacco-related injury, but to date, none has been validated since it has not been demonstrated that a change in the biomarker reliably predicts a difference in disease risk (Hatsukami et al., 2006).

Detecting adverse outcomes from tobacco use may take decades to develop. Biomarkers are measures that can be used in the early stages of tobacco use to assess exposure to tobacco toxins or to predict adverse health outcomes with which they are associated. Examples of biomarkers include specific chemical components of tobacco or their metabolites; early biochemical, histological, or physiological effects; and early health effects. Mechanistically relevant and quantitatively valid biomarkers are essential for assessing the ultimate impact of new products, treatments, preventive measures, and public health policies on tobacco-related disease. The tobacco industry's recent introduction of a variety of new tobacco products or devices with implied claims of reduced health risks highlights the need to develop methods for assessing their potential for benefit or harm. A wide variety of biomarkers for tobacco exposure or harm has been studied. Although many questions about their use remain unanswered, substantial data exist regarding their validity and utility. This conference reviewed both the general issues surrounding biomarker use and the current state of knowledge regarding the most widely studied and promising biomarkers.

The WHO Study Group on Tobacco Product Regulation recognizes that effective regulation of tobacco products, particularly products offered as reduced exposure or reduced risk products, can be greatly facilitated by the development of validated biomarkers of individual constituent exposure, biomarkers of exposure that are useful proxies for total tobacco emissions exposure, and biomarkers that can reliably predict differences in disease outcome. WHO has previously described the limitations of some of the present methods for making these assessments (WHO, 2003; WHO, 2004). A large number of potential biomarkers of exposure and effect have been identified in various research settings. A review (Hatsukami et al., 2006) of these potential biomarkers concluded that validated biomarkers exist for exposure to some tobacco emissions and for some biological processes such as inflammation and endothelial injury, but the review also came to the conclusion that "we have no valid biomarkers that serve as proxies for tobacco-related disease."

Biomarkers of tobacco toxin exposure are also related to heart disease (Cohn et al., 2004; Patel et al., 2004) and lung cancer (Hecht, 2001) which make them attractive surrogate end points to evaluate the effectiveness of smoking reduction interventions at improving health. For example, the tobacco-specific carcinogen 4-(methylnitrosamino)-1-(3-pyridyl)-1-butanone (NNK) and polycyclic aromatic hydrocarbons (PAH) present in tobacco smoke are believed to be the major causative agents for lung cancer in smokers (Jarvis et al., 2001; Joseph et al., 2005; Whincup et al., 2004). NNK metabolite 4-(methylnitrosamino)-1-(3-pyridyl)-1-butanol (NNAL), its glucuronide (NNAL-Gluc), and a PAH metabolite 1-hydroxypyrene (1-HOP) are biomarkers of carcinogen exposure (Boffetta et al., 2006; Stratton et al., 2001a). 1-HOP is the major metabolite of the abundant but noncarcinogenic PAH pyrene, which is always present in mixtures of PAH (Byrd et al., 1995).

12.2 MEASURING EXPOSURE

Biomarkers of exposure provide evidence of the presence of a tobacco toxicant and/ or its metabolites in the body. The most straightforward biomarkers directly measure the concentration of the toxicant or its metabolites in biological fluids or hair and nail. Ideal characteristics for a biomarker of exposure include tobacco smoke being the only source of the biomarker, with other sources of exposure being minor or nonexistent; the marker should be easily detectable; the analysis methods should be reproducible across laboratories; and the marker should reflect a specific toxic exposure or be a reliable surrogate of tobacco smoke toxicant exposure. Other issues of importance in using biomarkers as measures of exposure include how well the biomarker reflects long-term exposure to tobacco (the half-life of the biomarker, $t1/2$, indicates the period of time for which the biomarker reflects exposure and may vary from several hours to several weeks), what additional information is obtained by adding a particular biomarker to existing indicators, and how applicable the biomarker is to studying large populations in epidemiological studies.

The simplest use of biomarkers of exposure is to define tobacco use status. This is usually achieved by setting a value of the biomarker above which the individual is presumed to be a current user. Since current users include those who are very light or nondaily users, and since some nonsmokers are exposed to very high concentrations of secondhand smoke, there will be some overlap between light current users and heavily exposed nonsmokers in the constituent levels present in biological fluids, even for those constituents that are only present in tobacco smoke (e.g., nicotine or TSNAs). This overlap is greater for constituents with other sources of exposure (e.g., carbon monoxide [CO]), and some studies have used combinations of biomarkers to define tobacco use status. Nevertheless, it is generally accepted that biochemical verification of tobacco use status leads to a substantially more accurate definition of who is a current user.

A second use of biomarkers of exposure is to quantify the amount of exposure experienced by the individual user. This quantification may be specific to the constituent and the constituent's consequences, for example, quantifying nicotine levels in studies of dependence. The level of an individual constituent biomarker may also be used as a proxy to quantify whole smoke or total smokeless tobacco exposure. The relationship between biomarker levels of a single constituent and the total smoke or tobacco exposure may be influenced by individual characteristics, genetic and metabolic differences, patterns of use, and the presence of other sources of the constituent in the environment affecting the individual. When levels of a specific biomarker are used for whole exposure comparisons between products, differences in the composition of the emissions of the different products also need to be considered. For example, the toxicant burden from using smokeless tobacco may be well estimated by cotinine levels among users of a single smokeless product, but the toxicant burden at the same cotinine level will be very different among users of smokeless tobacco products in India when compared with smokeless tobacco users in Sweden because of the much higher concentrations of many toxic constituents present in the products sold in India.

A final form of exposure biomarker is one that measures the biologically effective dose of a single constituent or groups of constituents. These biomarkers attempt to quantify the exposure that has reached the tissue in ways that can result in injury and cellular or organ damage. Measurement of carcinogen–DNA adducts in lung tissue is one example of this effort to measure the biologically effective dose. The concept of a biologically effective dose is based on an understanding of the mechanism(s) by which constituents cause disease and attempts to quantify with precision the dose of the agent present in that mechanistic pathway. A limiting corollary of that mechanistic precision is that the biomarker may have less validity for organs or disease processes other than the one measured. For example, carcinogen–DNA adducts in lung tissue may define a biologically effective dose for the carcinogen(s) in the lung but may have less relevance to estimating the biologically effective dose for heart disease.

12.3   MEASURING BIOLOGICAL CHANGES

Cigarette smoke contains more than 4000 individual constituents and damages most of the organs in the body (HHS, 2004). Our understanding of the mechanistic relationships between exposure to these individual constituents and the diseases produced by inhalation of whole smoke is incomplete, and efforts to estimate individual disease risks from known constituent levels and toxicity underpredict the level of risk that occurs with smoking (Fowles and Dybing, 2003). Furthermore, reduction of any one chemical or class of chemicals will not necessarily lead to a reduction in disease risk, and the magnitude of chemical reduction needed to produce a substantive change in disease risk remains uncertain.

These limitations of tobacco chemistry and toxicology for estimating disease risks related to cigarette smoking have led to the development of measures of the cellular and physiological responses of subjects who are exposed to the integrated tobacco emissions of whole smoke. Such markers of biological changes do describe differences in biological response to the complex mixtures of tobacco emissions and can be used to characterize differences in that biological response with the use of different tobacco products and potential reduced exposure products (PREPs). Additional research will be required to demonstrate which of these biomarkers are valid predictors of disease risks and the levels of change in the validated biomarkers that are required to predict a meaningful difference in subsequent disease manifestation.

To date, none of the markers of biological change discussed in this report has been validated through the establishment of their independent predictive validity. Many of them are established as risk factors for disease in epidemiological studies, and some have been shown to decline with cessation or reduction in tobacco use, demonstrating a dose–response relationship with smoke exposure. The uncertainty that remains relates to whether individuals who have a change in exposure and who have a larger change in the biomarker have less subsequent disease risk than those who have the same change in exposure and a lesser change in the biomarker. It also remains to be established which of these biomarkers simply reflect biological responses to tobacco exposure and which reflect biological changes that are part of the critical pathways by which exposure progresses to disease. The research goal being pursued is the development of biomarkers for which a change in level of the biomarker can be reliably inferred to define a change in the likelihood of future disease. Many of the biomarkers discussed in this book offer considerable promise for achieving that goal. Markers of biological changes are not specific to tobacco use. However, the biomarkers discussed in this book are the ones where significant differences have been demonstrated between smokers and nonsmokers and where changes in smoke exposure through cessation or reduction in use have been associated with changes in the level of the biomarker, suggesting their potential value as biomarkers of injury and risk (Hatsukami et al., 2006).

12.4   MEASURING INJURY AND DISEASE

A substantial body of evidence exists on the mechanisms by which tobacco use causes various diseases, and a number of biochemical, cellular, and organ system measures exist that define the various mechanistic pathways either qualitatively or quantitatively. Similarly, a number of measures exist that can qualitatively or quantitatively predict the rate of disease occurring in a population and that are accepted as independent risk factors for disease. This is particularly true for cardiovascular disease for which a substantial number of risk factors have been identified. Measuring changes early in the mechanistic pathway of disease occurrence offers the promise of more rapid characterization of the risks that can result from the use of different tobacco products, and this promise has stimulated great interest in defining biomarkers for which a change in the level of the biomarker would accurately predict a change in disease risks.

Unfortunately, our understanding of the mechanisms by which smoking causes disease is not complete enough to identify with confidence the rate-limiting steps in the mechanistic pathways and therefore the changes that will reliably predict risk. We are also unsure which changes are markers of tobacco use, and therefore, their presence is associated with increased risk but not part of the pathway by which disease occurs and therefore, if altered, will not alter risk. These limitations mean that acceptance of a given biological change as a biomarker of injury and risk requires validation that a change in the biomarker independently predicts a change in the frequency of disease occurrence.

Biomarkers do exist that can measure the presence and extent of various systemic processes, including inflammation, which may play a mechanistic role in disease occurrence. However, the diseases caused by cigarette smoking involve multiple processes, and it remains unproven whether alteration of a single process (e.g., reduced inflammation) will reduce disease frequency.

12.5 RECOMMENDED USES FOR BIOMARKERS OF EXPOSURE AND EFFECT

Biomarkers can be useful tools to regulators in their efforts to understand tobacco products and reduce the public health harm of tobacco, and thus, biomarkers may be useful in several contexts.

12.5.1 IMPROVING THE ACCURACY OF THE DEFINITION OF CURRENT TOBACCO USE STATUS

For many purposes, self-report of tobacco use status provides information with sufficient accuracy to make judgments about the effects of tobacco control interventions or trends in tobacco use. However, with recent former smokers (IARC, 2006), those who may have a disproportionate incentive to misrepresent smoking status, or where the precision of definition of smoking status is critical (such as in clinical trials), biomarkers of exposure can play an important role. In the absence of nicotine replacement therapy or other tobacco use, cotinine levels are the best biomarker to define whether an individual is a current smoker. Cotinine levels are less useful for the definition of smoking status among adolescent smokers (where the definition of current smoking is any cigarette use in the past 30 days) (Kandel et al., 2006) and in populations where there are large numbers of nondaily cigarette smokers. In addition, while uncommon, very heavy secondhand smoke exposure can result in cotinine levels that overlap with those of light active smokers, particularly occasional smokers. When nicotine replacement therapy is being used, cotinine levels are not useful for defining smoking status, and other biomarkers such as levels of minor tobacco alkaloids or total NNAL in urine can be used. Similar utilities and constraints exist for using biomarkers of exposure to improve the accuracy of defining smokeless tobacco use status. Biomarkers of exposure may also be useful to insurance and regulatory authorities in monitoring the success rates of clinical cessation programs and other funded interventions in order to establish relative effectiveness of cessation programs and their cost-effectiveness. The prevalence of tobacco use at the population level can be investigated by population-based surveys that measure tobacco sales or self-reported consumption. The costs and response burden of obtaining biomarkers for representative samples of the population often make impractical the use of biomarkers to improve the accuracy of self-response measures of tobacco use status in surveillance tools. The intrusive nature of collection of many types of biomarkers may potentially decrease participation rates, and the increased cost of collection and analysis of biomarkers needs to be balanced against the information that could be gained by investing those costs in expanding the size of the population sample surveyed. Self-response data remain adequate for most indications, including examination of general population effects of tobacco control interventions and describing tobacco use over time. Improved accuracy in the definition of tobacco use status may be useful when monitoring or evaluating the effectiveness of smoking cessation interventions in health-care settings and other locations. Use for this purpose is complicated by the inability to use cotinine levels for establishing tobacco use status when former users are using nicotine replacement products. The cost of using minor alkaloids in tobacco or total NNAL may be justified in this setting by the improved confidence with which the results can be stated and by the potential reduction in the number of individuals who have to be examined to define a statistically significant result. The periodic collection of population-based sets of biomarkers would be an invaluable research tool. These data can be used to identify those settings where tobacco use status is less accurate with self-report and to describe individual genetic and metabolic characteristics and their influences on the relationships between exposure and biomarkers in biological samples of that exposure. The use of biomarkers to improve the accuracy of the definition of tobacco use status is recognized as an essential component of studies submitted to regulatory authorities seeking approval for smoking cessation therapies, and it is also the standard by which decisions about differential insurance rates or employment opportunities are related to tobacco use status. The use of biomarkers is also highly recommended as a component of studies evaluating or monitoring the effectiveness of tobacco control interventions for purposes of making public policy decisions about inclusion in programmatic efforts or funding support. In these settings, larger public policy decisions will hinge on the results of the evaluations, making accuracy in defining tobacco use status a substantive value.

12.5.2   Evaluating the Intensity of Exposure to Specific Constituents

There is no question that levels of nicotine or cotinine in urine, blood, nails, and hair are more accurate measures of the amount of recent nicotine intake for an individual than the number of cigarettes smoked per day or other self-reported measures of intensity of tobacco use. This is also true for the other biomarkers of carcinogen/mutagen intake such as, NNK, B[a]P, VOCs, and heavy metals relative to the specific carcinogen or family of carcinogens being measured.

This improved accuracy in quantifying exposure to individual constituents is of great value in experimental studies examining the mechanisms by which tobacco use causes addiction and disease. Improved accuracy in the measurement of intensity of exposure to nicotine or other constituents can also be of use to regulatory authorities for validating claims of reduced exposure with different products and assessing exposures that occur in different settings.

A number of PREPs have been developed and marketed by tobacco manufacturers. Some, such as the *low-tar* cigarette, do not result in reductions in exposure or risk of lung cancer (Hecht et al., 2005; NIH, 2001; Stratton et al., 2001a). For other tobacco products making reduced exposure claims, reductions in some carcinogens were verified, but there were no reductions in others (Hatsukami et al., 2004). The variability in these results makes it clear that verification of reduced exposure claims through experimental studies using biomarkers, rather than measurements of emissions, is essential for the regulation of these claims.

The assessment of exposure-reduction claims is complicated by the need to separate the differences due to individual characteristics of users and the factors that define self-selection of the product used from the differences due to changes in product design. At present, exposure-reduction claims can only be reliably examined in experimental investigations where groups of users are randomly assigned to the use of different products. Examining levels of biomarkers of exposure in self-selected users in the general population leads to confounding of the characteristics of the individual with the characteristics of the product, both of which can influence exposure levels. Therefore, differences in levels of biomarkers of exposure between users of different products found in the general population cannot be reliably ascribed to differences in the product.

Even with an experimental design, careful attention to selection of the control population is necessary (Hatsukami et al., 2003). Smokers who are experimentally switched to a different tobacco product often reduce the intensity of use of that product, particularly if they find it very different from their usual brand or unsatisfactory. Since individuals who find the product unsatisfactory would be unlikely to use it in a nonexperimental setting, including these individuals in a comparison of the exposures that occur with switching to a new product is likely to result in an underestimate of the exposure that would occur with self-selected use of the product by the general population. Comparing their exposure before and after switching without a control group could then result in a reduced exposure level that is due to the unacceptability of the product to a fraction of the population studied, rather than to differences in exposure due to differences in product design. For this reason, the recommended experimental approach is to have a control group and to have both the test and the control groups of users switch to different products. The challenge for this experimental design is to find a product that the control group can switch to that is substantively different in design from the product being tested but that has a similar level of unacceptability. Differences in satisfaction with the control and test products need to be considered when assessing differences in exposure observed in these experimental evaluations (Hatsukami et al., 2005).

The biomarkers of exposure presented can reliably quantify the level of exposure to individual constituents for individual users, and levels of cotinine in general population studies have been used to examine several important public policy questions. Biomarkers of exposure (cotinine) collected in conjunction with representative population surveys such as the National Health and Nutrition Examination Survey and the Health Survey for England have allowed population-based statements about the changing level of secondhand smoke exposure in the United States (Pirkle et al., 2006; Pirkle et al., 1996) and the consistency of nicotine exposure in smokers of cigarettes with different machine-measured yields in the United Kingdom (Jarvis et al., 2001). Both of these analyses have substantial regulatory significance and demonstrate the value of collecting biomarkers on representative samples of the population to support research on issues of regulatory importance. In particular, atmospheric measurements of smoke constituents in conjunction with biomarkers of exposure are useful tools in demonstrating changes in exposure following implementation of restrictions on where smoking is allowed. The clear demonstration of substantial differences in secondhand smoke exposure following implementation of restrictions on smoking in various locations has also helped to justify and build support for implementing and extending these restrictions.

Biomarker levels are influenced by individual characteristics including race, metabolic status, and genotype, and trends in tobacco use behavior may also vary according to some of these characteristics. For example, cotinine levels can vary according to individual characteristics such as ethnicity and genetic and metabolic status even for the same level of nicotine intake. Additional research using population-based surveys of biomarkers and smoking behaviors will help to identify the individual characteristics of tobacco users that influence the relationship between actual level of exposure and level of the biomarker in the biological fluids. A better understanding of these individual determinants of biomarker variability is needed before population-based surveys of biomarkers can reliably be substituted for self-reported behavior and per capita consumption data in evaluating trends in tobacco use at the general population level.

12.5.3 Evaluating the Intensity of Exposure as a Proxy for Total Tobacco Exposure

Biomarkers of individual tobacco smoke constituents are often also used as quantitative estimates of total smoke exposure. This use is based on the assumption that there is a fixed proportionality to the ratio of nicotine or other biomarker exposure and exposure to other smoke constituents or total smoke exposure. Similarly, single measures of tobacco emissions in smokeless tobacco users have been used as estimates of total smokeless tobacco exposure.

This assumption has some general validity. For example, smokers with high cotinine levels have higher levels of several other biomarkers including CO and those measuring nitrosamine and PAH exposures (Joseph et al., 2005), as well as having a higher self-reported number of cigarettes smoked per day. However, even when examined within individuals not changing brands, differences in levels of one biomarker do not reliably predict quantitative differences in other biomarkers (Joseph et al., 2005). Comparisons across brands are also complicated by the differences in composition of the smoke generated and presumably the exposures that result, with different brands of cigarettes (Counts et al., 2004). As an example, an experimental switching study to a PREP demonstrated that reductions in one biomarker do not necessarily predict that other biomarkers will also be reduced

(Hatsukami et al., 2004). These constraints limit the regulatory use of individual biomarkers as proxies for total tobacco exposure, and the very limited number of existing validated biomarkers of individual toxic constituents limits the regulatory use of groups of biomarkers as proxies for total toxicant exposure. As with active smoking, the use of a single biomarker of exposure as a proxy for exposure to all of the constituents of secondhand smoke depends on an assumption that there are fixed ratios between the many different constituents in smoke. If the question being asked is a general one about the relative levels of secondhand smoke exposure in different populations exposed to conventional cigarettes in a single country, this assumption is sufficiently valid to allow the use of biomarkers to assess the impact of changes in policy on levels of secondhand smoke exposure. However, if the question being asked relates to exposures to specific constituents, to exposures with PREPs, or to comparisons between countries, consideration of the emission characteristics of the products needs to be included in the judgment about differences in exposure levels.

12.6   MEASURING REDUCED INJURY OR HARM

Tobacco harm reduction is defined for this report as "minimizing harms and decreasing total morbidity and mortality, without completely eliminating tobacco and nicotine use" (Stratton et al., 2001a). Modifications of existing cigarettes, devices that heat rather than burn tobacco, and oral delivery of nicotine through a variety of products and devices that allow inhalation of nicotine have all been offered as harm-reduction products and collectively are known as PREPs (Boyle et al., 2006; Counts et al., 2004; Stratton et al., 2001a). Past efforts at such *reduced exposure products* include filtered cigarettes and *light* and *mild* cigarettes. Both of these product types are now known not to reduce either exposure or risk. This experience contributes to the need for regulators to evaluate PREPs carefully before allowing harm-reduction claims.

Biomarkers of exposure can be used in an experimental setting to evaluate the exposure-reduction claims made for these PREPs, but the absence of validated biomarkers for harm or risk makes it currently impossible to establish harm-reduction claims in the absence of outcome measures of actual disease frequencies (Hatsukami et al., 2005, 2006; Henningfield et al., 2005). In addition, the available biomarkers of exposure do not offer a comprehensive or reliable estimate of total toxicant exposure and cannot be used as summary estimates of total exposure or risk.

A limited number of biomarkers that measure early biological effects; alterations in morphology, structure, or function; and clinical symptoms consistent with harm do not offer scientifically valid estimates of disease risk for cancer or any of the other diseases caused by smoking (Hatsukami et al., 2006). Unfortunately, this absence of scientifically validated measures prevents regulatory authorities at the present time from being able to adequately evaluate harm-reduction claims based on these biomarkers alone and suggests that exposure (rather than risk or harm) reduction may be the limit of the claims that can be supported by biomarkers using existing science.

The WHO Study Group on Tobacco Product Regulation has previously recommended that regulatory authorities should not allow harm-reduction claims in the absence of evidence demonstrating actual reductions in disease risks

(WHO, 2004b). For some products, for example, smokeless tobacco and nicotine replacement therapy, a substantial body of outcome data exists from epidemiological and clinical studies, and this information can be used to supplement the evidence from biomarkers of exposure in evaluating the risks of these products. Effect biomarkers can be used to characterize differences in biological response with use of different types of tobacco products including cigarettes, water pipes, smokeless tobacco, and PREPs. Effect biomarkers may also be useful for regulators in characterizing the biological response that results from changes in product composition. These differences in biological response, coupled with measures of product emissions and exposure biomarkers, can provide useful inputs for expert panels tasked with advising regulators on the development of regulatory controls intended to reduce the harm of tobacco use.

The WHO Study Group on Tobacco Product Regulation recognizes the obligation for regulators to act, and that actions may need to be implemented even with limitations on or absence of scientific certainty. The description of biomarkers is intended to present the current level of scientific evidence supporting the use of biomarkers so that regulators can distinguish those questions that can be answered with scientific certainty from those where qualified extrapolation of existing science by expert panels is needed.

12.7   SUMMARY OF BIOMARKER RECOMMENDATIONS

Biomarkers of exposure should be required in studies submitted for regulatory approval of tobacco use cessation interventions, in support of exposure reduction claims, in studies defining the dependence potential of different products, and when evaluating or monitoring the effectiveness of individual-level tobacco cessation interventions. In addition, biomarkers of exposure have great utility in evaluating specific public policy questions about the effect of specific regulatory changes on exposures in the general population, notably whether restrictions on smoking in general or in specific locations reduce exposure among nonsmokers.

The biomarkers currently most useful for these purposes are measures of cotinine in urine, blood, saliva, nails, and hair. In settings where individuals may be using nicotine replacement therapy, combinations of CO and mercapturic acids have been used, and the minor tobacco alkaloids anabasine and anatabine or NNAL in the urine are highly specific for tobacco use if laboratory capacity for their accurate measurement exists.

Quantitative levels of biomarkers of exposure to nicotine can be used to differentiate more and less intense users of tobacco products, but when they are used to compare exposures from different tobacco products, they do not accurately define either levels of other toxicants or the total toxicant burden from tobacco use. Differences in nicotine biomarkers are not sufficient by themselves to support exposure-reduction claims for constituents other than nicotine. Validated biomarkers of a limited number of other tobacco toxicants, largely carcinogens, are available to assess differences in exposure to those constituents, but the existing scientific understanding of tobacco smoke and the mechanisms by which it causes disease is not sufficient to allow a battery of existing biomarkers of exposure to serve as a reliable measure of total toxicant burden or of the risk that will result from that toxicant burden.

Self-reported data on tobacco use status and frequency of use remain the currently recommended measure for estimating and evaluating trends in overall tobacco use behaviors in the general population. Nevertheless, the improvement in accuracy of smoking status ascertainment and quantitative exposures provided by biomarkers offers substantive value in the investigational assessment of changes in tobacco use status or intensity of use in response to public policy changes.

Biomarkers of exposure and biomarkers of biological effects can be used in controlled experimental studies to examine exposure and biological responses that result from the use of different tobacco products, including smokeless tobacco products, PREPs, and products making exposure-reduction claims.

Changes in existing biomarkers of biological effects have not been validated as predicting differences in tobacco-related injury or disease risk, either as individual measures or as panels of measures. No currently existing biomarkers, or panels of biomarkers, are sufficiently robust to support a risk- or harm-reduction claim in a regulatory setting. Validated biomarkers of some processes, notably inflammation, oxidative stress, and endothelial dysfunction, do exist and can provide information to guide regulatory authorities in examining the biological responses to different tobacco products that may be part of disease mechanisms. Information from these biological process biomarkers should be combined with chemical measurement of emissions, exposure biomarkers, design characteristics, and existing epidemiological and clinical data in forming assessments of the toxicities of different tobacco products. Such overall assessments will aid in the regulation of tobacco products with the aim of reducing tobacco-related injury and disease.

REFERENCES

Baltar, V. T., Xun, W. W., Chuang, S. C. et al. Smoking, secondhand smoke, and cotinine levels in a subset of EPIC cohort. *Cancer Epidemiology Biomarkers & Prevention*. 2011, 20, 869–875.

Benowitz, N. L., Hall, S. M., Herning, R. I. et al. Smokers of low-yield cigarettes do not consume less nicotine. *The New England Journal of Medicine*. 1983, 309, 139–142.

Boffetta, P., Clark, S., Shen, M. et al. Serum cotinine level as predictor of lung cancer risk. *Cancer Epidemiol Biomarkers and Prevention*. 2006, 15, 1184–1188.

Boyle, P., Ariyaratne, M. A., Barrington, R. et al. Tobacco: Deadly in any form or disguise. *Lancet*. 2006, 367, 1710–1712.

Byrd, G. D., Robinson, J. H., Caldwell, W. S. et al. Comparison of measured and FTC-predicted nicotine uptake in smokers. *Psychopharmacology* (Berl). 1995, 122, 95–103.

Cohn, J. N., Quyyumi, A. A., Hollenberg, N. K. et al. Surrogate markers for cardiovascular disease functional markers. *Circulation*. 2004, 109, IV-31–IV-46.

Counts, M. E., Hsu, F. S., Laffoon, S. W. et al. Mainstream smoke constituent yields and predicting relationships from a worldwide market sample of cigarette brands: ISO smoking conditions. *Regulatory Toxicology and Pharmacology*. 2004, 39, 111–134.

FCTC. The scientific basis of tobacco product regulation. WHO technical report series no. 945. WHO Press, Geneva, Switzerland, 2007.

Fowles, J., Dybing, E. Application of toxicological risk assessment principles to the chemical constituents of cigarette smoke. *Tobacco Control*. 2003, 12, 424–430.

Hatsukami, D. K., Benowitz, N. L., Rennard, S. et al. Biomarkers to assess the utility of potential reduced exposure tobacco products. *Nicotine & Tobacco Research*. 2006, 8, 599–622.

Hatsukami, D. K., Giovino, G. A., Eissenberg, T. et al. Methods to assess potential reduced exposure products. *Nicotine & Tobacco Research*. 2005, 7, 827–844.

Hatsukami, D. K., Hecht, S. S., Hennrikus, D. J. et al. Biomarkers of tobacco exposure or harm: Application to clinical and epidemiological studies. *Nicotine & Tobacco Research*. 2003, 5, 387–396.

Hatsukami, D. K., Lemmonds, C., Zhang, Y. et al. Evaluation of carcinogen exposure in people who used "reduced exposure" tobacco products. *Journal of the National Cancer Institute*. 2004, 96, 844–852.

Hecht, S. Carcinogen biomarkers for lung or oral cancer chemoprevention trials. IARC scientific publications, 154. 2001, p. 245.

Hecht, S. S., Murphy, S. E., Carmella, S. G. et al. Similar uptake of lung carcinogens by smokers of regular, light, and ultralight cigarettes. *Cancer Epidemiology Biomarkers & Prevention*. 2005, 14, 693–698.

Henningfield, J. E., Burns, D. M., Dybing, E. Guidance for research and testing to reduce tobacco toxicant exposure. *Nicotine and Tobacco Research*. 2005, 7, 821–826.

HHS. The health consequences of smoking: A report of the Surgeon General. United States Department of Health and Human Services, Centers for Disease Control and Prevention, National Center for Chronic Disease Prevention and Health Promotion, Office on Smoking and Health, Atlanta, GA, 2004.

IARC. *Tobacco Control: Reversal of Risk after Quitting Smoking* (IARC Handbooks of Cancer Prevention), IARC Handbook 11 International Agency for Research on Cancer, World Health Organization, Lyon, France, 2006.

Jarvis, M. J., Boreham, R., Primatesta, P. et al. Nicotine yield from machine-smoked cigarettes and nicotine intakes in smokers: Evidence from a representative population survey. *Journal of the National Cancer Institute*. 2001, 93, 134–138.

Joseph, A. M, Hecht, S. S., Murphy, S. E. et al. Relationships between cigarette consumption and biomarkers of tobacco toxin exposure. *Cancer Epidemiology, Biomarkers & Prevention*. 2005, 14, 2963–2968.

Kandel, D. B., Schaffran, C., Griesler, P. C. et al. Salivary cotinine concentration versus self-reported cigarette smoking: Three patterns of inconsistency in adolescence. *Nicotine & Tobacco Research*. 2006, 8, 525–537.

NIH. *Risks Associated with Smoking Cigarettes with Low Machine-Measured Yields of Tar and Nicotine*. Smoking and Tobacco Control Monograph No. 13. Bethesda, MD, United States Department of Health and Human Services, Public Health Service, National Institutes of Health, National Cancer Institute. 2001, NIH Publication No. 02-5074.

Patel, S. N., Rajaram, V., Pandya, S. et al. Emerging, noninvasive surrogate markers of atherosclerosis. *Current Atherosclerosis Reports*. 2004, 6, 60–68.

Pirkle, J. L., Bernert, J. T., Caudill, S. P. et al. Trends in the exposure of nonsmokers in the US population to secondhand smoke: 1988–2002. *Environmental Health Perspectives*. 2006, 114, 853.

Pirkle, J. L., Flegal, K. M., Bernert, J. T. et al. Exposure of the US population to environmental tobacco smoke: The Third National Health and Nutrition Examination Survey, 1988 to 1991. *JAMA*. 1996, 275, 1233–1240.

Stratton, K., Shetty, P., Wallace, R. et al. *Clearing the Smoke: Assessing the Science Base for Tobacco Harm Reduction*. National Academy Press, Washington, DC, 2001a.

Appendix A: Guideline Document for Ethics Review Concerning Research Involving Intervention on Humans by the European Commission

This document summarizes the discussions of a working group of experts in research ethics, established by the Ethics Review Sector of Unit L3 – Governance and Ethics of DG RTD, to address the issue of human intervention in FP7-supported research and the ethics review process to be used by the European Commission and its agencies. These guidance notes aim to provide some guidelines to those involved in the ethics screening and review of proposals submitted for funding by the European Commission. This document provides a general introduction and lists illustrative examples of research proposals involving human intervention that should be forwarded to the European Commission for Central Ethics Review or, alternatively, intervention where it is considered that the proposals need not be submitted to Central Ethics Review by the European Commission but, rather can be dealt with by appropriate, competent local/national ethics committees or authorities. A checklist table is provided for reference. It must be emphasised that any proposals submitted to the European Commission for funding involving actions or intervention with humans not explicitly included in this guidance document need to be carefully evaluated. These may require referral by the screener to Ethics Review for further investigation and discussion. It should be stressed that ethics screening and ethics review cannot substitute the national approvals and/or opinion of local ethics committees. Ethics review and ethics screening also cannot be used in any way to obtain national/local approvals and/or opinions.

INTRODUCTION

The purpose of this document is to provide guidance for the ethics review process and, particularly, the ethics screening of research proposals involving intervention in humans. The European Commission (EC) ethics review process entails the organisation of Ethics Review Panel Meetings among independent experts and is carried out by the Ethics Review Sector of DG RTD. Prior to ethics review, all proposals that raise ethical issues that have not been adequately dealt with by the applicant are screened by the Commission with the help of independent experts. Beyond the provided guidelines, each proposal must be screened and then viewed on its own merits, taking into account all aspects of the research that ethics screeners and evaluators are able to examine in order to make an informed decision as whether or not any particular proposal requires Central EC Review. Reviewers are encouraged to use their judgement and, in cases of doubt, refer the proposal for Central Review.

For the purpose of this procedure, "intervention" may be defined as any action on the human body that produces an effect and which is carried out to alter/monitor the course of a patho-physiological process or to collect material from human subjects for the purpose of research or to observe and document other phenomena, including brain activity.

As a general rule, there are three overriding criteria to establish when research proposals should be referred to EC Ethics Review. Proposals that include:

- research on vulnerable individuals or populations (e.g. Children);
- interventions that are deemed to be more than minimally invasive or intrusive (e.g. Experimental surgery);
- fast advancing innovative technologies where ethical and safety guidelines may not be fully established (e.g. Neuroscience);

For the purpose of this document, a vulnerable individual or population can be defined as one facing "a significant probability of incurring an identifiable harm while substantially lacking ability and/or means to protect oneself".

The definition of vulnerability needs to be broadly interpreted to include those not capable of providing informed consent or those individuals or groups that are in compromising situations. This may include individuals who are mentally, psychologically or psychiatrically impaired; individuals who are impaired by hearing, eyesight or physical disabilities/deficiencies; the illiterate; the elderly in care institutions; prisoners; asylum seekers; illegal immigrants; refugees; illicit drug users; commercial sex workers; ethnic minorities, etc.

Invasive and intrusive interventions are those that require entry into the living body and which may intrude significantly upon the privacy and dignity of the patient and/or research participant. Research involving the use of human cadavers should also be referred for touching upon ethical, philosophical, cultural and religious issues that might be raised.

There must be robust procedures described for recruitment of participants: fully-informed consent, confidentiality and data protection. There must be appropriate insurance coverage or protection against potential harm and a clearly defined and transparent policy for the handling of incidental findings, where relevant.

Furthermore, medical interventions on humans are rapidly changing, raising ethical concerns that are ever changing and that may not be captured in this document. Therefore, it is urged that all reviewers act on the side of caution when new treatments or technologies are developed. If in doubt, refer these to the European Commission for ethics review.

RATIONALE FOR AUTOMATIC EUROPEAN COMMISSION REVIEW

The need for safeguarding the privacy, integrity and dignity of the research participants as such and preventing constitutes the cornerstone of the Commission's Ethics Review procedure. Detailed review by the Commission is considered, in principle, as necessary in all those cases where detailed, explicit and robust informed consent and assent procedures need to be designed and followed. In all these cases, the risks and the benefits to the person participating in the research project must be clearly described and convincing justification for his/her involvement need to be provided so as to prevent the extortion or exploitation of the potential vulnerability of an individual or particular group of population.

- Research Involving Intervention with Children
  It is recommended that in case a research proposal involves intervention with children, this should be referred to the European Commission for Ethics Review.
- Research Involving Interventions with Vulnerable Adults
  It is recommended that any research proposals that include the involvement and interviewing of vulnerable adults should be referred to the European Commission for Ethical Review.
- Oocytes Collection
  Oocytes are unique cells that elicit strong philosophical and religious opinions from different members of society, views that differ widely across European and other cultures. The practice of the collection of oocytes may involve commercial interests and potential exploitation of donors.
  It is recommended that any research proposals that include the collection, manipulation, fertilization or implantation of oocytes from human donors should be referred to the European Commission for Ethical Review.
- Research Involving Innovative Surgical Techniques
  Such procedures are highly invasive and carry major risks. Even if insurance protection is in place, full review by the European Commission is necessary to ensure that the research complies with rigorous ethical standards and that the risk/benefit analysis is justifiable.
- Research Involving Innovative Intrusive Neuroscience
  Neuroethics encompasses a wide array of ethical issues emerging from different branches of clinical neuroscience (neurology, psychiatry, psychopharmacology) and basic neuroscience (cognitive neuroscience, affective neuroscience). These include ethical problems raised by advances in functional neuroimaging, brain implants and brain-machine interfaces and psychopharmacology, as well as by a growing understanding of the neural bases of behavior, personality, and consciousness. Research aimed at investigating brain functioning, predicting and modifying behavior, displaying conscious or unconscious attitudes raise numerous ethical, legal and social policy questions, including unknown risks. Therefore, it is recommended that any research proposals in this field should be referred to the European Commission for Ethical Review.

- Research Involving Pregnant Women
  It is recommended that any research studies involving pregnant women in which interventions are planned should be referred to the European Commission for Ethical Review and should include details of the proposed intervention and the informed consent procedures. Detailed review by the Commission is required to ensure that the health and welfare of the pregnant mother and her child are appropriately safeguarded, and this should be clearly noted in the application and that the informed consent procedures are rigorous and appropriate. Where necessary, Ethical Review will establish if counselling or support mechanisms need to be put in place.
- Interventions Involving ICT Implants
  Due to the rapid development of this field, issues of privacy and the potential for misuse of any contemplated intervention should be referred to the European Commission for Central Review.
- Research Involving Interventions with Participants from Developing Countries
  It is recommended that any research proposals that include the involvement of participants from developing countries should be referred to the European Commission for Ethics Review. Detailed Review by the Commission will be required to ensure that applicable legislation is complied with, that a robust informed consent procedure ensured and that benefit sharing, data protection and standard of care issues are appropriately addressed. Review by the Commission is required to ensure that there can be no danger of exploitation of a potentially vulnerable group, and of incentives that amount to undue inducement to participate.
- Interventions with Dual Use Applications
  It is recommended that any research proposal that involves human intervention which raises reasonable concerns about the potential dual use and/or misuse of its outcomes/products/collected data could be used, either directly or indirectly, or for dual use applications should be referred to the European Commission for a review of the possible ethical, implications. It is recommended that ethics reviewers make an effort to identify potential dual use aspects of the proposed research, making a distinction between dual use and misuse. Proposals raising dual use issue should be discussed with the Ethics Review Sector of DG RTD on a case-by-case basis.

RATIONALE FOR NATIONAL/LOCAL REVIEW ON THE BASIS OF EC ETHICS SCREENING

- Clinical trials
  For clinical trials involving medicinal products covered by EU Law local/national ethics reviews and approvals from the competent authorities should be sought by the applicants. Clinical trials involving vulnerable individuals and/or populations would generally come under central review as noted above. All other types of clinical trials should be justified on the basis of

the individual application before deciding on ER ethical review or national review

- The collection, processing, use and storage of biological samples for research and testing or bio-banking, including secondary uses

  Although collection of samples as such might not be intrusive, their storage might implicate issues of privacy, and there is the matter of secondary use - including whether and/or when blanket consent, or waiver of consent is ethical.

- Collection and the transfer of biological samples and materials such as saliva, buccal smears, dental plaque, nasal swabs, sweat, tears, blood, urine, faces, hair, small pieces of waste skin, seminal fluid and sperm (see below also), cervical smears, and breast milk, and the specific extraction of DNA

- Use of ionization radiation and MRI

  Research proposals that include the use of ionizing radiation (x-ray, CT, PET/SPECT) and MRI can be referred for review by local/national ethics committees, however, due to the possibility of health risk and safety issues, the applicants must demonstrate to possess the appropriate awareness and expertise.

- Psychometric or other basic cognitive testing

- Research involving secondary data that have been previously obtained following appropriate ethical approval

- Dictary interventions

- Use of "alternative medicine" (complementary) in treatment of a health problem

Appendix B: Declaration of Helsinki (V2008)

A. INTRODUCTION

1. The World Medical Association (WMA) has developed the Declaration of Helsinki as a statement of ethical principles for medical research involving human subjects, including research on identifiable human material and data. The Declaration is intended to be read as a whole and each of its constituent paragraphs should not be applied without consideration of all other relevant paragraphs.

2. Although the Declaration is addressed primarily to physicians, the WMA encourages other participants in medical research involving human subjects to adopt these principles.

3. It is the duty of the physician to promote and safeguard the health of patients, including those who are involved in medical research. The physician's knowledge and conscience are dedicated to the fulfilment of this duty.

4. The Declaration of Geneva of the WMA binds the physician with the words, "The health of my patient will be my first consideration," and the International Code of Medical Ethics declares that, "A physician shall act in the patient's best interest when providing medical care."

5. Medical progress is based on research that ultimately must include studies involving human subjects. Populations that are underrepresented in medical research should be provided appropriate access to participation in research.

6. In medical research involving human subjects, the well-being of the individual research subject must take precedence over all other interests.

7. The primary purpose of medical research involving human subjects is to understand the causes, development and effects of diseases and improve preventive, diagnostic and therapeutic interventions (methods, procedures and treatments). Even the best current interventions must be evaluated continually through research for their safety, effectiveness, efficiency, accessibility and quality.

8. In medical practice and in medical research, most interventions involve risks and burdens.

9. Medical research is subject to ethical standards that promote respect for all human subjects and protect their health and rights. Some research populations are particularly vulnerable and need special protection. These include those who cannot give or refuse consent for themselves and those who may be vulnerable to coercion or undue influence.

10. Physicians should consider the ethical, legal and regulatory norms and standards for research involving human subjects in their own countries as well as applicable international norms and standards. No national or international ethical, legal or regulatory requirement should reduce or eliminate any of the protections for research subjects set forth in this Declaration.

B. BASIC PRINCIPLES FOR ALL MEDICAL RESEARCH

11. It is the duty of physicians who participate in medical research to protect the life, health, dignity, integrity, right to self-determination, privacy, and confidentiality of personal information of research subjects.
12. Medical research involving human subjects must conform to generally accepted scientific principles, be based on a thorough knowledge of the scientific literature, other relevant sources of information, and adequate laboratory and, as appropriate, animal experimentation. The welfare of animals used for research must be respected.
13. Appropriate caution must be exercised in the conduct of medical research that may harm the environment.
14. The design and performance of each research study involving human subjects must be clearly described in a research protocol. The protocol should contain a statement of the ethical considerations involved and should indicate how the principles in this Declaration have been addressed. The protocol should include information regarding funding, sponsors, institutional affiliations, other potential conflicts of interest, incentives for subjects and provisions for treating and/or compensating subjects who are harmed as a consequence of participation in the research study. The protocol should describe arrangements for post-study access by study subjects to interventions identified as beneficial in the study or access to other appropriate care or benefits.
15. The research protocol must be submitted for consideration, comment, guidance and approval to a research ethics committee before the study begins. This committee must be independent of the researcher, the sponsor and any other undue influence. It must take into consideration the laws and regulations of the country or countries in which the research is to be performed as well as applicable international norms and standards but these must not be allowed to reduce or eliminate any of the protections for research subjects set forth in this Declaration. The committee must have the right to monitor ongoing studies. The researcher must provide monitoring information to the committee, especially information about any serious adverse events. No change to the protocol may be made without consideration and approval by the committee.
16. Medical research involving human subjects must be conducted only by individuals with the appropriate scientific training and qualifications. Research on patients or healthy volunteers requires the supervision of a competent and appropriately qualified physician or other health care professional. The responsibility for the protection of research subjects must always rest

with the physician or other health care professional and never the research subjects, even though they have given consent.

17. Medical research involving a disadvantaged or vulnerable population or community is only justified if the research is responsive to the health needs and priorities of this population or community and if there is a reasonable likelihood that this population or community stands to benefit from the results of the research.

18. Every medical research study involving human subjects must be preceded by careful assessment of predictable risks and burdens to the individuals and communities involved in the research in comparison with foreseeable benefits to them and to other individuals or communities affected by the condition under investigation.

19. Every clinical trial must be registered in a publicly accessible database before recruitment of the first subject.

20. Physicians may not participate in a research study involving human subjects unless they are confident that the risks involved have been adequately assessed and can be satisfactorily managed. Physicians must immediately stop a study when the risks are found to outweigh the potential benefits or when there is conclusive proof of positive and beneficial results.

21. Medical research involving human subjects may only be conducted if the importance of the objective outweighs the inherent risks and burdens to the research subjects.

22. Participation by competent individuals as subjects in medical research must be voluntary. Although it may be appropriate to consult family members or community leaders, no competent individual may be enrolled in a research study unless he or she freely agrees.

23. Every precaution must be taken to protect the privacy of research subjects and the confidentiality of their personal information and to minimize the impact of the study on their physical, mental and social integrity.

24. In medical research involving competent human subjects, each potential subject must be adequately informed of the aims, methods, sources of funding, any possible conflicts of interest, institutional affiliations of the researcher, the anticipated benefits and potential risks of the study and the discomfort it may entail, and any other relevant aspects of the study. The potential subject must be informed of the right to refuse to participate in the study or to withdraw consent to participate at any time without reprisal. Special attention should be given to the specific information needs of individual potential subjects as well as to the methods used to deliver the information. After ensuring that the potential subject has understood the information, the physician or another appropriately qualified individual must then seek the potential subject's freely-given informed consent, preferably in writing. If the consent cannot be expressed in writing, the non-written consent must be formally documented and witnessed.

25. For medical research using identifiable human material or data, physicians must normally seek consent for the collection, analysis, storage and/or reuse. There may be situations where consent would be impossible or

impractical to obtain for such research or would pose a threat to the validity of the research. In such situations the research may be done only after consideration and approval of a research ethics committee.

26. When seeking informed consent for participation in a research study the physician should be particularly cautious if the potential subject is in a dependent relationship with the physician or may consent under duress. In such situations the informed consent should be sought by an appropriately qualified individual who is completely independent of this relationship.

27. For a potential research subject who is incompetent, the physician must seek informed consent from the legally authorized representative. These individuals must not be included in a research study that has no likelihood of benefit for them unless it is intended to promote the health of the population represented by the potential subject, the research cannot instead be performed with competent persons, and the research entails only minimal risk and minimal burden.

28. When a potential research subject who is deemed incompetent is able to give assent to decisions about participation in research, the physician must seek that assent in addition to the consent of the legally authorized representative. The potential subject's dissent should be respected.

29. Research involving subjects who are physically or mentally incapable of giving consent, for example, unconscious patients, may be done only if the physical or mental condition that prevents giving informed consent is a necessary characteristic of the research population. In such circumstances the physician should seek informed consent from the legally authorized representative. If no such representative is available and if the research cannot be delayed, the study may proceed without informed consent provided that the specific reasons for involving subjects with a condition that renders them unable to give informed consent have been stated in the research protocol and the study has been approved by a research ethics committee. Consent to remain in the research should be obtained as soon as possible from the subject or a legally authorized representative.

30. Authors, editors and publishers all have ethical obligations with regard to the publication of the results of research. Authors have a duty to make publicly available the results of their research on human subjects and are accountable for the completeness and accuracy of their reports. They should adhere to accepted guidelines for ethical reporting. Negative and inconclusive as well as positive results should be published or otherwise made publicly available. Sources of funding, institutional affiliations and conflicts of interest should be declared in the publication. Reports of research not in accordance with the principles of this Declaration should not be accepted for publication.

31. The physician may combine medical research with medical care only to the extent that the research is justified by its potential preventive, diagnostic or therapeutic value and if the physician has good reason to believe that participation in the research study will not adversely affect the health of the patients who serve as research subjects.

32. The benefits, risks, burdens and effectiveness of a new intervention must be tested against those of the best current proven intervention, except in the following circumstances: The use of placebo, or no treatment, is acceptable in studies where no current proven intervention exists; or Where for compelling and scientifically sound methodological reasons the use of placebo is necessary to determine the efficacy or safety of an intervention and the patients who receive placebo or no treatment will not be subject to any risk of serious or irreversible harm. Extreme care must be taken to avoid abuse of this option.

33. At the conclusion of the study, patients entered into the study are entitled to be informed about the outcome of the study and to share any benefits that result from it, for example, access to interventions identified as beneficial in the study or to other appropriate care or benefits.

34. The physician must fully inform the patient which aspects of the care are related to the research. The refusal of a patient to participate in a study or the patient's decision to withdraw from the study must never interfere with the patient-physician relationship.

35. In the treatment of a patient, where proven interventions do not exist or have been ineffective, the physician, after seeking expert advice, with informed consent from the patient or a legally authorized representative, may use an unproven intervention if in the physician's judgement it offers hope of saving life, re-establishing health or alleviating suffering. Where possible, this intervention should be made the object of research, designed to evaluate its safety and efficacy. In all cases, new information should be recorded and, where appropriate, made publicly available. (22.10.2008)

Appendix C: Protection of Animals Used for Scientific Purposes

THE EUROPEAN PARLIAMENT AND THE COUNCIL OF THE EUROPEAN UNION,

Having regard to the Treaty on the Functioning of the European Union, and in particular Article 114 thereof,

Having regard to the proposal from the European Commission,

Having regard to the opinion of the European Economic and Social Committee (1),

After consulting the Committee of the Regions,

Acting in accordance with the ordinary legislative procedure (2),

Whereas:

(1) On 24 November 1986 the Council adopted Directive 86/609/EEC (3) in order to eliminate disparities between laws, regulations and administrative provisions of the Member States regarding the protection of animals used for experimental and other scientific purposes. Since the adoption of that Directive, further disparities between Member States have emerged. Certain Member States have adopted national implementing measures that ensure a high level of protection of animals used for scientific purposes, while others only apply the minimum requirements laid down in Directive 86/609/EEC. These disparities are liable to constitute barriers to trade in products and substances the development of which involves experiments on animals. Accordingly, this Directive should provide for more detailed rules in order to reduce such disparities by approximating the rules applicable in that area and to ensure a proper functioning of the internal market.

(2) Animal welfare is a value of the Union that is enshrined in Article 13 of the Treaty on the Functioning of the European Union (TFEU).

(3) On 23 March 1998 the Council adopted Decision 1999/575/EC concerning the conclusion by the Community of the European Convention for the protection of vertebrate animals used for experimental and other scientific purposes (4). By becoming party to that Convention, the Community acknowledged the importance of the protection and welfare of animals used for scientific purposes at international level.

(4) The European Parliament in its resolution of 5 December 2002 on Directive 86/609/EEC called for the Commission to come forward with a proposal for

a revision of that Directive with more stringent and transparent measures in the area of animal experimentation.

(5) On 15 June 2006, the Fourth Multilateral Consultation of Parties to the European Convention for the protection of vertebrate animals used for experimental and other scientific purposes adopted a revised Appendix A to that Convention, which set out guidelines for the accommodation and care of experimental animals. Commission Recommendation 2007/526/EC of 18 June 2007 on guidelines for the accommodation and care of animals used for experimental and other scientific purposes (5) incorporated those guidelines.

(6) New scientific knowledge is available in respect of factors influencing animal welfare as well as the capacity of animals to sense and express pain, suffering, distress and lasting harm. It is therefore necessary to improve the welfare of animals used in scientific procedures by raising the minimum standards for their protection in line with the latest scientific developments.

(7) Attitudes towards animals also depend on national perceptions, and there is a demand in certain Member States to maintain more extensive animal-welfare rules than those agreed upon at the level of the Union. In the interests of the animals, and provided it does not affect the functioning of the internal market, it is appropriate to allow the Member States certain flexibility to maintain national rules aimed at more extensive protection of animals in so far as they are compatible with the TFEU.

(8) In addition to vertebrate animals including cyclostomes, cephalopods should also be included in the scope of this Directive, as there is scientific evidence of their ability to experience pain, suffering, distress and lasting harm.

(9) This Directive should also cover foetal forms of mammals, as there is scientific evidence showing that such forms in the last third of the period of their development are at an increased risk of experiencing pain, suffering and distress, which may also affect negatively their subsequent development. Scientific evidence also shows that procedures carried out on embryonic and foetal forms at an earlier stage of development could result in pain, suffering, distress or lasting harm, should the developmental forms be allowed to live beyond the first two thirds of their development.

(10) While it is desirable to replace the use of live animals in procedures by other methods not entailing the use of live animals, the use of live animals continues to be necessary to protect human and animal health and the environment. However, this Directive represents an important step towards achieving the final goal of full replacement of procedures on live animals for scientific and educational purposes as soon as it is scientifically possible to do so. To that end, it seeks to facilitate and promote the advancement of alternative approaches. It also seeks to ensure a high level of protection for animals that still need to be used in procedures. This Directive should be reviewed regularly in light of evolving science and animal-protection measures.

(11) The care and use of live animals for scientific purposes is governed by internationally established principles of replacement, reduction and refinement. To ensure that the way in which animals are bred, cared for and used in procedures within the Union is in line with that of the other international and national standards applicable outside the Union, the principles of replacement, reduction and refinement should be considered systematically when implementing this Directive. When choosing methods, the principles of replacement, reduction and refinement should be implemented through a strict hierarchy of the requirement to use alternative methods. Where no alternative method is recognised by the legislation of the Union, the numbers of animals used may be reduced by resorting to other methods and by implementing testing strategies, such as the use of in vitro and other methods that would reduce and refine the use of animals.

(12) Animals have an intrinsic value which must be respected. There are also the ethical concerns of the general public as regards the use of animals in procedures. Therefore, animals should always be treated as sentient creatures and their use in procedures should be restricted to areas which may ultimately benefit human or animal health, or the environment. The use of animals for scientific or educational purposes should therefore only be considered where a non-animal alternative is unavailable. Use of animals for scientific procedures in other areas under the competence of the Union should be prohibited.

(13) The choice of methods and the species to be used have a direct impact on both the numbers of animals used and their welfare. The choice of methods should therefore ensure the selection of the method that is able to provide the most satisfactory results and is likely to cause the minimum pain, suffering or distress. The methods selected should use the minimum number of animals that would provide reliable results and require the use of species with the lowest capacity to experience pain, suffering, distress or lasting harm that are optimal for extrapolation into target species.

(14) The methods selected should avoid, as far as possible, death as an end-point due to the severe suffering experienced during the period before death. Where possible, it should be substituted by more humane end-points using clinical signs that determine the impending death, thereby allowing the animal to be killed without any further suffering.

(15) The use of inappropriate methods for killing an animal can cause significant pain, distress and suffering to the animal. The level of competence of the person carrying out this operation is equally important. Animals should therefore be killed only by a competent person using a method that is appropriate to the species.

(16) It is necessary to ensure that the use of animals in procedures does not pose a threat to biodiversity. Therefore, the use of endangered species in procedures should be limited to a strict minimum.

(17) Having regard to the present state of scientific knowledge, the use of non-human primates in scientific procedures is still necessary in biomedical research. Due to their genetic proximity to human beings and to their

highly developed social skills, the use of non-human primates in scientific procedures raises specific ethical and practical problems in terms of meeting their behavioural, environmental and social needs in a laboratory environment. Furthermore, the use of non-human primates is of the greatest concern to the public. Therefore the use of non-human primates should be permitted only in those biomedical areas essential for the benefit of human beings, for which no other alternative replacement methods are yet available. Their use should be permitted only for basic research, the preservation of the respective non-human primate species or when the work, including xenotransplantation, is carried out in relation to potentially life-threatening conditions in humans or in relation to cases having a substantial impact on a person's day-to-day functioning, i.e. debilitating conditions.

(18) The use of great apes, as the closest species to human beings with the most advanced social and behavioural skills, should be permitted only for the purposes of research aimed at the preservation of those species and where action in relation to a life-threatening, debilitating condition endangering human beings is warranted, and no other species or alternative method would suffice in order to achieve the aims of the procedure. The Member State claiming such a need should provide information necessary for the Commission to take a decision.

(19) The capture of non-human primates from the wild is highly stressful for the animals concerned and carries an elevated risk of injury and suffering during capture and transport. In order to end the capturing of animals from the wild for breeding purposes, only animals that are the offspring of an animal which has been bred in captivity, or that are sourced from self-sustaining colonies, should be used in procedures after an appropriate transition period. A feasibility study should be carried out to that effect and the transition period adopted if necessary. The feasibility of moving towards sourcing non-human primates only from self-sustaining colonies as an ultimate goal should also be examined.

(20) There is a need for certain species of vertebrate animals used in procedures to be bred specifically for that purpose so that their genetic, biological and behavioural background is well-known to persons undertaking the procedures. Such knowledge both increases the scientific quality and reliability of the results and decreases the variability, ultimately resulting in fewer procedures and reduced animal use. Furthermore, for reasons of animal welfare and conservation, the use of animals taken from the wild in procedures should be limited to cases where the purpose of the procedures cannot be achieved using animals bred specifically for use in procedures.

(21) Since the background of stray and feral animals of domestic species is not known, and since capture and placement into establishments increases distress for such animals, they should not, as a general rule, be used in procedures.

(22) To enhance transparency, facilitate the project authorisation, and provide tools for monitoring compliance, a severity classification of procedures

should be introduced on the basis of estimated levels of pain, suffering, distress and lasting harm that is inflicted on the animals.

(23) From an ethical standpoint, there should be an upper limit of pain, suffering and distress above which animals should not be subjected in scientific procedures. To that end, the performance of procedures that result in severe pain, suffering or distress, which is likely to be long-lasting and cannot be ameliorated, should be prohibited.

(24) When developing a common format for reporting purposes, the actual severity of the pain, suffering, distress or lasting harm experienced by the animal should be taken into account rather than the predicted severity at the time of the project evaluation.

(25) The number of animals used in procedures could be reduced by performing procedures on animals more than once, where this does not detract from the scientific objective or result in poor animal welfare. However, the benefit of reusing animals should be balanced against any adverse effects on their welfare, taking into account the lifetime experience of the individual animal. As a result of this potential conflict, the reuse of animals should be considered on a case-by-case basis.

(26) At the end of the procedure, the most appropriate decision should be taken as regards the future of the animal on the basis of animal welfare and potential risks to the environment. The animals whose welfare would be compromised should be killed. In some cases, animals should be returned to a suitable habitat or husbandry system or animals such as dogs and cats should be allowed to be rehomed in families as there is a high level of public concern as to the fate of such animals. Should Member States allow rehoming, it is essential that the breeder, supplier or user has a scheme in place to provide appropriate socialisation to those animals in order to ensure successful rehoming as well as to avoid unnecessary distress to the animals and to guarantee public safety.

(27) Animal tissue and organs are used for the development of in vitro methods. To promote the principle of reduction, Member States should, where appropriate, facilitate the establishment of programmes for sharing the organs and tissue of animals that are killed.

(28) The welfare of the animals used in procedures is highly dependent on the quality and professional competence of the personnel supervising procedures, as well as of those performing procedures or supervising those taking care of the animals on a daily basis. Member States should ensure through authorisation or by other means that staff are adequately educated, trained and competent. Furthermore, it is important that staff are supervised until they have obtained and demonstrated the requisite competence. Non-binding guidelines at the level of the Union concerning educational requirements would, in the long run, promote the free movement of personnel.

(29) The establishments of breeders, suppliers and users should have adequate installations and equipment in place to meet the accommodation requirements of the animal species concerned and to allow the procedures to be performed efficiently and with the least distress to the animals. The breeders,

suppliers and users should operate only if they are authorised by the competent authorities.

(30) To ensure the ongoing monitoring of animal-welfare needs, appropriate veterinary care should be available at all times and a staff member should be made responsible for the care and welfare of animals in each establishment.

(31) Animal-welfare considerations should be given the highest priority in the context of animal keeping, breeding and use. Breeders, suppliers and users should therefore have an animal-welfare body in place with the primary task of focusing on giving advice on animal-welfare issues. The body should also follow the development and outcome of projects at establishment level, foster a climate of care and provide tools for the practical application and timely implementation of recent technical and scientific developments in relation to the principles of replacement, reduction and refinement, in order to enhance the life-time experience of the animals. The advice given by the animal-welfare body should be properly documented and open to scrutiny during inspections.

(32) In order to enable competent authorities to monitor compliance with this Directive, each breeder, supplier and user should maintain accurate records of the numbers of animals, their origins and fate.

(33) Non-human primates, dogs and cats should have a personal history file from birth covering their lifetimes in order to be able to receive the care, accommodation and treatment that meet their individual needs and characteristics.

(34) The accommodation and care of animals should be based on the specific needs and characteristics of each species.

(35) There are differences in the requirements for the accommodation and care of animals between Member States, which contribute to the distortion of the internal market. Furthermore, some of those requirements no longer reflect the most recent knowledge on the impacts of accommodation and care conditions on both the animal welfare and the scientific results of procedures. It is therefore necessary to establish in this Directive harmonised requirements for accommodation and care. These requirements should be updated on the basis of scientific and technical development.

(36) To monitor compliance with this Directive, Member States should carry out regular inspections of breeders, suppliers and users on a risk basis. To ensure public confidence and promote transparency, an appropriate proportion of the inspections should be carried out without prior warning.

(37) To assist the Member States in the enforcement of this Directive and on the basis of the findings in the reports on the operation of the national inspections, the Commission should, when there is reason for concern, carry out controls of the national inspection systems. Member States should address any weaknesses identified in the findings of these controls.

(38) Comprehensive project evaluation, taking into account ethical considerations in the use of animals, forms the core of project authorisation and should ensure the implementation of principles of replacement, reduction and refinement in those projects.

(39) It is also essential, both on moral and scientific grounds, to ensure that each use of an animal is carefully evaluated as to the scientific or educational validity, usefulness and relevance of the expected result of that use. The likely harm to the animal should be balanced against the expected benefits of the project. Therefore, an impartial project evaluation independent of those involved in the study should be carried out as part of the authorisation process of projects involving the use of live animals. Effective implementation of a project evaluation should also allow for an appropriate assessment of the use of any new scientific experimental techniques as they emerge.

(40) Due to the nature of the project, the type of species used and the likelihood of achieving the desired objectives of the project, it might be necessary to carry out a retrospective assessment. Since projects may vary significantly in terms of complexity, length, and the time period for obtaining the results, it is necessary that the decision on retrospective assessment should be made taking those aspects fully into account.

(41) To ensure that the public is informed, it is important that objective information concerning projects using live animals is made publicly available. This should not violate proprietary rights or expose confidential information. Therefore, users should provide anonymous non-technical summaries of those projects, which Member States should publish. The published details should not breach the anonymity of the users.

(42) To manage risks to human and animal health and the environment, the legislation of the Union provides that substances and products can be marketed only after appropriate safety and efficacy data have been submitted. Some of those requirements can be fulfilled only by resorting to animal testing, hereinafter referred to as 'regulatory testing'. It is necessary to introduce specific measures in order to increase the use of alternative approaches and to eliminate unnecessary duplication of regulatory testing. For that purpose Member States should recognise the validity of test data produced using test methods provided for under the legislation of the Union.

(43) To reduce the administrative workload and enhance the competitiveness of research and industry in the Union, it should be possible to authorise multiple generic projects when carried out using established methods for testing, diagnostic or production purposes under one group authorisation, albeit without exempting any of these procedures from the project evaluation.

(44) To ensure effective examination of authorisation applications and to enhance the competitiveness of research and industry in the Union, a time-limit should be set for the competent authorities to evaluate project proposals and take decisions on the authorisation of such projects. In order not to compromise the quality of the project evaluation, additional time might be required for more complex project proposals due to the number of disciplines involved, the novel characteristics and more complex techniques of the proposed project. However, the extension of deadlines for project evaluation should remain the exception.

(45) Given the routine or repetitive nature of certain procedures, it is appropriate to provide for a regulatory option whereby the Member States could

introduce a simplified administrative procedure for the evaluation of projects containing such procedures, provided certain requirements laid down in this Directive are complied with.

(46) The availability of alternative methods is highly dependent on the progress of the research into the development of alternatives. The Community Framework Programmes for Research and Technological Development provided increasing funding for projects which aim to replace, reduce and refine the use of animals in procedures. In order to increase competitiveness of research and industry in the Union and to replace, reduce and refine the use of animals in procedures, the Commission and the Member States should contribute through research and by other means to the development and validation of alternative approaches.

(47) The European Centre for the Validation of Alternative Methods, a policy action within the Joint Research Centre of the Commission, has coordinated the validation of alternative approaches in the Union since 1991. However, there is an increasing need for new methods to be developed and proposed for validation, which requires a reference laboratory of the Union for the validation of alternative methods to be established formally. This laboratory should be referred to as the European Centre for the Validation of Alternative Methods (ECVAM). It is necessary for the Commission to cooperate with the Member States when setting priorities for validation studies. The Member States should assist the Commission in identifying and nominating suitable laboratories to carry out such validation studies. For validation studies that are similar to previously validated methods and in respect of which a validation represents a significant competitive advantage, ECVAM should be able to collect charges from those who submit their methods for validation. Such charges should not be prohibitive of healthy competition in the testing industry.

(48) There is a need to ensure a coherent approach to project evaluation and review strategies at national level. Member States should establish national committees for the protection of animals used for scientific purposes to give advice to the competent authorities and animal-welfare bodies in order to promote the principles of replacement, reduction and refinement. A network of national committees should play a role in the exchange of best practice at the level of the Union.

(49) Technical and scientific advancements in biomedical research can be rapid, as can the increase in knowledge of factors influencing animal welfare. It is therefore necessary to provide for a review of this Directive. Such review should examine the possible replacement of the use of animals, and in particular non-human primates, as a matter of priority where it is possible, taking into account the advancement of science. The Commission should also conduct periodic thematic reviews concerning the replacement, reduction and refinement of the use of animals in procedures.

(50) In order to ensure uniform conditions for implementation, implementing powers should be conferred on the Commission to adopt guidelines at the level of the Union on the requirements with regard to education, training

and competence of breeders', suppliers' and users' staff, to adopt detailed rules regarding the Union Reference Laboratory, its duties and tasks and the charges it may collect, to establish a common format for submitting the information by Member States to the Commission on the implementation of this Directive, statistical information and other specific information, and for the application of safeguard clauses. According to Article 291 TFEU, rules and general principles concerning mechanisms for the control by Member States of the Commission's exercise of implementing powers shall be laid down in advance by a regulation adopted in accordance with the ordinary legislative procedure. Pending the adoption of that new regulation, Council Decision 1999/468/EC of 28 June 1999 laying down the procedures for the exercise of implementing powers conferred on the Commission (6) continues to apply, with the exception of the regulatory procedure with scrutiny, which is not applicable.

(51) The Commission should be empowered to adopt delegated acts in accordance with Article 290 TFEU in respect of the following: modifications of the list of species falling under the obligation of being specifically bred for use in procedures; modifications of the care and accommodation standards; modifications of methods of killing, including their specifications; modifications of the elements to be used for the establishment by Member States of requirements with regard to education, training and competence of breeders', suppliers' and users' personnel; modifications of certain obligatory elements of the application for authorisation; modifications regarding the Union Reference Laboratory, its duties and tasks; as well as modifications of examples of different types of procedures assigned to each of the severity categories on the basis of factors related to the type of procedure. It is of particular importance that the Commission carry out appropriate consultation during its preparatory work, including at expert level.

(52) Member States should lay down rules on penalties applicable to infringements of the provisions of this Directive and ensure that they are implemented. Those penalties should be effective, proportionate and dissuasive.

(53) Directive 86/609/EEC should therefore be repealed. Certain modifications introduced by this Directive have a direct impact on the application of Regulation (EC) No 1069/2009 of the European Parliament and of the Council of 21 October 2009 laying down health rules as regards animal by-products and derived products not intended for human consumption (7). It is therefore appropriate to amend a provision of that Regulation accordingly.

(54) Benefits to animal welfare from applying project authorisation retrospectively, and the related administrative costs, can only be justified for long term ongoing projects. Therefore, it is necessary to include transitional measures for ongoing short and medium term projects to avoid the need for retrospective authorisation with only limited benefits.

(55) In accordance with paragraph 34 of the Interinstitutional Agreement on better law-making, Member States are encouraged to draw up, for themselves

and in the interests of the Union, their own tables, which will, as far as possible, illustrate the correlation between this Directive and the transposition measures, and to make them public.

(56) Since the objective of this Directive, namely the harmonisation of legislation concerning the use of animals for scientific purposes, cannot be sufficiently achieved by the Member States and can therefore, by reason of its scale and effects, be better achieved at Union level, the Union may adopt measures, in accordance with the principle of subsidiarity as set out in Article 5 of the Treaty on European Union. In accordance with the principle of proportionality, as set out in that Article, this Directive does not go beyond what is necessary in order to achieve that objective,

HAVE ADOPTED THIS DIRECTIVE:

Chapter I
General Provisions

Article 1
Subject Matter and Scope

1. This Directive establishes measures for the protection of animals used for scientific or educational purposes.
   To that end, it lays down rules on the following:
   (a) the replacement and reduction of the use of animals in procedures and the refinement of the breeding, accommodation, care and use of animals in procedures;
   (b) the origin, breeding, marking, care and accommodation and killing of animals;
   (c) the operations of breeders, suppliers and users;
   (d) the evaluation and authorisation of projects involving the use of animals in procedures.

2. This Directive shall apply where animals are used or intended to be used in procedures, or bred specifically so that their organs or tissues may be used for scientific purposes.
   This Directive shall apply until the animals referred to in the first subparagraph have been killed, rehomed or returned to a suitable habitat or husbandry system.
   The elimination of pain, suffering, distress or lasting harm by the successful use of anaesthesia, analgesia or other methods shall not exclude the use of an animal in procedures from the scope of this Directive.

3. This Directive shall apply to the following animals:
   (a) live non-human vertebrate animals, including:
       (i)   independently feeding larval forms; and
       (ii) foetal forms of mammals as from the last third of their normal development;
   (b) live cephalopods.

4. This Directive shall apply to animals used in procedures, which are at an earlier stage of development than that referred to in point (a) of paragraph 3, if the animal is to be allowed to live beyond that stage of development and, as a result of the procedures performed, is likely to experience pain, suffering, distress or lasting harm after it has reached that stage of development.

5. This Directive shall not apply to the following:
   (a) non-experimental agricultural practices;
   (b) non-experimental clinical veterinary practices;
   (c) veterinary clinical trials required for the marketing authorisation of a veterinary medicinal product;
   (d) practices undertaken for the purposes of recognised animal husbandry;
   (e) practices undertaken for the primary purpose of identification of an animal;
   (f) practices not likely to cause pain, suffering, distress or lasting harm equivalent to, or higher than, that caused by the introduction of a needle in accordance with good veterinary practice.

6. This Directive shall apply without prejudice to Council Directive 76/768/ EEC of 27 July 1976 on the approximation of the laws of the Member States relating to cosmetic products (8).

Article 2

Stricter National Measures

1. Member States may, while observing the general rules laid down in the TFEU, maintain provisions in force on 9 November 2010, aimed at ensuring more extensive protection of animals falling within the scope of this Directive than those contained in this Directive.

    Before 1 January 2013 Member States shall inform the Commission about such national provisions. The Commission shall bring them to the attention of other Member States.

2. When acting pursuant to paragraph 1, a Member State shall not prohibit or impede the supply or use of animals bred or kept in another Member State in accordance with this Directive, nor shall it prohibit or impede the placing on the market of products developed with the use of such animals in accordance with this Directive.

Article 3

Definitions

For the purposes of this Directive the following definitions shall apply:

1. 'procedure' means any use, invasive or non-invasive, of an animal for experimental or other scientific purposes, with known or unknown outcome, or educational purposes, which may cause the animal a level of pain, suffering, distress or lasting harm equivalent to, or higher than, that caused by the introduction of a needle in accordance with good veterinary practice.

This includes any course of action intended, or liable, to result in the birth or hatching of an animal or the creation and maintenance of a genetically modified animal line in any such condition, but excludes the killing of animals solely for the use of their organs or tissues;

2. 'project' means a programme of work having a defined scientific objective and involving one or more procedures;

3. 'establishment' means any installation, building, group of buildings or other premises and may include a place that is not wholly enclosed or covered and mobile facilities;

4. 'breeder' means any natural or legal person breeding animals referred to in Annex I with a view to their use in procedures or for the use of their tissue or organs for scientific purposes, or breeding other animals primarily for those purposes, whether for profit or not;

5. 'supplier' means any natural or legal person, other than a breeder, supplying animals with a view to their use in procedures or for the use of their tissue or organs for scientific purposes, whether for profit or not;

6. 'user' means any natural or legal person using animals in procedures, whether for profit or not;

7. 'competent authority' means an authority or authorities or bodies designated by a Member State to carry out the obligations arising from this Directive.

Article 4
Principle of Replacement, Reduction and Refinement

1. Member States shall ensure that, wherever possible, a scientifically satisfactory method or testing strategy, not entailing the use of live animals, shall be used instead of a procedure.

2. Member States shall ensure that the number of animals used in projects is reduced to a minimum without compromising the objectives of the project.

3. Member States shall ensure refinement of breeding, accommodation and care, and of methods used in procedures, eliminating or reducing to the minimum any possible pain, suffering, distress or lasting harm to the animals.

4. This Article shall, in the choice of methods, be implemented in accordance with Article 13.

Article 5
Purposes of Procedures

Procedures may be carried out for the following purposes only:

(a) basic research;
(b) translational or applied research with any of the following aims:
    (i) the avoidance, prevention, diagnosis or treatment of disease, ill-health or other abnormality or their effects in human beings, animals or plants.

    (ii)   the assessment, detection, regulation or modification of physiological conditions in human beings, animals or plants; or

   (iii)   the welfare of animals and the improvement of the production conditions for animals reared for agricultural purpose.

(c) for any of the aims in point (b) in the development, manufacture or testing of the quality, effectiveness and safety of drugs, foodstuffs and feed-stuffs and other substances or products.

(d) protection of the natural environment in the interests of the health or welfare of human beings or animals.

(e) research aimed at preservation of the species.

(f) higher education, or training for the acquisition, maintenance or improvement of vocational skills.

(g) forensic inquiries.

Article 6

Methods of Killing

1. Member States shall ensure that animals are killed with minimum pain, suffering and distress.
2. Member States shall ensure that animals are killed in the establishment of a breeder, supplier or user, by a competent person.

    However, in the case of a field study an animal may be killed by a competent person outside of an establishment.
3. In relation to the animals covered by Annex IV, the appropriate method of killing as set out in that Annex shall be used.
4. Competent authorities may grant exemptions from the requirement in paragraph 3:
   (a) to allow the use of another method provided that, on the basis of scientific evidence, the method is considered to be at least as humane; or
   (b) when, on the basis of scientific justification, the purpose of the procedure cannot be achieved by the use of a method of killing set out in Annex IV.
5. Paragraphs 2 and 3 shall not apply where an animal has to be killed in emergency circumstances for animal-welfare, public-health, public-security, animal-health or environmental reasons.

Chapter II

Provisions on the Use of Certain Animals in Procedures

Article 7

Endangered Species

1. Specimens of those endangered species listed in Annex A to Council Regulation (EC) No 338/97 of 9 December 1996 on the protection of species of wild fauna and flora by regulating trade therein (9), which do not fall within the scope of Article 7(1) of that Regulation, shall not be used in

procedures, with the exception of those procedures meeting the following conditions:

(a) the procedure has one of the purposes referred to in points (b)(i), (c) or (e) of Article 5 of this Directive; and

(b) there is scientific justification to the effect that the purpose of the procedure cannot be achieved by the use of species other than those listed in that Annex.

2. Paragraph 1 shall not apply to any species of non-human primates.

Article 8
Non-Human Primates

1. Subject to paragraph 2, specimens of non-human primates shall not be used in procedures, with the exception of those procedures meeting the following conditions:

(a) the procedure has one of the purposes referred to in

(i) points (b)(i) or (c) of Article 5 of this Directive and is undertaken with a view to the avoidance, prevention, diagnosis or treatment of debilitating or potentially life-threatening clinical conditions in human beings; or

(ii) points (a) or (e) of Article 5;

and

(b) there is scientific justification to the effect that the purpose of the procedure cannot be achieved by the use of species other than non-human primates.

A debilitating clinical condition for the purposes of this Directive means a reduction in a person's normal physical or psychological ability to function.

2. Specimens of non-human primates listed in Annex A to Regulation (EC) No 338/97, which do not fall within the scope of Article 7(1) of that Regulation, shall not be used in procedures, with the exception of those procedures meeting the following conditions:

(a) the procedure has one of the purposes referred to in:

(i) points (b)(i) or (c) of Article 5 of this Directive and is undertaken with a view to the avoidance, prevention, diagnosis or treatment of debilitating or potentially life-threatening clinical conditions in human beings; or

(ii) Article 5(e);

and

(b) there is scientific justification to the effect that the purpose of the procedure cannot be achieved by the use of species other than non-human primates and by the use of species not listed in that Annex.

3. Notwithstanding paragraphs 1 and 2, great apes shall not be used in procedures, subject to the use of the safeguard clause in Article 55(2).

Article 9
Animals Taken from the Wild

1. Animals taken from the wild shall not be used in procedures.
2. Competent authorities may grant exemptions from paragraph 1 on the basis of scientific justification to the effect that the purpose of the procedure cannot be achieved by the use of an animal which has been bred for use in procedures.
3. The capture of animals in the wild shall be carried out only by competent persons using methods which do not cause the animals avoidable pain, suffering, distress or lasting harm.

Any animal found, at or after capture, to be injured or in poor health shall be examined by a veterinarian or another competent person and action shall be taken to minimise the suffering of the animal. Competent authorities may grant exemptions from the requirement of taking action to minimise the suffering of the animal if there is scientific justification.

Article 10
Animals Bred for Use in Procedures

1. Member States shall ensure that animals belonging to the species listed in Annex I may only be used in procedures where those animals have been bred for use in procedures.

   However, from the dates set out in Annex II, Member States shall ensure that non-human primates listed therein may be used in procedures only where they are the offspring of non-human primates which have been bred in captivity or where they are sourced from self-sustaining colonies.

   For the purposes of this Article a 'self-sustaining colony' means a colony in which animals are bred only within the colony or sourced from other colonies but not taken from the wild, and where the animals are kept in a way that ensures that they are accustomed to humans.

   The Commission shall, in consultation with the Member States and stakeholders, conduct a feasibility study, which shall include an animal health and welfare assessment, of the requirement laid down in the second subparagraph. The study shall be published no later than 10 November 2017. It shall be accompanied, where appropriate, by proposals for amendments to Annex II.
2. The Commission shall keep under review the use of sourcing non-human primates from self-sustaining colonies and, in consultation with the Member States and stakeholders, conduct a study to analyse the feasibility of sourcing animals only from self-sustaining colonies.

   The study shall be published no later than 10 November 2022.
3. Competent authorities may grant exemptions from paragraph 1 on the basis of scientific justification.

Article 11

Stray and Feral Animals of Domestic Species

1. Stray and feral animals of domestic species shall not be used in procedures.
2. The competent authorities may only grant exemptions from paragraph 1 subject to the following conditions:
   (a) there is an essential need for studies concerning the health and welfare of the animals or serious threats to the environment or to human or animal health; and
   (b) there is scientific justification to the effect that the purpose of the procedure can be achieved only by the use of a stray or a feral animal.

Chapter III

Procedures

Article 12

Procedures

1. Member States shall ensure that procedures are carried out in a user's establishment.

   The competent authority may grant an exemption from the first subparagraph on the basis of scientific justification.
2. Procedures may be carried out only within the framework of a project.

Article 13

Choice of Methods

1. Without prejudice to national legislation prohibiting certain types of methods, Member States shall ensure that a procedure is not carried out if another method or testing strategy for obtaining the result sought, not entailing the use of a live animal, is recognised under the legislation of the Union.
2. In choosing between procedures, those which to the greatest extent meet the following requirements shall be selected:
   (a) use the minimum number of animals;
   (b) involve animals with the lowest capacity to experience pain, suffering, distress or lasting harm;
   (c) cause the least pain, suffering, distress or lasting harm;
   and are most likely to provide satisfactory results.
3. Death as the end-point of a procedure shall be avoided as far as possible and replaced by early and humane end-points. Where death as the end-point is unavoidable, the procedure shall be designed so as to:
   (a) result in the deaths of as few animals as possible; and
   (b) reduce the duration and intensity of suffering to the animal to the minimum possible and, as far as possible, ensure a painless death.

Article 14

Anaesthesia

1. Member States shall ensure that, unless it is inappropriate, procedures are carried out under general or local anaesthesia, and that analgesia or another appropriate method is used to ensure that pain, suffering and distress are kept to a minimum.

   Procedures that involve serious injuries that may cause severe pain shall not be carried out without anaesthesia.
2. When deciding on the appropriateness of using anaesthesia, the following shall be taken into account:
   (a) whether anaesthesia is judged to be more traumatic to the animal than the procedure itself; and
   (b) whether anaesthesia is incompatible with the purpose of the procedure.
3. Member States shall ensure that animals are not given any drug to stop or restrict their showing pain without an adequate level of anaesthesia or analgesia.

   In these cases, a scientific justification shall be provided, accompanied by the details of the anaesthetic or analgesic regimen.
4. An animal, which may suffer pain once anaesthesia has worn off, shall be treated with pre-emptive and post-operative analgesics or other appropriate pain-relieving methods provided that it is compatible with the purpose of the procedure.
5. As soon as the purpose of the procedure has been achieved appropriate action shall be taken to minimise the suffering of the animal.

Article 15

Classification of Severity of Procedures

1. Member States shall ensure that all procedures are classified as 'non-recovery', 'mild', 'moderate', or 'severe' on a case-by-case basis using the assignment criteria set out in Annex VIII.
2. Subject to the use of the safeguard clause in Article 55(3), Member States shall ensure that a procedure is not performed if it involves severe pain, suffering or distress that is likely to be long-lasting and cannot be ameliorated.

Article 16

Reuse

1. Member States shall ensure that an animal already used in one or more procedures, when a different animal on which no procedure has previously been carried out could also be used, may only be reused in a new procedure provided that the following conditions are met:
   (a) the actual severity of the previous procedures was 'mild' or 'moderate';
   (b) it is demonstrated that the animal's general state of health and well-being has been fully restored;

(c) the further procedure is classified as 'mild', 'moderate' or 'non-recovery'; and

(d) it is in accordance with veterinary advice, taking into account the lifetime experience of the animal.

2. In exceptional circumstances, by way of derogation from point (a) of paragraph 1 and after a veterinary examination of the animal, the competent authority may allow reuse of an animal, provided the animal has not been used more than once in a procedure entailing severe pain, distress or equivalent suffering.

Article 17
End of the Procedure

1. A procedure shall be deemed to end when no further observations are to be made for that procedure or, as regards new genetically modified animal lines, when the progeny are no longer observed or expected to experience pain, suffering, distress or lasting harm equivalent to, or higher than, that caused by the introduction of a needle.

2. At the end of a procedure, a decision to keep an animal alive shall be taken by a veterinarian or by another competent person. An animal shall be killed when it is likely to remain in moderate or severe pain, suffering, distress or lasting harm.

3. Where an animal is to be kept alive, it shall receive care and accommodation appropriate to its state of health.

Article 18
Sharing Organs and Tissues

Member States shall facilitate, where appropriate, the establishment of programmes for the sharing of organs and tissues of animals killed.

Article 19
Setting Free of Animals and Rehoming

Member States may allow animals used or intended to be used in procedures to be rehomed, or returned to a suitable habitat or husbandry system appropriate to the species, provided that the following conditions are met:

(a) the state of health of the animal allows it;

(b) there is no danger to public health, animal health or the environment; and

(c) appropriate measures have been taken to safeguard the well-being of the animal.

Chapter IV
Authorisation

Section 1
Requirements for Breeders, Suppliers and Users

Article 20
Authorisation of Breeders, Suppliers and Users

1. Member States shall ensure that all breeders, suppliers and users are authorised by, and registered with, the competent authority. Such authorisation
   may be granted for a limited period.
       Authorisation shall be granted only if the breeder, supplier or user and
   its establishment is in compliance with the requirements of this Directive.
2. The authorisation shall specify the person responsible for ensuring compliance with the provisions of this Directive and the person or persons referred
   to in Article 24(1) and in Article 25.
3. Renewal of the authorisation shall be required for any significant change to
   the structure or the function of an establishment of a breeder, supplier or
   user that could negatively affect animal welfare.
4. Member States shall ensure that the competent authority is notified of any
   changes of the person or persons referred to in paragraph 2.

Article 21
Suspension and Withdrawal of Authorisation

1. Where a breeder, supplier or user no longer complies with the requirements
   set out in this Directive, the competent authority shall take appropriate
   remedial action, or require such action to be taken, or suspend or withdraw
   its authorisation.
2. Member States shall ensure that, where the authorisation is suspended or
   withdrawn, the welfare of the animals housed in the establishment is not
   adversely affected.

Article 22
Requirements for Installations and Equipment

1. Member States shall ensure that all establishments of a breeder, supplier
   or user have installations and equipment suited to the species of animals
   housed and, where procedures are carried out, to the performance of the
   procedures.
2. The design, construction and method of functioning of the installations and
   equipment referred to in paragraph 1 shall ensure that the procedures are

carried out as effectively as possible, and aim at obtaining reliable results using the minimum number of animals and causing the minimum degree of pain, suffering, distress or lasting harm.

3. For the purposes of implementation of paragraphs 1 and 2, Member States shall ensure that the relevant requirements as set out in Annex III are complied with.

Article 23
Competence of Personnel

1. Member States shall ensure that each breeder, supplier and user has sufficient staff on site.
2. The staff shall be adequately educated and trained before they perform any of the following functions:
   (a) carrying out procedures on animals;
   (b) designing procedures and projects;
   (c) taking care of animals; or
   (d) killing animals.

   Persons carrying out the functions referred to in point (b) shall have received instruction in a scientific discipline relevant to the work being undertaken and shall have species-specific knowledge.

   Staff carrying out functions referred to in points (a), (c) or (d) shall be supervised in the performance of their tasks until they have demonstrated the requisite competence.

   Member States shall ensure, through authorisation or by other means, that the requirements laid down in this paragraph are fulfilled.
3. Member States shall publish, on the basis of the elements set out in Annex V, minimum requirements with regard to education and training and the requirements for obtaining, maintaining and demonstrating requisite competence for the functions set out in paragraph 2.
4. Non-binding guidelines at the level of the Union on the requirements laid down in paragraph 2 may be adopted in accordance with the advisory procedure referred to in Article 56(2).

Article 24
Specific Requirements for Personnel

1. Member States shall ensure that each breeder, supplier and user has one or several persons on site who shall:
   (a) be responsible for overseeing the welfare and care of the animals in the establishment;
   (b) ensure that the staff dealing with animals have access to information specific to the species housed in the establishment;
   (c) be responsible for ensuring that the staff are adequately educated, competent and continuously trained and that they are supervised until they have demonstrated the requisite competence.

2. Member States shall ensure that persons specified in Article 40(2)(b) shall:
   (a) ensure that any unnecessary pain, suffering, distress or lasting harm that is being inflicted on an animal in the course of a procedure is stopped; and
   (b) ensure that the projects are carried out in accordance with the project authorisation or, in the cases referred to in Article 42, in accordance with the application sent to the competent authority or any decision taken by the competent authority, and ensure that in the event of non-compliance, the appropriate measures to rectify it are taken and recorded.

Article 25
Designated Veterinarian

Member States shall ensure that each breeder, supplier and user has a designated veterinarian with expertise in laboratory animal medicine, or a suitably qualified expert where more appropriate, charged with advisory duties in relation to the well-being and treatment of the animals.

Article 26
Animal-Welfare Body

1. Member States shall ensure that each breeder, supplier and user sets up an animal-welfare body.
2. The animal-welfare body shall include at least the person or persons responsible for the welfare and care of the animals and, in the case of a user, a scientific member. The animal-welfare body shall also receive input from the designated veterinarian or the expert referred to in Article 25.
3. Member States may allow small breeders, suppliers and users to fulfil the tasks laid down in Article 27(1) by other means.

Article 27
Tasks of the Animal-Welfare Body

1. The animal-welfare body shall, as a minimum, carry out the following tasks:
   (a) advise the staff dealing with animals on matters related to the welfare of animals, in relation to their acquisition, accommodation, care and use;
   (b) advise the staff on the application of the requirement of replacement, reduction and refinement, and keep it informed of technical and scientific developments concerning the application of that requirement;
   (c) establish and review internal operational processes as regards monitoring, reporting and follow-up in relation to the welfare of animals housed or used in the establishment;

(d) follow the development and outcome of projects, taking into account the effect on the animals used, and identify and advise as regards elements that further contribute to replacement, reduction and refinement; and

(e) advise on rehoming schemes, including the appropriate socialisation of the animals to be rehomed.

2. Member States shall ensure that the records of any advice given by the animal-welfare body and decisions taken regarding that advice are kept for at least 3 years.

The records shall be made available to the competent authority upon request.

Article 28

Breeding Strategy for Non-Human Primates

Member States shall ensure that breeders of non-human primates have a strategy in place for increasing the proportion of animals that are the offspring of non-human primates that have been bred in captivity.

Article 29

Scheme for Rehoming or Setting Free of Animals

Where Member States allow rehoming, the breeders, suppliers and users from which animals are intended to be rehomed shall have a rehoming scheme in place that ensures socialisation of the animals that are rehomed. In the case of wild animals, where appropriate, a programme of rehabilitation shall be in place before they are returned to their habitat.

Article 30

Animal Records

1. Member States shall ensure that all breeders, suppliers and users keep records of at least the following:
   (a) the number and the species of animals bred, acquired, supplied, used in procedures, set-free or rehomed;
   (b) the origin of the animals, including whether they are bred for use in procedures;
   (c) the dates on which the animals are acquired, supplied, released or rehomed;
   (d) from whom the animals are acquired;
   (e) the name and address of the recipient of animals;
   (f) the number and species of animals which died or were killed in each establishment. For animals that have died, the cause of death shall, when known, be noted; and
   (g) in the case of users, the projects in which animals are used.

2. The records referred to in paragraph 1 shall be kept for a minimum of 5 years and made available to the competent authority upon request.

Article 31

Information on Dogs, Cats and Non-Human Primates

1. Member States shall ensure that all breeders, suppliers and users keep the following information on each dog, cat and non-human primate:
   (a) identity;
   (b) place and date of birth, when available;
   (c) whether it is bred for use in procedures; and
   (d) in the case of a non-human primate, whether it is the offspring of non-human primates that have been bred in captivity.
2. Each dog, cat and non-human primate shall have an individual history file, which follows the animal as long as it is kept for the purposes of this Directive.

   The file shall be established at birth or as soon as possible thereafter and shall cover any relevant reproductive, veterinary and social information on the individual animal and the projects in which it has been used.
3. The information referred to in this Article shall be kept for a minimum of 3 years after the death or rehoming of the animal and shall be made available to the competent authority upon request.

   In the case of rehoming, relevant veterinary care and social information from the individual history file referred to in paragraph 2 shall accompany the animal.

Article 32

Marking and Identification of Dogs, Cats and Non-Human Primates

1. Each dog, cat or non-human primate shall be provided, at the latest at the time of weaning, with a permanent individual identification mark in the least painful manner possible.
2. Where a dog, cat or non-human primate is transferred from one breeder, supplier or user to another before it is weaned, and it is not practicable to mark it beforehand, a record, specifying in particular its mother, must be maintained by the receiver until it is marked.
3. Where an unmarked dog, cat or non-human primate, which is weaned, is received by a breeder, supplier or user it shall be permanently marked as soon as possible and in the least painful manner possible.
4. The breeder, supplier and user shall provide, at the request of the competent authority, reasons for which the animal is unmarked.

Article 33

Care and Accommodation

1. Member States shall, as far as the care and accommodation of animals is concerned, ensure that:
   (a) all animals are provided with accommodation, an environment, food, water and care which are appropriate to their health and well-being;

   (b) any restrictions on the extent to which an animal can satisfy its physiological and ethological needs are kept to a minimum;

   (c) the environmental conditions in which animals are bred, kept or used
are checked daily;

   (d) arrangements are made to ensure that any defect or avoidable pain, suffering, distress or lasting harm discovered is eliminated as quickly as
possible; and

   (e) animals are transported under appropriate conditions.

2. For the purposes of paragraph 1, Member States shall ensure that the care
and accommodation standards set out in Annex III are applied from the
dates provided for therein.

3. Member States may allow exemptions from the requirements of paragraph
1(a) or paragraph 2 for scientific, animal-welfare or animal-health reasons.

Section 2

Inspections

Article 34

Inspections by the Member States

1. Member States shall ensure that the competent authorities carry out regular
inspections of all breeders, suppliers and users, including their establishments, to verify compliance with the requirements of this Directive.

2. The competent authority shall adapt the frequency of inspections on the
basis of a risk analysis for each establishment, taking account of:

   (a) the number and species of animals housed;

   (b) the record of the breeder, supplier or user in complying with the requirements of this Directive;

   (c) the number and types of projects carried out by the user in question;
and

   (d) any information that might indicate non-compliance.

3. Inspections shall be carried out on at least one third of the users each year
in accordance with the risk analysis referred to in paragraph 2. However,
breeders, suppliers and users of non-human primates shall be inspected at
least once a year.

4. An appropriate proportion of the inspections shall be carried out without
prior warning.

5. Records of all inspections shall be kept for at least 5 years.

Article 35

Controls of Member State Inspections

1. The Commission shall, when there is due reason for concern, taking into
account, inter alia, the proportion of inspections carried out without prior
warning, undertake controls of the infrastructure and operation of national
inspections in Member States.

2. The Member State in the territory of which the control referred to in paragraph 1 is being carried out shall give all necessary assistance to the experts of the Commission in carrying out their duties. The Commission shall inform the competent authority of the Member State concerned of the results of the control.

3. The competent authority of the Member State concerned shall take measures to take account of the results of the control referred to in paragraph 1.

Section 3

Requirements for Projects

Article 36

Project Authorisation

1. Member States shall ensure, without prejudice to Article 42, that projects are not carried out without prior authorisation from the competent authority, and that projects are carried out in accordance with the authorisation or, in the cases referred to in Article 42, in accordance with the application sent to the competent authority or any decision taken by the competent authority.

2. Member States shall ensure that no project is carried out unless a favourable project evaluation by the competent authority has been received in accordance with Article 38.

Article 37

Application for Project Authorisation

1. Member States shall ensure that an application for project authorisation is submitted by the user or the person responsible for the project. The application shall include at least the following:
   (a) the project proposal;
   (b) a non-technical project summary; and
   (c) information on the elements set out in Annex VI.

2. Member States may waive the requirement in paragraph 1(b) for projects referred to in Article 42(1).

Article 38

Project Evaluation

1. The project evaluation shall be performed with a degree of detail appropriate for the type of project and shall verify that the project meets the following criteria:
   (a) the project is justified from a scientific or educational point of view or required by law;
   (b) the purposes of the project justify the use of animals; and
   (c) the project is designed so as to enable procedures to be carried out in the most humane and environmentally sensitive manner possible.

2. The project evaluation shall consist in particular of the following:
   (a) an evaluation of the objectives of the project, the predicted scientific benefits or educational value;
   (b) an assessment of the compliance of the project with the requirement of replacement, reduction and refinement;
   (c) an assessment and assignment of the classification of the severity of procedures;
   (d) a harm-benefit analysis of the project, to assess whether the harm to the animals in terms of suffering, pain and distress is justified by the expected outcome taking into account ethical considerations, and may ultimately benefit human beings, animals or the environment;
   (e) an assessment of any justification referred to in Articles 6 to 12, 14, 16 and 33; and
   (f) a determination as to whether and when the project should be assessed retrospectively.
3. The competent authority carrying out the project evaluation shall consider expertise in particular in the following areas:
   (a) the areas of scientific use for which animals will be used including replacement, reduction and refinement in the respective areas;
   (b) experimental design, including statistics where appropriate;
   (c) veterinary practice in laboratory animal science or wildlife veterinary practice where appropriate;
   (d) animal husbandry and care, in relation to the species that are intended to be used.
4. The project evaluation process shall be transparent.

Subject to safeguarding intellectual property and confidential information, the project evaluation shall be performed in an impartial manner and may integrate the opinion of independent parties.

Article 39

Retrospective Assessment

1. Member States shall ensure that when determined in accordance with Article 38(2)(f), the retrospective assessment shall be carried out by the competent authority which shall, on the basis of the necessary documentation submitted by the user, evaluate the following:
   (a) whether the objectives of the project were achieved;
   (b) the harm inflicted on animals, including the numbers and species of animals used, and the severity of the procedures; and
   (c) any elements that may contribute to the further implementation of the requirement of replacement, reduction and refinement.
2. All projects using non-human primates and projects involving procedures classified as 'severe', including those referred to in Article 15(2), shall undergo a retrospective assessment.

3. Without prejudice to paragraph 2 and by way of derogation from Article 38(2)(f), Member States may exempt projects involving only procedures classified as 'mild' or 'non-recovery' from the requirement for a retrospective assessment.

Article 40
Granting of Project Authorisation

1. The project authorisation shall be limited to procedures which have been subject to:
   (a) a project evaluation; and
   (b) the severity classifications assigned to those procedures.
2. The project authorisation shall specify the following:
   (a) the user who undertakes the project;
   (b) the persons responsible for the overall implementation of the project and its compliance with the project authorisation;
   (c) the establishments in which the project will be undertaken, where applicable; and
   (d) any specific conditions following the project evaluation, including whether and when the project shall be assessed retrospectively.
3. Project authorisations shall be granted for a period not exceeding 5 years.
4. Member States may allow the authorisation of multiple generic projects carried out by the same user if such projects are to satisfy regulatory requirements or if such projects use animals for production or diagnostic purposes with established methods.

Article 41
Authorisation Decisions

1. Member States shall ensure that the decision regarding authorisation is taken and communicated to the applicant 40 working days at the latest from the receipt of the complete and correct application. This period shall include the project evaluation.
2. When justified by the complexity or the multi-disciplinary nature of the project, the competent authority may extend the period referred to in paragraph 1 once, by an additional period not exceeding 15 working days. The extension and its duration shall be duly motivated and shall be notified to the applicant before the expiry of the period referred to in paragraph 1.
3. Competent authorities shall acknowledge to the applicant all applications for authorisations as quickly as possible, and shall indicate the period referred to in paragraph 1 within which the decision is to be taken.
4. In the case of an incomplete or incorrect application, the competent authority shall, as quickly as possible, inform the applicant of the need to supply any additional documentation and of any possible effects on the running of the applicable time period.

Article 42

Simplified Administrative Procedure

1. Member States may decide to introduce a simplified administrative procedure for projects containing procedures classified as 'non-recovery', 'mild' or 'moderate' and not using non-human primates, that are necessary to satisfy regulatory requirements, or which use animals for production or diagnostic purposes with established methods.
2. When introducing a simplified administrative procedure, Member States shall ensure that the following provisions are met:
   (a) the application specifies elements referred to in Article 40(2)(a), (b) and (c);
   (b) a project evaluation is performed in accordance with Article 38; and
   (c) that the period referred to in Article 41(1) is not exceeded.
3. If a project is changed in a way that may have a negative impact on animal welfare, Member States shall require an additional project evaluation with a favourable outcome.
4. Article 40(3) and (4), Article 41(3) and Article 44(3), (4) and (5) shall apply mutatis mutandis to projects that are allowed to be carried out in accordance with this Article.

Article 43

Non-Technical Project Summaries

1. Subject to safeguarding intellectual property and confidential information, the non-technical project summary shall provide the following:
   (a) information on the objectives of the project, including the predicted harm and benefits and the number and types of animals to be used;
   (b) a demonstration of compliance with the requirement of replacement, reduction and refinement.
   The non-technical project summary shall be anonymous and shall not contain the names and addresses of the user and its personnel.
2. Member States may require the non-technical project summary to specify whether a project is to undergo a retrospective assessment and by what deadline. In such a case, Member States shall ensure that the non-technical project summary is updated with the results of any retrospective assessment.
3. Member States shall publish the non-technical project summaries of authorised projects and any updates thereto.

Article 44

Amendment, Renewal and Withdrawal of a Project Authorisation

1. Member States shall ensure that amendment or renewal of the project authorisation is required for any change of the project that may have a negative impact on animal welfare.

2. Any amendment or renewal of a project authorisation shall be subject to a further favourable outcome of the project evaluation.
3. The competent authority may withdraw the project authorisation where the project is not carried out in accordance with the project authorisation.
4. Where a project authorisation is withdrawn, the welfare of the animals used or intended to be used in the project must not be adversely affected.
5. Member States shall establish and publish conditions for amendment and renewal of project authorisations.

Article 45

Documentation

1. Member States shall ensure that all relevant documentation, including project authorisations and the result of the project evaluation is kept for at least 3 years from the expiry date of the authorisation of the project or from the expiry of the period referred to in Article 41(1) and shall be available to the competent authority.
2. Without prejudice to paragraph 1, the documentation for projects which have to undergo retrospective assessment shall be kept until the retrospective assessment has been completed.

Chapter V

Avoidance of Duplication and Alternative Approaches

Article 46

Avoidance of Duplication of Procedures

Each Member State shall accept data from other Member States that are generated by procedures recognised by the legislation of the Union, unless further procedures need to be carried out regarding that data for the protection of public health, safety or the environment.

Article 47

Alternative Approaches

1. The Commission and the Member States shall contribute to the development and validation of alternative approaches which could provide the same or higher levels of information as those obtained in procedures using animals, but which do not involve the use of animals or use fewer animals or which entail less painful procedures, and they shall take such other steps as they consider appropriate to encourage research in this field.
2. Member States shall assist the Commission in identifying and nominating suitable specialised and qualified laboratories to carry out such validation studies.
3. After consulting the Member States, the Commission shall set the priorities for those validation studies and allocate the tasks between the laboratories for carrying out those studies.

4. Member States shall, at national level, ensure the promotion of alternative approaches and the dissemination of information thereon.
5. Member States shall nominate a single point of contact to provide advice on the regulatory relevance and suitability of alternative approaches proposed for validation.
6. The Commission shall take appropriate action with a view to obtaining international acceptance of alternative approaches validated in the Union.

Article 48
Union Reference Laboratory

1. The Union Reference Laboratory and its duties and tasks shall be those referred to in Annex VII.
2. The Union Reference Laboratory may collect charges for the services it provides that do not directly contribute to the further advancement of replacement, reduction and refinement.
3. Detailed rules necessary for the implementation of paragraph 2 of this Article and Annex VII may be adopted in accordance with the regulatory procedure referred to in Article 56(3).

Article 49
National Committees for the Protection of Animals Used for Scientific Purposes

1. Each Member State shall establish a national committee for the protection of animals used for scientific purposes. It shall advise the competent authorities and animal-welfare bodies on matters dealing with the acquisition, breeding, accommodation, care and use of animals in procedures and ensure sharing of best practice.
2. The national committees referred to in paragraph 1 shall exchange information on the operation of animal-welfare bodies and project evaluation and share best practice within the Union.

Chapter VI
Final Provisions

Article 50
Adaptation of Annexes to Technical Progress

In order to ensure that the provisions of Annexes I and III to VIII reflect the state of technical or scientific progress, taking into account the experience gained in the implementation of this Directive, in particular through the reporting referred to in Article 54(1), the Commission may adopt, by means of delegated acts in accordance with Article 51 and subject to the conditions laid down in Articles 52 and 53, modifications of those Annexes, with the exception of provisions of Sections I and II of Annex VIII. The dates referred to in Section B of Annex III shall not be brought forward. When adopting such delegated acts, the Commission shall act in accordance with the relevant provisions of this Directive.

Article 51
Exercise of the Delegation

1. The power to adopt delegated acts referred to in Article 50 shall be conferred on the Commission for a period of 8 years beginning on 9 November 2010. The Commission shall make a report in respect of the delegated power at the latest 12 months before the end of the 8-year period. The delegation of power shall be automatically extended for periods of an identical duration, unless the European Parliament or the Council revokes it in accordance with Article 52.
2. As soon as it adopts a delegated act, the Commission shall notify it simultaneously to the European Parliament and to the Council.
3. The power to adopt delegated acts is conferred on the Commission subject to the conditions laid down in Articles 52 and 53.

Article 52
Revocation of the Delegation

1. The delegation of power referred to in Article 50 may be revoked at any time by the European Parliament or by the Council.
2. The institution which has commenced an internal procedure for deciding whether to revoke the delegation of power shall endeavour to inform the other institution and the Commission within a reasonable time before the final decision is taken, indicating the delegated power which could be subject to revocation and possible reasons for a revocation.
3. The decision of revocation shall put an end to the delegation of the power specified in that decision. It shall take effect immediately or at a later date specified therein. It shall not affect the validity of the delegated acts already in force. It shall be published in the *Official Journal of the European Union*.

Article 53
Objections to Delegated Acts

1. The European Parliament or the Council may object to a delegated act within a period of 2 months from the date of notification.
   At the initiative of the European Parliament or the Council this period shall be extended by 2 months.
2. If, on expiry of that period, neither the European Parliament nor the Council has objected to the delegated act, it shall be published in the *Official Journal of the European Union* and shall enter into force at the date stated therein.
   The delegated act may be published in the *Official Journal of the European Union* and enter into force before the expiry of that period if the

European Parliament and the Council have both informed the Commission of their intention not to raise objections.
3. If the European Parliament or the Council objects to a delegated act, it shall not enter into force. The institution which objects shall state the reasons for objecting to the delegated act.

Article 54
Reporting

1. Member States shall by 10 November 2018, and every 5 years thereafter, send the information on the implementation of this Directive and in particular Articles 10(1), 26, 28, 34, 38, 39, 43 and 46 to the Commission.
2. Member States shall collect and make publicly available, on an annual basis, statistical information on the use of animals in procedures, including information on the actual severity of the procedures and on the origin and species of non-human primates used in procedures.

   Member States shall submit that statistical information to the Commission by 10 November 2015 and every year thereafter.
3. Member States shall submit to the Commission, on annual basis, detailed information on exemptions granted under Article 6(4)(a).
4. The Commission shall by 10 May 2012 establish a common format for submitting the information referred to in paragraphs 1, 2, and 3 of this Article in accordance with the regulatory procedure referred to in Article 56(3).

Article 55
Safeguard Clauses

1. Where a Member State has scientifically justifiable grounds for believing it is essential to use non-human primates for the purposes referred to in Article 8(1)(a)(i) with regard to human beings, but where the use is not undertaken with a view to the avoidance, prevention, diagnosis or treatment of debilitating or potentially life-threatening clinical conditions, it may adopt a provisional measure allowing such use, provided the purpose cannot be achieved by the use of species other than non-human primates.
2. Where a Member State has justifiable grounds for believing that action is essential for the preservation of the species or in relation to an unexpected outbreak of a life-threatening or debilitating clinical condition in human beings, it may adopt a provisional measure allowing the use of great apes in procedures having one of the purposes referred to in points (b)(i), (c) or (e) of Article 5; provided that the purpose of the procedure cannot be achieved by the use of species other than great apes or by the use of alternative methods. However, the reference to Article 5(b)(i) shall not be taken to include the reference to animals and plants.
3. Where, for exceptional and scientifically justifiable reasons, a Member State deems it necessary to allow the use of a procedure involving severe pain, suffering or distress that is likely to be long-lasting and cannot be

ameliorated, as referred to in Article 15(2), it may adopt a provisional measure to allow such procedure. Member States may decide not to allow the use of non-human primates in such procedures.

4. A Member State which has adopted a provisional measure in accordance with paragraph 1, 2 or 3 shall immediately inform the Commission and the other Member States thereof, giving reasons for its decision and submitting evidence of the situation as described in paragraphs 1, 2 and 3 on which the provisional measure is based.

   The Commission shall put the matter before the Committee referred to in Article 56(1) within 30 days of receipt of the information from the Member State and shall, in accordance with the regulatory procedure referred to in Article 56(3), either:

   (a) authorise the provisional measure for a time period defined in the decision; or

   (b) require the Member State to revoke the provisional measure.

Article 56

Committee

1. The Commission shall be assisted by a Committee.
2. Where reference is made to this paragraph, Articles 3 and 7 of Decision 1999/468/EC shall apply, having regard to the provisions of Article 8 thereof.
3. Where reference is made to this paragraph, Articles 5 and 7 of Decision 1999/468/EC shall apply, having regard to the provisions of Article 8 thereof.

The period laid down in Article 5(6) of Decision 1999/468/EC shall be set at 3 months.

Article 57

Commission Report

1. By 10 November 2019 and every 5 years thereafter, the Commission shall, based on the information received from the Member States under Article 54(1), submit to the European Parliament and the Council a report on the implementation of this Directive.
2. By 10 November 2019 and every 3 years thereafter, the Commission shall, based on the statistical information submitted by Member States under Article 54(2), submit to the European Parliament and the Council a summary report on that information.

Article 58

Review

The Commission shall review this Directive by 10 November 2017, taking into account advancements in the development of alternative methods not entailing the use of animals, in particular of non-human primates, and shall propose amendments, where appropriate.

The Commission shall, where appropriate, and in consultation with the Member States and stakeholders, conduct periodic thematic reviews of the replacement, reduction and refinement of the use of animals in procedures, paying specific attention to non-human primates, technological developments, and new scientific and animal-welfare knowledge.

Article 59
Competent Authorities

1. Each Member State shall designate one or more competent authorities responsible for the implementation of this Directive.

   Member States may designate bodies other than public authorities for the implementation of specific tasks laid down in this Directive only if there is proof that the body:
   (a) has the expertise and infrastructure required to carry out the tasks; and
   (b) is free of any conflict of interests as regards the performance of the tasks.

   Bodies thus designated shall be considered competent authorities for the purposes of this Directive.
2. Each Member State shall communicate details of a national authority serving as contact point for the purposes of this Directive to the Commission by 10 February 2011, as well as any update to such data.

The Commission shall make publicly available the list of those contact points.

Article 60
Penalties

Member States shall lay down the rules on penalties applicable to infringements of the national provisions adopted pursuant to this Directive and shall take all measures necessary to ensure that they are implemented. The penalties provided for must be effective, proportionate and dissuasive. The Member States shall notify those provisions to the Commission by 10 February 2013, and shall notify the Commission without delay of any subsequent amendment affecting them.

Article 61
Transposition

1. Member States shall adopt and publish, by 10 November 2012, the laws, regulations and administrative provisions necessary to comply with this Directive. They shall forthwith communicate to the Commission the text of those provisions.

   They shall apply those provisions from 1 January 2013.

   When Member States adopt those provisions, they shall contain a reference to this Directive or be accompanied by such reference on the occasion
   of their official publication. The method of making such reference shall be
   laid down by Member States.

2. Member States shall communicate to the Commission the text of the main
   provisions of national law which they adopt in the field covered by this
   Directive.

Article 62
Repeal

1. Directive 86/609/EEC is repealed with effect from 1 January 2013 with
   the exception of Article 13, which shall be repealed with effect from
   10 May 2013.
2. References to the repealed Directive shall be construed as references to this
   Directive.

Article 63
Amendment of Regulation (EC) No 1069/2009

Point (a)(iv) of Article 8 of Regulation (EC) No 1069/2009 is replaced by the
following:

'(iv) animals used in a procedure or procedures defined in Article 3 of Directive
2010/63/EU of the European Parliament and of the Council of 22 September 2010
on the protection of animals used for scientific purposes (10), in cases where the
competent authority decides that such animals or any of their body parts have the
potential to pose serious health risks to humans or to other animals, as a result of
that procedure or those procedures without prejudice to Article 3(2) of Regulation
(EC) No 1831/2003;

Article 64
Transitional Provisions

1. Member States shall not apply laws, regulations and administrative provisions adopted in accordance with Articles 36 to 45 to projects which have
   been approved before 1 January 2013 and the duration of which does not
   extend beyond 1 January 2018.
2. Projects which have been approved before 1 January 2013 and the duration
   of which extends beyond 1 January 2018 shall obtain project authorisation
   by 1 January 2018.

Article 65
Entry into Force

This Directive shall enter into force on the 20th day following its publication in the
Official Journal of the European Union.

Article 66

Addressees

This Directive is addressed to the Member States.
Done at Strasbourg, 22 September 2010.

> *For the European Parliament*
> *The President*
> J. BUZEK
> *For the Council*
> *The President*
> O. CHASTEL

(1) OJC277,17.11.2009,p.51.
(2) Position of the European Parliament of 5 May 2009 (OJC212E,5.8.2010,p.170), position of the Council of 13 September 2010 (not yet published in the Official Journal) and position of the European Parliament of 8 September 2010 (not yet published in the Official Journal).
(3) OJL358,18.12.1986,p.1.
(4) OJL222,24.8.1999,p.29.
(5) OJL197,30.7.2007,p.1.
(6) OJL184,17.7.1999,p.23.
(7) OJL300,14.11.2009,p.1.
(8) OJL262,27.9.1976,p.169. Directive recast by Regulation (EC) No 1223/2009 of the European Parliament and the Council of 30 November 2009 on cosmetic products (OJL342,22.12.2009,p.59), which applies as from 11 July 2013.
(9) OJL61,3.3.1997,p.1.
(10) OJL276,20.10.2010,p.33'.

Annex I

List of Animals Referred to in Article 10

1. Mouse (*Mus musculus*)
2. Rat (*Rattus norvegicus*)
3. Guinea pig (*Cavia porcellus*)
4. Syrian (golden) hamster (*Mesocricetus auratus*)
5. Chinese hamster (*Cricetulus griseus*)
6. Mongolian gerbil (*Meriones unguiculatus*)
7. Rabbit (*Oryctolagus cuniculus*)
8. Dog (*Canis familiaris*)
9. Cat (*Felis catus*)
10. All species of non-human primates
11. Frog (*Xenopus* (*laevis*, *tropicalis*), Rana (*temporaria*, *pipiens*))
12. Zebra fish (*Danio rerio*)

Annex II

List of Non-Human Primates and Dates Referred to in the Second Subparagraph of Article 10(1)

Species	Dates
Marmoset (*Callithrix jacchus*)	1 January 2013
Cynomolgus monkey (*Macaca fascicularis*)	5 years after the publication of the feasibility study referred to in Article 10(1), fourth subparagraph, provided the study does not recommend an extended period
Rhesus monkey (*Macaca mulatta*)	5 years after the publication of the feasibility study referred to in Article 10(1), fourth subparagraph, provided the study does not recommend an extended period
Other species of non-human primates	5 years after the publication of the feasibility study referred to in Article 10(1), fourth subparagraph, provided the study does not recommend an extended period

Annex III

Requirements for Establishments and for the Care and Accommodation of Animals

SECTION A: GENERAL SECTION

1   THE PHYSICAL FACILITIES

1.1   FUNCTIONS AND GENERAL DESIGN

(a) All facilities shall be constructed so as to provide an environment which takes into account the physiological and ethological needs of the species kept in them. Facilities shall also be designed and managed to prevent access by unauthorised persons and the ingress or escape of animals.

(b) Establishments shall have an active maintenance programme to prevent and remedy any defect in buildings or equipment.

1.2   HOLDING ROOMS

(a) Establishments shall have a regular and efficient cleaning schedule for the rooms and shall maintain satisfactory hygienic standards.

(b) Walls and floors shall be surfaced with a material resistant to the heavy wear and tear caused by the animals and the cleaning process. The material shall not be detrimental to the health of the animals and shall be such that the animals cannot hurt themselves. Additional protection shall be given to any equipment or fixtures so that they are not damaged by the animals nor do they cause injury to the animals themselves.

(c) Species that are incompatible, for example predator and prey, or animals requiring different environmental conditions, shall not be housed in the same room nor, in the case of predator and prey, within sight, smell or sound of each other.

1.3   GENERAL AND SPECIAL PURPOSE PROCEDURE ROOMS

(a) Establishments shall, where appropriate, have available laboratory facilities for the carrying out of simple diagnostic tests, post-mortem examinations, and/or the collection of samples that are to be subjected to more extensive laboratory investigations elsewhere. General and special purpose procedure rooms shall be available for situations where it is undesirable to carry out the procedures or observations in the holding rooms.
(b) Facilities shall be provided to enable newly-acquired animals to be isolated until their health status can be determined and the potential health risk to established animals assessed and minimised.
(c) There shall be accommodation for the separate housing of sick or injured animals.

1.4   SERVICE ROOMS

(a) Store-rooms shall be designed, used and maintained to safeguard the quality of food and bedding. These rooms shall be vermin and insect-proof, as far as possible. Other materials, which may be contaminated or present a hazard to animals or staff, shall be stored separately.
(b) The cleaning and washing areas shall be large enough to accommodate the installations necessary to decontaminate and clean used equipment. The cleaning process shall be arranged so as to separate the flow of clean and dirty equipment to prevent the contamination of newly-cleaned equipment.
(c) Establishments shall provide for the hygienic storage and safe disposal of carcasses and animal waste.
(d) Where surgical procedures under aseptic conditions are required there shall be provision for one or more than one suitably equipped room, and facilities provided for postoperative recovery.

2   THE ENVIRONMENT AND CONTROL THEREOF

2.1   VENTILATION AND TEMPERATURE

(a) Insulation, heating and ventilation of the holding room shall ensure that the air circulation, dust levels, and gas concentrations are kept within limits that are not harmful to the animals housed.
(b) Temperature and relative humidity in the holding rooms shall be adapted to the species and age groups housed. The temperature shall be measured and logged on a daily basis.
(c) Animals shall not be restricted to outdoor areas under climatic conditions which may cause them distress.

2.2 LIGHTING

(a) Where natural light does not provide an appropriate light/dark cycle, controlled lighting shall be provided to satisfy the biological requirements of the animals and to provide a satisfactory working environment.
(b) Illumination shall satisfy the needs for the performance of husbandry procedures and inspection of the animals.
(c) Regular photoperiods and intensity of light adapted to the species shall be provided.
(d) When keeping albino animals, the lighting shall be adjusted to take into account their sensitivity to light.

2.3 NOISE

(a) Noise levels including ultrasound, shall not adversely affect animal welfare.
(b) Establishments shall have alarm systems that sound outside the sensitive hearing range of the animals, where this does not conflict with their audibility to human beings.
(c) Holding rooms shall where appropriate be provided with noise insulation and absorption materials.

2.4 ALARM SYSTEMS

(a) Establishments relying on electrical or mechanical equipment for environmental control and protection, shall have a stand-by system to maintain essential services and emergency lighting systems as well as to ensure that alarm systems themselves do not fail to operate.
(b) Heating and ventilation systems shall be equipped with monitoring devices and alarms.
(c) Clear instructions on emergency procedures shall be prominently displayed.

3 CARE OF ANIMALS

3.1 HEALTH

(a) Establishments shall have a strategy in place to ensure that a health status of the animals is maintained that safeguards animal welfare and meets scientific requirements. This strategy shall include regular health monitoring, a microbiological surveillance programme and plans for dealing with health breakdowns and shall define health parameters and procedures for the introduction of new animals.
(b) Animals shall be checked at least daily by a competent person. These checks shall ensure that all sick or injured animals are identified and appropriate action is taken.

3.2 ANIMALS TAKEN FROM THE WILD

(a) Transport containers and means of transport adapted to the species concerned shall be available at capture sites, in case animals need to be moved for examination or treatment.

(b) Special consideration shall be given and appropriate measures taken for the acclimatisation, quarantine, housing, husbandry, care of animals taken from the wild and, as appropriate, provisions for setting them free at the end of procedures.

3.3 HOUSING AND ENRICHMENT

(a) Housing

Animals, except those which are naturally solitary, shall be socially housed in stable groups of compatible individuals. In cases where single housing is allowed in accordance with article 33(3) the duration shall be limited to the minimum period necessary and visual, auditory, olfactory and/or tactile contact shall be maintained. The introduction or re-introduction of animals to established groups shall be carefully monitored to avoid problems of incompatibility and disrupted social relationships.

(b) Enrichment

All animals shall be provided with space of sufficient complexity to allow expression of a wide range of normal behaviour. They shall be given a degree of control and choice over their environment to reduce stress-induced behaviour. Establishments shall have appropriate enrichment techniques in place, to extend the range of activities available to the animals and increase their coping activities including physical exercise, foraging, manipulative and cognitive activities, as appropriate to the species. Environmental enrichment in animal enclosures shall be adapted to the species and individual needs of the animals concerned. The enrichment strategies in establishments shall be regularly reviewed and updated.

(c) Animal enclosures

Animal enclosures shall not be made out of materials detrimental to the health of the animals. Their design and construction shall be such that no injury to the animals is caused. Unless they are disposable, they shall be made from materials that will withstand cleaning and decontamination techniques. The design of animal enclosure floors shall be adapted to the species and age of the animals and be designed to facilitate the removal of excreta.

3.4 FEEDING

(a) The form, content and presentation of the diet shall meet the nutritional and behavioural needs of the animal.

(b) The animals' diet shall be palatable and non-contaminated. In the selection of raw materials, production, preparation and presentation of feed, establishments shall take measures to minimise chemical, physical and microbiological contamination.

(c) Packing, transport and storage shall be such as to avoid contamination, deterioration or destruction. All feed hoppers, troughs or other utensils used for feeding shall be regularly cleaned and, if necessary, sterilised.

(d) Each animal shall be able to access the food, with sufficient feeding space provided to limit competition.

3.5 WATERING

(a) Uncontaminated drinking water shall always be available to all animals.

(b) When automatic watering systems are used, they shall be regularly checked, serviced and flushed to avoid accidents. If solid-bottomed cages are used, care shall be taken to minimise the risk of flooding.

(c) Provision shall be made to adapt the water supply for aquaria and tanks to the needs and tolerance limits of the individual fish, amphibian and reptile species.

3.6 RESTING AND SLEEPING AREAS

(a) Bedding materials or sleeping structures adapted to the species shall always be provided, including nesting materials or structures for breeding animals.

(b) Within the animal enclosure, as appropriate to the species, a solid, comfortable resting area for all animals shall be provided. All sleeping areas shall be kept clean and dry.

3.7 HANDLING

Establishments shall set up habituation and training programmes suitable for the animals, the procedures and length of the project.

SECTION B: SPECIES-SPECIFIC SECTION

1 MICE, RATS, GERBILS, HAMSTERS AND GUINEA PIGS

In this and subsequent tables for mice, rats, gerbils, hamsters and guinea pigs, 'enclosure height' means the vertical distance between the enclosure floor and the top of the enclosure and this height applies over more than 50% of the minimum enclosure floor area prior to the addition of enrichment devices.

When designing procedures, consideration shall be given to the potential growth of the animals to ensure adequate space is provided (as detailed in Tables 1.1 to 1.5) for the duration of the study.

TABLE 1.1
Mice

	Body Weight (g)	Minimum Enclosure Size (cm²)	Floor Area per Animal (cm²)	Minimum Enclosure Height (cm)	Date Referred to in Article 33(2)
In stock and during procedures	Up to 20	330	60	12	1 January 2017
	Over 20 to 25	330	70	12	
	Over 25 to 30	330	80	12	
	Over 30	330	100	12	
Breeding		330 For a monogamous pair (outbred/inbred) or a trio (inbred). For each additional female plus litter 180 cm² shall be added.		12	
Stock at breeders (1) Enclosure size 950 cm²	Less than 20	950	40	12	
Enclosure size 1 500 cm²	Less than 20	1 500	30	12	

TABLE 1.2
Rats

	Body Weight(g)	Minimum Enclosure Size (cm²)	Floor Area per Animal (cm²)	Minimum Enclosure Height (cm)	Date Referred to in Article 33(2)
In stock and during procedures (2)	Up to 200	800	200	18	1 January 2017
	Over 200 to 300	800	250	18	
	Over 300 to 400	800	350	18	
	Over 400 to 600	800	450	18	
	Over 600	1 500	600	18	

(*Continued*)

TABLE 1.2 (*Continued*)
Rats

	Body Weight(g)	Minimum Enclosure Size (cm²)	Floor Area per Animal (cm²)	Minimum Enclosure Height (cm)	Date Referred to in Article 33(2)
Breeding		800 Mother and litter. For each additional adult animal permanently added to the enclosure add 400 cm²		18	
Stock at breeders (3) Enclosure size 1 500 cm²	Up to 50	1 500	100	18	
	Over 50 to 100	1 500	125	18	
	Over 100 to 150	1 500	150	18	
	Over 150 to 200	1 500	175	18	
Stock at breeders (3) Enclosure size 2 500 cm²	Up to 100	2 500	100	18	
	Over 100 to 150	2 500	125	18	
	Over 150 to 200	2 500	150	18	

TABLE 1.3
Gerbils

	Body Weight (g)	Minimum Enclosure Size (cm²)	Floor Area per Animal (cm²)	Minimum Enclosure Height (cm)	Date Referred to in Article 33(2)
In stock and during procedures	Up to 40	1 200	150	18	1 January 2017
	Over 40	1 200	250	18	
Breeding		1 200 Monogamous pair or trio with offspring		18	

TABLE 1.4
Hamsters

	Body Weight (g)	Minimum Enclosure Size (cm²)	Floor Area per Animal (cm²)	Minimum Enclosure Height (cm)	Date Referred to in Article 33(2)
In stock and during procedures	Up to 60	800	150	14	1 January 2017
	Over 60 to 100	800	200	14	
	Over 100	800	250	14	
Breeding		800 Mother or monogamous pair with litter		14	
Stock at breeders	Less than 60	1 500	100	14	

TABLE 1.5
Guinea Pigs

	Body Weight (g)	Minimum Enclosure Size (cm²)	Floor Area per Animal (cm²)	Minimum Enclosure Height (cm)	Date Referred to in Article 33(2)
In stock and during procedures	Up to 200	1 800	200	23	1 January 2017
	Over 200 to 300	1 800	350	23	
	Over 300 to 450	1 800	500	23	
	Over 450 to 700	2 500	700	23	
	Over 700	2 500	900	23	
Breeding		2 500 Pair with litter. For each additional breeding female add 1 000 cm²		23	

2 RABBITS

During agricultural research, when the aim of the project requires that the animals are kept under similar conditions to those under which commercial farm animals are kept, the keeping of the animals shall at least follow the standards laid down in Directive 98/58/EC (5).

A raised area shall be provided within the enclosure. This raised area must allow the animal to lie and sit and easily move underneath, and shall not cover more than 40% of the floor space. When for scientific or veterinary reasons a raised area cannot be used, the enclosure shall be 33% larger for a single rabbit and 60% larger for two rabbits. Where a raised area is provided for rabbits of less than 10 weeks of age, the size of the raised area shall be at least of 55 cm by 25 cm and the height above the floor shall be such that the animals can make use of it.

TABLE 2.1
Rabbits over 10 Weeks of Age

Table 2.1 is to be used for both cages and pens. The additional floor area is as a minimum 3 000 cm² per rabbit for the third, the fourth, the fifth and the sixth rabbit, while 2 500 cm² as a minimum shall be added for each additional rabbit above a number of six.

Final Body Weight (kg)	Minimum Floor Area for One or Two Socially Harmonious Animals (cm²)	Minimum Height (cm)	Date Referred to in Article 33(2)
Less than 3	3 500	45	1 January 2017
From 3 to 5	4 200	45	
Over 5	5 400	60	

TABLE 2.2
Doe plus Litter

Doe Weight (kg)	Minimum Enclosure Size (cm²)	Addition for Nest Boxes (cm²)	Minimum Height (cm)	Date Referred to in Article 33(2)
Less than 3	3 500	1 000	45	1 January 2017
From 3 to 5	4 200	1 200	45	
Over 5	5 400	1 400	60	

TABLE 2.3
Rabbits Less than 10 Weeks of Age

Table 2.3 is to be used for both cages and pens.

Age	Minimum Enclosure Size (cm²)	Minimum Floor Area per Animal (cm²)	Minimum Height (cm)	Date Referred to in Article 33(2)
Weaning to 7 weeks	4 000	800	40	1 January 2017
From 7 to 10 weeks	4 000	1200	40	

TABLE 2.4

Rabbits: Optimal Dimensions for Raised Areas for Enclosures Having the Dimensions Indicated in Table 2.1.

Age in Weeks	Final Body Weight (kg)	Optimum Size (cm × cm)	Optimum Height from the Enclosure Floor (cm)	Date Referred to in Article 33(2)
Over 10	Less than 3	55×25	25	1 January 2017
	From 3 to 5	55×30	25	
	Over 5	60×35	30	

3 CATS

Cats shall not be single-housed for more than 24 hours at a time. Cats that are repeatedly aggressive towards other cats shall be housed singly only if a compatible companion cannot be found. Social stress in all pair- or group-housed individuals shall be monitored at least weekly. Females with kittens under four weeks of age or in the last two weeks of pregnancy may be housed singly.

TABLE 3

Cats

The minimum space in which a queen and litter may be held is the space for a single cat, which shall be gradually increased so that by 4 months of age litters have been rehoused following the space requirements for adults.

Areas for feeding and for litter trays shall not be less than 0,5 metres apart and shall not be interchanged.

	Floor(6) (m²)	Shelves (m²)	Height (m)	Date Referred to in Article 33(2)
Minimum for one adult animal	1,5	0,5	2	1 January 2017
For each additional animal add	0,75	0,25	—	

4 DOGS

Dogs shall where possible be provided with outside runs. Dogs shall not be single-housed for more than 4 hours at a time.

The internal enclosure shall represent at least 50% of the minimum space to be made available to the dogs, as detailed in Table 4.1.

The space allowances detailed below are based on the requirements of beagles, but giant breeds such as St Bernards or Irish wolfhounds shall be provided with allowances significantly in excess of those detailed in Table 4.1. For breeds other than the laboratory beagle, space allowances shall be determined in consultation with veterinary staff.

TABLE 4.1
Dogs

Dogs that are pair or group housed may each be constrained to half the total space provided (2 m^2 for a dog under 20 kg, 4 m^2 for a dog over 20 kg) while they are undergoing procedures as defined in this Directive, if this separation is essential for scientific purposes. The period for which a dog is so constrained shall not exceed 4 hours at a time.

A nursing bitch and litter shall have the same space allowance as a single bitch of equivalent weight. The whelping pen shall be designed so that the bitch can move to an additional compartment or raised area away from the puppies.

Weight (kg)	Minimum Enclosure Size (m^2)	Minimum Floor Area for One or Two Animals (m^2)	For Each Additional Animal Add a Minimum of (m^2)	Minimum Height (m)	Date Referred to in Article 33(2)
Up to 20	4	4	2	2	1 January 2017
Over 20	8	8	4	2	

TABLE 4.2
Dogs—Post-Weaned Stock

Weight of Dog(kg)	Minimum Enclosure Size (m^2)	Minimum Floor Area/Animal (m^2)	Minimum Height (m)	Date Referred to in Article 33(2)
Up to 5	4	0,5	2	1 January 2017
Over 5 to 10	4	1,0	2	
Over 10 to 15	4	1,5	2	
Over 15 to 20	4	2	2	
Over 20	8	4	2	

5   FERRETS

TABLE 5
Ferrets

	Minimum Enclosure Size (cm^2)	Minimum Floor Area per Animal (cm^2)	Minimum Height (cm)	Date Referred to in Article 33(2)
Animals up to 600 g	4 500	1 500	50	1 January 2017
Animals over 600 g	4 500	3 000	50	
Adult males	6 000	6 000	50	
Jill and litter	5 400	5 400	50	

6 NON-HUMAN PRIMATES

Young non-human primates shall not be separated from their mothers until they are, depending on the species, 6 to 12 months old.

The environment shall enable non-human primates to carry out a complex daily programme of activity. The enclosure shall allow non-human primates to adopt as wide a behavioural repertoire as possible, provide it with a sense of security, and a suitably complex environment to allow the animal to run, walk, climb and jump.

TABLE 6.1

Marmosets and Tamarins

	Minimum Floor Area of Enclosures for 1 (7) or 2 Animals Plus Offspring up to 5 Months Old (m²)	Minimum Volume per Additional Animal over 5 Months (m³)	Minimum Enclosure Height (m) (8)	Date Referred to in Article 33(2)
Marmosets	0,5	0,2	1,5	1 January 2017
Tamarins	1,5	0,2	1,5	

For marmosets and tamarins, separation from the mother shall not take place before 8 months of age.

TABLE 6.2

Squirrel Monkeys

Minimum Floor Area for 1 (9) or 2 Animals (m²)	Minimum Volume per Additional Animal over 6 Months of Age (m³)	Minimum Enclosure Height (m)	Date Referred to in Article 33(2)
2,0	0,5	1,8	1 January 2017

For squirrel monkeys, separation from the mother shall not take place before 6 months of age.

TABLE 6.3

Macaques and Vervets (10)

	Minimum Enclosure Size (m²)	Minimum Enclosure Volume (m³)	Minimum Volume per Animal (m³)	Minimum Enclosure Height (m)	Date Referred to in Article 33(2)
Animals less than 3 yrs of age (11)	2,0	3,6	1,0	1,8	1 January 2017
Animals from 3 yrs of age (12)	2,0	3,6	1,8	1,8	
Animals held for breeding purposes (13)			3,5	2,0	

For macaques and vervets, separation from the mother shall not take place before 8 months of age.

TABLE 6.4
Baboons (14)

	Minimum Enclosure Size (m²)	Minimum Enclosure Volume (m³)	Minimum Volume per Animal (m³)	Minimum Enclosure Height (m)	Date Referred to in Article 33(2)
Animals less than 4 yrs of age (15)	4,0	7,2	3,0	1,8	1 January 2017
Animals from 4 yrs of age (15)	7,0	12,6	6,0	1,8	
Animals held for breeding purposes (16)			12,0	2,0	

For baboons, separation from the mother shall not take place before 8 months of age.

7   FARM ANIMALS

During agricultural research, when the aim of the project requires that the animals are kept under similar conditions to those under which commercial farm animals are kept, the keeping of the animals shall comply at least with the standards laid down in Directives 98/58/EC, 91/629/EEC (17) and 91/630/EEC (18).

TABLE 7.1
Cattle

Body Weight (kg)	Minimum Enclosure Size (m²)	Minimum Floor Area/ Animal (m²/Animal)	Trough Space for Ad-Libitum Feeding of Polled Cattle (m/Animal)	Trough Space for Restricted Feeding of Polled Cattle (m/Animal)	Date Referred to in Article 33(2)
Up to 100	2,50	2,30	0,10	0,30	1 January 2017
Over 100 to 200	4,25	3,40	0,15	0,50	
Over 200 to 400	6,00	4,80	0,18	0,60	
Over 400 to 600	9,00	7,50	0,21	0,70	
Over 600 to 800	11,00	8,75	0,24	0,80	
Over 800	16,00	10,00	0,30	1,00	

TABLE 7.2
Sheep and Goats

Body Weight (kg)	Minimum Enclosure Size (m²)	Minimum Floor Area/ Animal (m²/ Animal)	Minimum Partition Height (m)	Trough Space for Ad-Libitum Feeding (m/Animal)	Trough Space for Restricted Feeding (m/Animal)	Date Referred to in Article 33(2)
Less than 20	1,0	0,7	1,0	0,10	0,25	1 January 2017
Over 20 to 35	1,5	1,0	1,2	0,10	0,30	
Over 35 to 60	2,0	1,5	1,2	0,12	0,40	
Over 60	3,0	1,8	1,5	0,12	0,50	

TABLE 7.3
Pigs and Minipigs

Live Weight (kg)	Minimum Enclosure Size (19) (m²)	Minimum Floor Area per Animal (m²/animal)	Minimum Lying Space per Animal (in, Thermoneutral Conditions) (m²/animal)	Date Referred to in Article 33(2)
Up to 5	2,0	0,20	0,10	1 January 2017
Over 5 to 10	2,0	0,25	0,11	
Over 10 to 20	2,0	0,35	0,18	
Over 20 to 30	2,0	0,50	0,24	
Over 30 to 50	2,0	0,70	0,33	
Over 50 to 70	3,0	0,80	0,41	
Over 70 to 100	3,0	1,00	0,53	
Over 100 to 150	4,0	1,35	0,70	
Over 150	5,0	2,50	0,95	
Adult (conventional) boars	7,5		1,30	

TABLE 7.4
Equines

The shortest side shall be a minimum of 1,5 times the wither height of the animal. The height of indoor enclosures shall allow animals to rear to their full height.

Wither Height(m)	Minimum Floor Area/Animal (m²/animal)		Foaling Box/Mare with Foal	Minimum Enclosure Height(m)	Date Referred to in Article 33(2)
	For Each Animal Held Singly or in Groups of up to 3 Animals	For Each Animal Held in Groups of 4 or More Animals			
1,00 to 1,40	9,0	6,0	16	3,00	1 January
Over 1,40 to 1,60	12,0	9,0	20	3,00	2017
Over 1,60	16,0	$(2 \times WH)^2$ (20)	20	3,00	

8   BIRDS

During agricultural research, when the aim of the project requires that the animals are kept under similar conditions to those under which commercial farm animals are kept, the keeping of the animals shall comply at least with the standards laid down in Directives 98/58/EC, 1999/74/EC (21) and 2007/43/EC (22).

TABLE 8.1
Domestic Fowl

Where these minimum enclosure sizes cannot be provided for scientific reasons, the duration of the confinement shall be justified by the experimenter in consultation with veterinary staff. In such circumstances, birds can be housed in smaller enclosures containing appropriate enrichment and with a minimum floor area of 0,75 m².

Body Mass (g)	Minimum Enclosure Size (m²)	Minimum Area per Bird (m²)	Minimum Height (cm)	Minimum Length of Feed Trough per Bird (cm)	Date Referred to in Article 33(2)
Up to 200	1,00	0,025	30	3	1 January 2017
Over 200 to 300	1,00	0,03	30	3	
Over 300 to 600	1,00	0,05	40	7	
Over 600 to 1 200	2,00	0,09	50	15	
Over 1 200 to 1 800	2,00	0,11	75	15	
Over 1 800 to 2 400	2,00	0,13	75	15	
Over 2 400	2,00	0,21	75	15	

TABLE 8.2
Domestic Turkeys

All enclosure sides shall be at least 1,5 m long. Where these minimum enclosures sizes cannot be provided for scientific reasons, the duration of the confinement shall be justified by the experimenter in consultation with veterinary staff. In such circumstances, birds can be housed in smaller enclosures containing appropriate enrichment and with a minimum floor area of 0,75 m² and a minimum height of 50 cm for birds below 0,6 kg, 75 cm for birds below 4 kg, and 100 cm for birds over 4 kg. These can be used to house small groups of birds in accordance with the space allowances given in Table 8.2.

Body Mass (kg)	Minimum Enclosure Size (m²)	Minimum Area per Bird (m²)	Minimum Height (cm)	Minimum Length of Feed Trough per Bird (cm)	Date Referred to in Article 33(2)
Up to 0,3	2,00	0,13	50	3	1 January 2017
Over 0,3 to 0,6	2,00	0,17	50	7	
Over 0,6 to 1	2,00	0,30	100	15	
Over 1 to 4	2,00	0,35	100	15	
Over 4 to 8	2,00	0,40	100	15	
Over 8 to 12	2,00	0,50	150	20	
Over 12 to 16	2,00	0,55	150	20	
Over 16 to 20	2,00	0,60	150	20	
Over 20	3,00	1,00	150	20	

TABLE 8.3
Quails

Body Mass (g)	Minimum Enclosure Size (m²)	Area per Bird Pair-Housed (m²)	Area per Additional Bird Group-Housed (m²)	Minimum Height (cm)	Minimum Length of Trough per Bird (cm)	Date Referred to in Article 33(2)
Up to 150	1,00	0,5	0,10	20	4	1 January 2017
Over 150	1,00	0,6	0,15	30	4	

TABLE 8.4
Ducks and Geese

Where these minimum enclosures sizes cannot be provided for scientific reasons, the duration of the confinement shall be justified by the experimenter in consultation with veterinary staff. In such circumstances, birds can be housed in smaller enclosures containing appropriate enrichment and with a minimum floor area of 0,75 m². These can be used to house small groups of birds in accordance with the space allowances given in Table 8.4.

Body Mass (g)	Minimum Enclosure Size (m²)	Area per Bird (m²) (23)	Minimum Height (cm)	Minimum Length of Feed Trough per Bird (cm)	Date Referred to in Article 33(2)
Ducks					1 January 2017
Up to 300	2,00	0,10	50	10	
Over 300 to 1 200 (24)	2,00	0,20	200	10	
Over 1 200 to 3 500	2,00	0,25	200	15	
Over 3 500	2,00	0,50	200	15	
Geese					
Up to 500	2,00	0,20	200	10	
Over 500 to 2 000	2,00	0,33	200	15	
Over 2 000	2,00	0,50	200	15	

TABLE 8.5
Ducks and Geese: Minimum Pond Sizes (25)

	Area (m²)	Depth (cm)
Ducks	0,5	30
Geese	0,5	From 10 to 30

TABLE 8.6
Pigeons

Enclosures shall be long and narrow (for example 2 m by 1 m) rather than square to allow birds to perform short flights.

Group Size	Minimum Enclosure Size (m²)	Minimum Height (cm)	Minimum Length of Food Trough per Bird (cm)	Minimum Length of Perch per Bird (cm)	Date Referred to in Article 33(2)
Up to 6	2	200	5	30	1 January 2017
From 7 to 12	3	200	5	30	
For each additional bird above 12	0,15		5	30	

TABLE 8.7
Zebra Finches

Enclosures shall be long and narrow (for example 2 m by 1 m) to enable birds to perform short flights. For breeding studies, pairs may be housed in smaller enclosures containing appropriate enrichment with a minimum floor area of 0,5 m² and a minimum height of 40 cm. The duration of the confinement shall be justified by the experimenter in consultation with veterinary staff.

Group Size	Minimum Enclosure Size (m²)	Minimum Height (cm)	Minimum Number of Feeders	Date Referred to in Article 33(2)
Up to 6	1,0	100	2	1 January 2017
7 to 12	1,5	200	2	
13 to 20	2,0	200	3	
For each additional bird above 20	0,05		1 per 6 birds	

9   AMPHIBIANS

TABLE 9.1
Aquatic Urodeles

Body Length (26) (cm)	Minimum Water Surface Area (cm²)	Minimum Water Surface Area for Each Additional Animal in Group-Holding (cm²)	Minimum Water Depth (cm)	Date Referred to in Article 33(2)
Up to 10	262,5	50	13	1 January 2017
Over 10 to 15	525	110	13	
Over 15 to 20	875	200	15	
Over 20 to 30	1 837,5	440	15	
Over 30	3 150	800	20	

TABLE 9.2
Aquatic Anurans (27)

Body Length (28) (cm)	Minimum Water Surface Area (cm²)	Minimum Water Surface Area for Each Additional Animal in Group-Holding (cm²)	Minimum Water Depth (cm)	Date Referred to in Article 33(2)
Less than 6	160	40	6	1 January 2017
From 6 to 9	300	75	8	
Over 9 to 12	600	150	10	
Over 12	920	230	12,5	

TABLE 9.3
Semi-Aquatic Anurans

Body Length (29) (cm)	Minimum Enclosure Size (30) (cm²)	Minimum Area for Each Additional Animal in Group Holding (cm²)	Minimum Enclosure Height (31) (cm)	Minimum Water Depth (cm)	Date Referred to in Article 33(2)
Up to 5,0	1500	200	20	10	1 January
Over 5,0 to 7,5	3500	500	30	10	2017
Over 7,5	4000	700	30	15	

TABLE 9.4
Semi-Terrestrial Anurans

Body Length (32) (cm)	Minimum Enclosure Size (33) (cm²)	Minimum Area for Each Additional Animal in Group-Holding (cm²)	Minimum Enclosure Height (34) (cm)	Minimum Water Depth (cm)	Date Referred to in Article 33(2)
Up to 5,0	1500	200	20	10	1 January 2017
Over 5,0 to 7,5	3500	500	30	10	
Over 7,5	4000	700	30	15	

TABLE 9.5
Arboreal Anurans

Body Length (35) (cm)	Minimum Enclosure Size (36) (cm²)	Minimum Area for Each Additional Animal in Group-Holding (cm²)	Minimum Enclosure Height (37) (cm)	Date Referred to in Article 33(2)
Up to 3,0	900	100	30	1 January 2017
Over 3,0	1500	200	30	

10   REPTILES

TABLE 10.1
Aquatic Chelonians

Body Length (38) (cm)	Minimum Water Surface Area (cm²)	Minimum Water Surface Area for Each Additional Animal in Group Holding (cm²)	Minimum Water Depth (cm)	Date Referred to in Article 33(2)
Up to 5	600	100	10	1 January 2017
Over 5 to 10	1600	300	15	
Over 10 to 15	3500	600	20	
Over 15 to 20	6000	1200	30	
Over 20 to 30	10000	2000	35	
Over 30	20000	5000	40	

TABLE 10.2
Terrestrial Snakes

Body Length (39) (cm)	Minimum Floor Area (cm²)	Minimum Area for Each Additional Animal in Group-Holding (cm²)	Minimum Enclosure Height (40) (cm)	Date Referred to in Article 33(2)
Up to 30	300	150	10	1 January 2017
Over 30 to 40	400	200	12	
Over 40 to 50	600	300	15	
Over 50 to 75	1200	600	20	
Over 75	2500	1200	28	

11   FISH

11.1   WATER SUPPLY AND QUALITY

Adequate water supply of suitable quality shall be provided at all times. Water flow in re-circulatory systems or filtration within tanks shall be sufficient to ensure that water quality parameters are maintained within acceptable levels. Water supply shall be filtered or treated to remove substances harmful to fish, where necessary. Water-quality parameters shall at all times be within the acceptable range that sustains normal activity and physiology for a given species and stage of development. The water flow shall be appropriate to enable fish to swim correctly and to maintain normal behaviour. Fish shall be given an appropriate time for acclimatisation and adaptation to changes in water-quality conditions.

11.2   Oxygen, Nitrogen Compounds, pH, and Salinity

Oxygen concentration shall be appropriate to the species and to the context in which the fish are held. Where necessary, supplementary aeration of tank water shall be provided. The concentrations of nitrogen compounds shall be kept low.

The pH level shall be adapted to the species and kept as stable as possible. The salinity shall be adapted to the requirements of the fish species and to the life stage of the fish. Changes in salinity shall take place gradually.

11.3   Temperature, Lighting, Noise

Temperature shall be maintained within the optimal range for the fish species concerned and kept as stable as possible. Changes in temperature shall take place gradually. Fish shall be maintained on an appropriate photoperiod. Noise levels shall be kept to a minimum and, where possible, equipment causing noise or vibration, such as power generators or filtration systems, shall be separate from the fish-holding tanks.

11.4   Stocking Density and Environmental Complexity

The stocking density of fish shall be based on the total needs of the fish in respect of environmental conditions, health and welfare. Fish shall have sufficient water volume for normal swimming, taking account of their size, age, health and feeding method. Fish shall be provided with an appropriate environmental enrichment, such as hiding places or bottom substrate, unless behavioural traits suggest none is required.

11.5   Feeding and Handling

Fish shall be fed a diet suitable for the fish at an appropriate feeding rate and frequency. Particular attention shall be given to feeding of larval fish during any transition from live to artificial diets. Handling of fish shall be kept to a minimum.

(1) Post-weaned mice may be kept at these higher stocking densities for the short period after weaning until issue, provided that the animals are housed in larger enclosures with adequate enrichment, and these housing conditions do not cause any welfare deficit such as increased levels of aggression, morbidity or mortality, stereotypes and other behavioural deficits, weight loss, or other physiological or behavioural stress responses.

(2) In long-term studies, if space allowances per individual animal fall below those indicated above towards the end of such studies, priority shall be given to maintaining stable social structures.

(3) Post-weaned rats may be kept at these higher stocking densities for the short period after weaning until issue, provided that the animals are housed in larger enclosures with adequate enrichment, and these housing conditions do not cause any welfare deficit such as increased levels of aggression, morbidity or mortality, stereotypes and other behavioural deficits, weight loss, or other physiological or behavioural stress responses.

(4) Post-weaned hamsters may be kept at these higher stocking densities, for the short period after weaning until issue provided that the animals are housed

in larger enclosures with adequate enrichment, and these housing conditions do not cause any welfare deficit such as increased levels of aggression, morbidity or mortality, stereotypes and other behavioural deficits, weight loss, or other physiological or behavioural stress responses.

(5) Council Directive 98/58/EC of 20 July 1998 concerning the protection of animals kept for farming purposes (OJL221,8.8.1998,p.23).

(6) Floor area excluding shelves.

(7) Animals shall be kept singly only in exceptional circumstances.

(8) The top of the enclosure shall be at least 1,8 m from the floor.

(9) Animals shall be kept singly only in exceptional circumstances.

(10) Animals shall be kept singly only in exceptional circumstances.

(11) An enclosure of minimum dimensions may hold up to three animals.

(12) An enclosure of minimum dimensions may hold up to two animals.

(13) In breeding colonies no additional space/volume allowance is required for young animals up to 2 years of age housed with their mother.

(14) Animals shall be kept singly only in exceptional circumstances.

(15) An enclosure of minimum dimensions may hold up to 2 animals.

(16) In breeding colonies no additional space/volume allowance is required for young animals up to 2 years of age housed with their mothers.

(17) Council Directive 91/629/EEC of 19 November 1991 laying down minimum standards for the protection of calves (OJL340,11.12.1991,p.28).

(18) Council Directive 91/630/EEC of 19 November 1991 laying down minimum standards for the protection of pigs (OJL340,11.12.1991,p.33).

(19) Pigs may be confined in smaller enclosures for short periods of time, for example by partitioning the main enclosure using dividers, when justified on veterinary or experimental grounds, for example where individual food consumption is required.

(20) To ensure adequate space is provided, space allowances for each individual animal shall be based on height to withers (WH).

(21) Council Directive 1999/74/EC of 19 July 1999 laying down minimum standards for the protection of laying hens (OJL203,3.8.1999,p.53).

(22) Council Directive 2007/43/EC of 28 June 2007 laying down minimum rules for the protection of chickens kept for meat production (OJL182,12.7.2007,p.19).

(23) This shall include a pond of minimum area 0,5 m^2 per 2 m^2 enclosure with a minimum depth of 30 cm. The pond may contribute up to 50% of the minimum enclosure size.

(24) Pre-fledged birds may be held in enclosures with a minimum height of 75 cm.

(25) Pond sizes are per 2 m^2 enclosure. The pond may contribute up to 50% of the minimum enclosure size.

(26) Measured from snout to vent.

(27) These conditions apply to holding (i.e. husbandry) tanks but not to those tanks used for natural mating and super-ovulation for reasons of efficiency, as the latter procedures require smaller individual tanks. Space requirements determined for adults in the indicated size categories; juveniles and tadpoles shall either be excluded, or dimensions altered according to the scaling principle.

(28) Measured from snout to vent.
(29) Measured from snout to vent.
(30) One-third land division, two-thirds water division sufficient for animals to submerge.
(31) Measured from the surface of the land division up to the inner part of the top of the terrarium; furthermore, the height of the enclosures shall be adapted to the interior design.
(32) Measured from snout to vent.
(33) Two-thirds land division, one-third water division sufficient for animals to submerge.
(34) Measured from the surface of the land division up to the inner part of the top of the terrarium; furthermore, the height of the enclosures shall be adapted to the interior design.
(35) Measured from snout to vent.
(36) Two-thirds land division, one-third pool division sufficient for animals to submerge.
(37) Measured from the surface of the land division up to the inner part of the top of the terrarium; furthermore, the height of the enclosures shall be adapted to the interior design.
(38) Measured in a straight line from the front edge to the back edge of the shell.
(39) Measured from snout to tail.
(40) Measured from the surface of the land division up to the inner part of the top of the terrarium; furthermore, the height of the enclosure shall be adapted to the interior design.

Annex IV

Methods of Killing Animals

1. In the process of killing animals, methods listed in the table below shall be used.
   Methods other than those listed in the table may be used:
   (a) on unconscious animals, providing the animal does not regain consciousness before death;
   (b) on animals used in agricultural research, when the aim of the project requires that the animals are kept under similar conditions to those under which commercial farm animals are kept; these animals may be killed in accordance with the requirements laid down in Annex I to Council Regulation (EC) No 1099/2009 of 24 September 2009 on the protection of animals at the time of killing(1).
2. The killing of animals shall be completed by one of the following methods:
   (a) confirmation of permanent cessation of the circulation;
   (b) destruction of the brain;
   (c) dislocation of the neck;
   (d) exsanguination; or
   (e) confirmation of the onset of *rigor mortis*.

3. Table

Animals-Remarks/Methods	Fish	Amphibians	Reptiles	Birds	Rodents	Rabbits	Dogs, Cats, Ferrets and Foxes	Large Mammals	Non-Human Primates
Anaesthetic overdose	(1)	(1)	(1)	(1)	(1)	(1)	(1)	(1)	(1)
Captive bolt			(2)						
Carbon dioxide					(3)				
Cervical dislocation				(4)	(5)	(6)			
Concussion/percussive blow to the head				(7)	(8)	(9)	(10)		
Decapitation				(11)	(12)				
Electrical stunning	(13)	(13)		(13)		(13)	(13)	(13)	
Inert gases (Ar, N$_2$)								(14)	
Shooting with a free bullet with appropriate rifles, guns and ammunition			(15)				(16)	(15)	

Requirements

1. Shall, where appropriate, be used with prior sedation.
2. Only to be used on large reptiles.
3. Only to be used in gradual fill. Not to be used for foetal and neonate rodents.
4. Only to be used for birds under 1 kg. Birds over 250 g shall be sedated.
5. Only to be used for rodents under 1 kg. Rodents over 150 g shall be sedated.
6. Only to be used for rabbits under 1 kg. Rabbits over 150 g shall be sedated.
7. Only to be used for birds under 5 kg.
8. Only to be used for rodents under 1 kg.
9. Only to be used for rabbits under 5 kg.
10. Only to be used on neonates.
11. Only to be used for birds under 250 g.
12. Only to be used if other methods are not possible.
13. Specialised equipment required.
14. Only to be used on pigs.
15. Only to be used in field conditions by experienced marksmen.
16. Only to be used in field conditions by experienced marksmen when other methods are not possible.

Annex V
List of Elements Referred to in Article 23(3)

1. National legislation in force relevant to the acquisition, husbandry, care and use of animals for scientific purposes.
2. Ethics in relation to human-animal relationship, intrinsic value of life and arguments for and against the use of animals for scientific purposes.
3. Basic and appropriate species-specific biology in relation to anatomy, physiological features, breeding, genetics and genetic alteration.
4. Animal behaviour, husbandry and enrichment.
5. Species-specific methods of handling and procedures, where appropriate.
6. Animal health management and hygiene.
7. Recognition of species-specific distress, pain and suffering of most common laboratory species.
8. Anaesthesia, pain relieving methods and killing.
9. Use of humane end-points.
10. Requirement of replacement, reduction and refinement.
11. Design of procedures and projects, where appropriate.

Annex VI
List of Elements Referred to in Article 37(1)(c)

1. Relevance and justification of the following:
   (a) use of animals including their origin, estimated numbers, species and life stages;
   (b) procedures.

2. Application of methods to replace, reduce and refine the use of animals in procedures.
3. The planned use of anaesthesia, analgesia and other pain relieving methods.
4. Reduction, avoidance and alleviation of any form of animal suffering, from birth to death where appropriate.
5. Use of humane end-points.
6. Experimental or observational strategy and statistical design to minimise animal numbers, pain, suffering, distress and environmental impact where appropriate.
7. Reuse of animals and the accumulative effect thereof on the animals.
8. The proposed severity classification of procedures.
9. Avoidance of unjustified duplication of procedures where appropriate.
10. Housing, husbandry and care conditions for the animals.
11. Methods of killing.
12. Competence of persons involved in the project.

Annex VII

Duties and Tasks of the Union Reference Laboratory

1. The Union Reference Laboratory referred to in Article 48 is the Commission's Joint Research Centre.
2. The Union Reference Laboratory shall be responsible, in particular, for:
    (a) coordinating and promoting the development and use of alternatives to procedures including in the areas of basic and applied research and regulatory testing;
    (b) coordinating the validation of alternative approaches at Union level;
    (c) acting as a focal point for the exchange of information on the development of alternative approaches;
    (d) setting up, maintaining and managing public databases and information systems on alternative approaches and their state of development;
    (e) promoting dialogue between legislators, regulators, and all relevant stakeholders, in particular, industry, biomedical scientists, consumer organisations and animal-welfare groups, with a view to the development, validation, regulatory acceptance, international recognition, and application of alternative approaches.
3. The Union Reference Laboratory shall participate in the validation of alternative approaches.

Annex VIII

Severity Classification of Procedures

The severity of a procedure shall be determined by the degree of pain, suffering, distress or lasting harm expected to be experienced by an individual animal during the course of the procedure.

SECTION I: SEVERITY CATEGORIES

Non-Recovery

Procedures which are performed entirely under general anaesthesia from which the animal shall not recover consciousness shall be classified as 'non-recovery'.

Mild

Procedures on animals as a result of which the animals are likely to experience short-term mild pain, suffering or distress, as well as procedures with no significant impairment of the well-being or general condition of the animals shall be classified as 'mild'.

Moderate

Procedures on animals as a result of which the animals are likely to experience short-term moderate pain, suffering or distress, or long-lasting mild pain, suffering or distress as well as procedures that are likely to cause moderate impairment of the well-being or general condition of the animals shall be classified as 'moderate'.

Severe

Procedures on animals as a result of which the animals are likely to experience severe pain, suffering or distress, or long-lasting moderate pain, suffering or distress as well as procedures, that are likely to cause severe impairment of the well-being or general condition of the animals shall be classified as 'severe'.

SECTION II: ASSIGNMENT CRITERIA

The assignment of the severity category shall take into account any intervention or manipulation of an animal within a defined procedure. It shall be based on the most severe effects likely to be experienced by an individual animal after applying all appropriate refinement techniques.

When assigning a procedure to a particular category, the type of procedure and a number of other factors shall be taken into account. All these factors shall be considered on a case-by-case basis.

The factors related to the procedure shall include:

- type of manipulation, handling,
- nature of pain, suffering, distress or lasting harm caused by (all elements of) the procedure, and its intensity, the duration, frequency and multiplicity of techniques employed,
- cumulative suffering within a procedure,
- prevention from expressing natural behaviour including restrictions on the housing, husbandry and care standards.

Examples are given in Section III of procedures assigned to each of the severity categories on the basis of factors related to the type of the procedure alone. They shall provide the first indication as to what classification would be the most appropriate for a certain type of procedure.

However, for the purposes of the final severity classification of the procedure, the following additional factors, assessed on a case-by-case basis, shall also be taken into account:

- type of species and genotype,
- maturity, age and gender of the animal,
- training experience of the animal with respect to the procedure,
- if the animal is to be reused, the actual severity of the previous procedures,
- the methods used to reduce or eliminate pain, suffering and distress, including refinement of housing, husbandry and care conditions,
- humane end-points.

Section III

Examples of different types of procedure assigned to each of the severity categories on the basis of factors related to the type of the procedure

1. Mild

(a) administration of anaesthesia except for the sole purpose of killing;

(b) pharmacokinetic study where a single dose is administered and a limited number of blood samples are taken (totalling < 10% of circulating volume) and the substance is not expected to cause any detectable adverse effect;

(c) non-invasive imaging of animals (e.g. MRI) with appropriate sedation or anaesthesia;

(d) superficial procedures, e.g. ear and tail biopsies, non-surgical subcutaneous implantation of mini-pumps and transponders;

(e) application of external telemetry devices that cause only minor impairment to the animals or minor interference with normal activity and behaviour;

(f) administration of substances by subcutaneous, intramuscular, intraperitoneal routes, gavage and intravenously via superficial blood vessels, where the substance has no more than mild impact on the animal, and the volumes are within appropriate limits for the size and species of the animal;

(g) induction of tumours, or spontaneous tumours, that cause no detectable clinical adverse effects (e.g. small, subcutaneous, non-invasive nodules);

(h) breeding of genetically altered animals, which is expected to result in a phenotype with mild effects;

(i) feeding of modified diets, that do not meet all of the animals' nutritional needs and are expected to cause mild clinical abnormality within the timescale of the study;

(j) short-term (<24 h) restraint in metabolic cages;

(k) studies involving short-term deprivation of social partners, short-term solitary caging of adult rats or mice of sociable strains;

(l)   models which expose animals to noxious stimuli which are briefly associated with mild pain, suffering or distress, and which the animals can successfully avoid;
(m) a combination or accumulation of the following examples may result in classification as 'mild':
   (i)   assessing body composition by non-invasive measures and with minimal restraint;
   (ii)   monitoring ECG with non-invasive techniques with minimal or no restraint of habituated animals;
   (iii) application of external telemetry devices that are expected to cause no impairment to socially adapted animals and do not interfere with normal activity and behaviour;
   (iv) breeding genetically altered animals which are expected to have no clinically detectable adverse phenotype;
   (v)   adding inert markers in the diet to follow passage of digesta;
   (vi) withdrawal of food for <24 h in adult rats;
   (vii) open field testing.

2. Moderate

(a) frequent application of test substances which produce moderate clinical effects, and withdrawal of blood samples (>10% of circulating volume) in a conscious animal within a few days without volume replacement;
(b) acute dose-range finding studies, chronic toxicity/carcinogenicity tests, with non-lethal end-points;
(c) surgery under general anaesthesia and appropriate analgesia, associated with post surgical pain, suffering or impairment of general condition. Examples include: thoracotomy, craniotomy, laparotomy, orchidectomy, lymphadenectomy, thyroidectomy, orthopaedic surgery with effective stabilisation and wound management, organ transplantation with effective management of rejection, surgical implantation of catheters, or biomedical devices (e.g. telemetry transmitters, minipumps etc.);
(d) models of induction of tumours, or spontaneous tumours, that are expected to cause moderate pain or distress or moderate interference with normal behaviour;
(e) irradiation or chemotherapy with a sublethal dose, or with an otherwise lethal dose but with reconstitution of the immune system. Adverse effects would be expected to be mild or moderate and would be short-lived (<5 days);
(f) breeding of genetically altered animals which are expected to result in a phenotype with moderate effects;
(g) creation of genetically altered animals through surgical procedures;
(h) use of metabolic cages involving moderate restriction of movement over a prolonged period (up to 5 days);

(i) studies with modified diets that do not meet all of the animals' nutritional needs and are expected to cause moderate clinical abnormality within the time-scale of the study;
(j) withdrawal of food for 48 hours in adult rats;
(k) evoking escape and avoidance reactions where the animal is unable to escape or avoid the stimulus, and are expected to result in moderate distress.

3. Severe

(a) toxicity testing where death is the end-point, or fatalities are to be expected and severe pathophysiological states are induced. For example, single dose acute toxicity testing (see OECD testing guidelines);
(b) testing of device where failure may cause severe pain, distress or death of the animal (e.g. cardiac assist devices);
(c) vaccine potency testing characterised by persistent impairment of the animal's condition, progressive disease leading to death, associated with long-lasting moderate pain, distress or suffering;
(d) irradiation or chemotherapy with a lethal dose without reconstitution of the immune system, or reconstitution with production of graft versus host disease;
(e) models with induction of tumours, or with spontaneous tumours, that are expected to cause progressive lethal disease associated with long-lasting moderate pain, distress or suffering. For example tumours causing cachexia, invasive bone tumours, tumours resulting in metastatic spread, and tumours that are allowed to ulcerate;
(f) surgical and other interventions in animals under general anaesthesia which are expected to result in severe or persistent moderate postoperative pain, suffering or distress or severe and persistent impairment of the general condition of the animals. Production of unstable fractures, thoracotomy without adequate analgesia, or trauma to produce multiple organ failure;
(g) organ transplantation where organ rejection is likely to lead to severe distress or impairment of the general condition of the animals (e.g. xenotransplantation);
(h) breeding animals with genetic disorders that are expected to experience severe and persistent impairment of general condition, for example Huntington's disease, Muscular dystrophy, chronic relapsing neuritis models;
(i) use of metabolic cages involving severe restriction of movement over a prolonged period;
(j) inescapable electric shock (e.g. to produce learned helplessness);
(k) complete isolation for prolonged periods of social species e.g. dogs and non-human primates;
(l) immobilisation stress to induce gastric ulcers or cardiac failure in rats;
(m) forced swim or exercise tests with exhaustion as the end-point.

Appendix D: Guidance for Industry Bioanalytical Method Validation by the Food and Drug Administration

I. INTRODUCTION

This guidance provides assistance to sponsors of investigational new drug applications (INDs), new drug applications (NDAs), abbreviated new drug applications (ANDAs), and supplements in developing bioanalytical method validation information used in human clinical pharmacology, bioavailability (BA), and bioequivalence (BE) studies requiring pharmacokinetic (PK) evaluation. This guidance also applies to bioanalytical methods used for non-human pharmacology/toxicology studies and preclinical studies. For studies related to the veterinary drug approval process, this guidance applies only to blood and urine BA, BE, and PK studies.

The information in this guidance generally applies to bioanalytical procedures such as gas chromatography (GC), high-pressure liquid chromatography (LC), combined GC and LC mass spectrometric (MS) procedures such as LC-MS, LC-MS-MS, GC-MS, and GC-MS-MS performed for the quantitative determination of drugs and/or metabolites in biological matrices such as blood, serum, plasma, or urine. This guidance also applies to other bioanalytical methods, such as immunological and microbiological procedures, and to other biological matrices, such as tissue and skin samples.

This guidance provides general recommendations for bioanalytical method validation. The recommendations can be adjusted or modified depending on the specific type of analytical method used.

II. BACKGROUND

This guidance has been developed based on the deliberations of two workshops: (1) Analytical Methods Validation: Bioavailability, Bioequivalence, and Pharmacokinetic Studies (held on December 3-5, 1990) and (2) Bioanalytical Method Validation: A Revisit With a Decade of Progress (held on January 12-14, 2003).

Selective and sensitive analytical methods for the quantitative evaluation of drugs and their metabolites (analytes) are critical for the successful conduct of preclinical and/or biopharmaceutics and clinical pharmacology studies. Bioanalytical method validation includes all of the procedures that demonstrate that a particular method used for quantitative measurement of analytes in a given biological matrix, such as

247

blood, plasma, serum, or urine, is reliable and reproducible for the intended use. The fundamental parameters for this validation include (1) accuracy, (2) precision, (3) selectivity, (4) sensitivity, (5) reproducibility, and (6) stability. Validation involves documenting, through the use of specific laboratory investigations, that the performance characteristics of the method are suitable and reliable for the intended analytical applications. The acceptability of analytical data corresponds directly to the criteria used to validate the method.

Published methods of analysis are often modified to suit the requirements of the laboratory performing the assay. These modifications should be validated to ensure suitable performance of the analytical method. When changes are made to a previously validated method, the analyst should exercise judgment as to how much additional validation is needed. During the course of a typical drug development program, a defined bioanalytical method undergoes many modifications. The evolutionary changes to support specific studies and different levels of validation demonstrate the validity of an assay's performance. Different types and levels of validation are defined and characterized as follows:

A. FULL VALIDATION

- Full validation is important when developing and implementing a bioanalytical method for the first time.
- Full validation is important for a new drug entity.
- A full validation of the revised assay is important if metabolites are added to an existing assay for quantification.

B. PARTIAL VALIDATION

Partial validations are modifications of already validated bioanalytical methods. Partial validation can range from as little as one intra-assay accuracy and precision determination to a nearly full validation. Typical bioanalytical method changes that fall into this category include, but are not limited to:

- Bioanalytical method transfers between laboratories or analysts
- Change in analytical methodology (e.g., change in detection systems)
- Change in anticoagulant in harvesting biological fluid
- Change in matrix within species (e.g., human plasma to human urine)
- Change in sample processing procedures
- Change in species within matrix (e.g., rat plasma to mouse plasma)
- Change in relevant concentration range
- Changes in instruments and/or software platforms
- Limited sample volume (e.g., pediatric study)
- Rare matrices
- Selectivity demonstration of an analyte in the presence of concomitant medications
- Selectivity demonstration of an analyte in the presence of specific metabolites

C. CROSS-VALIDATION

Cross-validation is a comparison of validation parameters when two or more bio-analytical methods are used to generate data within the same study or across different studies. An example of cross-validation would be a situation where an original validated bioanalytical method serves as the reference and the revised bioanalytical method is the comparator. The comparisons should be done both ways.

When sample analyses within a single study are conducted at more than one site or more than one laboratory, cross-validation with spiked matrix standards and subject samples should be conducted at each site or laboratory to establish interlaboratory reliability. Cross-validation should also be considered when data generated using different analytical techniques (e.g., LC-MS-MS vs. ELISA[4]) in different studies are included in a regulatory submission.

All modifications should be assessed to determine the recommended degree of validation. The analytical laboratory conducting pharmacology/toxicology and other preclinical studies for regulatory submissions should adhere to FDA's Good Laboratory Practices (GLPs)[5] (21 CFR part 58) and to sound principles of quality assurance throughout the testing process. The bioanalytical method for human BA, BE, PK, and drug interaction studies must meet the criteria in 21 CFR 320.29. The analytical laboratory should have a written set of standard operating procedures (SOPs) to ensure a complete system of quality control and assurance. The SOPs should cover all aspects of analysis from the time the sample is collected and reaches the laboratory until the results of the analysis are reported. The SOPs also should include record keeping, security and chain of sample custody (accountability systems that ensure integrity of test articles), sample preparation, and analytical tools such as methods, reagents, equipment, instrumentation, and procedures for quality control and verification of results.

The process by which a specific bioanalytical method is developed, validated, and used in routine sample analysis can be divided into (1) reference standard preparation, (2) bioanalytical method development and establishment of assay procedure, and (3) application of validated bioanalytical method to routine drug analysis and acceptance criteria for the analytical run and/or batch. These three processes are described in the following sections of this guidance.

III. REFERENCE STANDARD

Analysis of drugs and their metabolites in a biological matrix is carried out using samples spiked with calibration (reference) standards and using quality control (QC) samples. The purity of the reference standard used to prepare spiked samples can affect study data. For this reason, an authenticated analytical reference standard of known identity and purity should be used to prepare solutions of known concentrations. If possible, the reference standard should be identical to the analyte. When this is not possible, an established chemical form (free base or acid, salt or ester) of known purity can be used.

Three types of reference standards are usually used: (1) certified reference standards (e.g., USP compendial standards); (2) commercially supplied reference standards obtained from a reputable commercial source; and/or (3) other materials of documented purity custom-synthesized by an analytical laboratory or other non-commercial establishment. The source and lot number, expiration date, certificates of analyses when available, and/or internally or externally generated evidence of identity and purity should be furnished for each reference standard.

IV. METHOD DEVELOPMENT: CHEMICAL ASSAY

The method development and establishment phase defines the chemical assay. The fundamental parameters for a bioanalytical method validation are accuracy, precision, selectivity, sensitivity, reproducibility, and stability. Measurements for each analyte in the biological matrix should be validated. In addition, the stability of the analyte in spiked samples should be determined. Typical method development and establishment for a bioanalytical method include determination of (1) selectivity, (2) accuracy, precision, recovery, (3) calibration curve, and (4) stability of analyte in spiked samples.

A. SELECTIVITY

Selectivity is the ability of an analytical method to differentiate and quantify the analyte in the presence of other components in the sample. For selectivity, analyses of blank samples of the appropriate biological matrix (plasma, urine, or other matrix) should be obtained from at least six sources. Each blank sample should be tested for interference, and selectivity should be ensured at the lower limit of quantification (LLOQ).

Potential interfering substances in a biological matrix include endogenous matrix components, metabolites, decomposition products, and in the actual study, concomitant medication and other exogenous xenobiotics. If the method is intended to quantify more than one analyte, each analyte should be tested to ensure that there is no interference.

B. ACCURACY, PRECISION, AND RECOVERY

The accuracy of an analytical method describes the closeness of mean test results obtained by the method to the true value (concentration) of the analyte. Accuracy is determined by replicate analysis of samples containing known amounts of the analyte. Accuracy should be measured using a minimum of five determinations per concentration. A minimum of three concentrations in the range of expected concentrations is recommended. The mean value should be within 15% of the actual value except at LLOQ, where it should not deviate by more than 20%. The deviation of the mean from the true value serves as the measure of accuracy.

The precision of an analytical method describes the closeness of individual measures of an analyte when the procedure is applied repeatedly to multiple aliquots of a single homogeneous volume of biological matrix. Precision should be measured using a minimum of five determinations per concentration. A minimum of three concentrations in the range of expected concentrations is recommended. The precision determined at each concentration level should not exceed 15% of the coefficient of variation (CV) except for the LLOQ, where it should not exceed 20% of the CV. Precision is further subdivided into within-run, intra-batch precision or repeatability, which assesses precision during a single analytical run, and between-run, interbatch precision or repeatability, which measures precision with time, and may involve different analysts, equipment, reagents, and laboratories.

The recovery of an analyte in an assay is the detector response obtained from an amount of the analyte added to and extracted from the biological matrix, compared to the detector response obtained for the true concentration of the pure authentic standard. Recovery pertains to the extraction efficiency of an analytical method within the limits of variability. Recovery of the analyte need not be 100%, but the extent of recovery of an analyte and of the internal standard should be consistent, precise, and reproducible. Recovery experiments should be performed by comparing the analytical results for extracted samples at three concentrations (low, medium, and high) with unextracted standards that represent 100% recovery.

C.   CALIBRATION/STANDARD CURVE

A calibration (standard) curve is the relationship between instrument response and known concentrations of the analyte. A calibration curve should be generated for each analyte in the sample. A sufficient number of standards should be used to adequately define the relationship between concentration and response. A calibration curve should be prepared in the same biological matrix as the samples in the intended study by spiking the matrix with known concentrations of the analyte. The number of standards used in constructing a calibration curve will be a function of the anticipated range of analytical values and the nature of the analyte/response relationship. Concentrations of standards should be chosen on the basis of the concentration range expected in a particular study. A calibration curve should consist of a blank sample (matrix sample processed without internal standard), a zero sample (matrix sample processed with internal standard), and six to eight non-zero samples covering the expected range, including LLOQ.

1.   Lower Limit of Quantification (LLOQ)

The lowest standard on the calibration curve should be accepted as the limit of quantification if the following conditions are met:

- The analyte response at the LLOQ should be at least 5 times the response compared to blank response.
- Analyte peak (response) should be identifiable, discrete, and reproducible with a precision of 20% and accuracy of 80–120%.

2.   Calibration Curve/Standard Curve/Concentration-Response

The simplest model that adequately describes the concentration-response relationship should be used. Selection of weighting and use of a complex regression equation should be justified. The following conditions should be met in developing a calibration curve:

- 20% deviation of the LLOQ from nominal concentration
- 15% deviation of standards other than LLOQ from nominal concentration

At least four out of six non-zero standards should meet the above criteria, including the LLOQ and the calibration standard at the highest concentration. Excluding the standards should not change the model used.

D.   STABILITY

Drug stability in a biological fluid is a function of the storage conditions, the chemical properties of the drug, the matrix, and the container system. The stability of an analyte in a particular matrix and container system is relevant only to that matrix and container system and should not be extrapolated to other matrices and container systems. Stability procedures should evaluate the stability of the analytes during sample collection and handling, after long-term (frozen at the intended storage temperature) and short-term (bench top, room temperature) storage, and after going through freeze and thaw cycles and the analytical process. Conditions used in stability experiments should reflect situations likely to be encountered during actual sample handling and analysis. The procedure should also include an evaluation of analyte stability in stock solution.

All stability determinations should use a set of samples prepared from a freshly made stock solution of the analyte in the appropriate analyte-free, interference-free biological matrix. Stock solutions of the analyte for stability evaluation should be prepared in an appropriate solvent at known concentrations.

1.   Freeze and Thaw Stability

Analyte stability should be determined after three freeze and thaw cycles. At least three aliquots at each of the low and high concentrations should be stored at the intended storage temperature for 24 hours and thawed unassisted at room temperature. When completely thawed, the samples should be refrozen for 12 to 24 hours under the same conditions. The freeze–thaw cycle should be repeated two more times, then analyzed on the third cycle. If an analyte is unstable at the intended storage temperature, the stability sample should be frozen at $-70°C$.

2.   Short-Term Temperature Stability

Three aliquots of each of the low and high concentrations should be thawed at room temperature and kept at this temperature from 4 to 24 hours (based on the expected duration that samples will be maintained at room temperature in the intended study) and analyzed.

3.   Long-Term Stability

The storage time in a long-term stability evaluation should exceed the time between the date of first sample collection and the date of last sample analysis. Long-term stability should be determined by storing at least three aliquots of each of the low and high concentrations under the same conditions as the study samples. The volume of samples should be sufficient for analysis on three separate occasions. The concentrations of all the stability samples should be compared to the mean of back-calculated values for the standards at the appropriate concentrations from the first day of long-term stability testing.

4. Stock Solution Stability

The stability of stock solutions of drug and the internal standard should be evaluated at room temperature for at least 6 hours. If the stock solutions are refrigerated or frozen for the relevant period, the stability should be documented. After completion of the desired storage time, the stability should be tested by comparing the instrument response with that of freshly prepared solutions.

5. Post-Preparative Stability

The stability of processed samples, including the resident time in the autosampler, should be determined. The stability of the drug and the internal standard should be assessed over the anticipated run time for the batch size in validation samples by determining concentrations on the basis of original calibration standards.

Although the traditional approach of comparing analytical results for stored samples with those for freshly prepared samples has been referred to in this guidance, other statistical approaches based on confidence limits for evaluation of an analyte's stability in a biological matrix can be used. SOPs should clearly describe the statistical method and rules used. Additional validation may include investigation of samples from dosed subjects.

E. PRINCIPLES OF BIOANALYTICAL METHOD VALIDATION AND ESTABLISHMENT

The fundamental parameters to ensure the acceptability of the performance of a bioanalytical method validation are accuracy, precision, selectivity, sensitivity, reproducibility, and stability.

- A specific, detailed description of the bioanalytical method should be written. This can be in the form of a protocol, study plan, report, and/or SOP.
- Each step in the method should be investigated to determine the extent to which environmental, matrix, material, or procedural variables can affect the estimation of analyte in the matrix from the time of collection of the material up to and including the time of analysis.
- It may be important to consider the variability of the matrix due to the physiological nature of the sample. In the case of LC-MS-MS-based procedures, appropriate steps should be taken to ensure the lack of matrix effects throughout the application of the method, especially if the nature of the matrix changes from the matrix used during method validation.
- A bioanalytical method should be validated for the intended use or application. All experiments used to make claims or draw conclusions about the validity of the method should be presented in a report (method validation report).

- Whenever possible, the same biological matrix as the matrix in the intended samples should be used for validation purposes. (For tissues of limited availability, such as bone marrow, physiologically appropriate proxy matrices can be substituted.)
- The stability of the analyte (drug and/or metabolite) in the matrix during the collection process and the sample storage period should be assessed, preferably prior to sample analysis.
- For compounds with potentially labile metabolites, the stability of analyte in matrix from dosed subjects (or species) should be confirmed.
- The accuracy, precision, reproducibility, response function, and selectivity of the method for endogenous substances, metabolites, and known degradation products should be established for the biological matrix. For selectivity, there should be evidence that the substance being quantified is the intended analyte.
- The concentration range over which the analyte will be determined should be defined in the bioanalytical method, based on the evaluation of actual standard samples over the range, including their statistical variation. This defines the *standard curve*.
- A sufficient number of standards should be used to adequately define the relationship between concentration and response. The relationship between response and concentration should be demonstrated to be continuous and reproducible. The number of standards used should be a function of the dynamic range and nature of the concentration-response relationship. In many cases, six to eight concentrations (excluding blank values) can define the standard curve. More standard concentrations may be recommended for nonlinear than for linear relationships.
- The ability to dilute samples originally above the upper limit of the standard curve should be demonstrated by accuracy and precision parameters in the validation.
- When considering high-throughput analyses, including but not limited to multiplexing, multicolumn, and parallel systems, a sufficient number of QC samples should be used to ensure control of the assay. This number should be determined based on the run size. The placement of QC samples should be judiciously considered in the run.
- For a bioanalytical method to be considered valid, specific acceptance criteria should be set in advance and achieved for accuracy and precision for the validation of QC samples over the range of the standards.

F. Specific Recommendations for Method Validation

- The matrix-based standard curve should consist of a minimum of six standard points, excluding blanks, using single or replicate samples. The standard curve should cover the entire range of expected concentrations.
- Standard curve fitting is determined by applying the simplest model that adequately describes the concentration-response relationship using appropriate weighting and statistical tests for goodness of fit.

- LLOQ is the lowest concentration of the standard curve that can be measured with acceptable accuracy and precision. The LLOQ should be established using at least five samples independent of standards and determining the coefficient of variation and/or appropriate confidence interval. The LLOQ should serve as the lowest concentration on the standard curve and should not be confused with the limit of detection and/or the low QC sample. The highest standard will define the upper limit of quantification (ULOQ) of an analytical method.
- For validation of the bioanalytical method, accuracy and precision should be determined using a minimum of five determinations per concentration level (excluding blank samples). The mean value should be within ±15% of the theoretical value, except at LLOQ, where it should not deviate by more than ±20%. The precision around the mean value should not exceed 15% of the CV, except for LLOQ, where it should not exceed 20% of the CV. Other methods of assessing accuracy and precision that meet these limits may be equally acceptable.
- The accuracy and precision with which known concentrations of analyte in biological matrix can be determined should be demonstrated. This can be accomplished by analysis of replicate sets of analyte samples of known concentrations—QC samples—from an equivalent biological matrix. At a minimum, three concentrations representing the entire range of the standard curve should be studied: one within 3× the lower limit of quantification (LLOQ) (low QC sample), one near the center (middle QC), and one near the upper boundary of the standard curve (high QC).
- Reported method validation data and the determination of accuracy and precision should include all outliers; however, calculations of accuracy and precision excluding values that are statistically determined as outliers can also be reported.
- The stability of the analyte in biological matrix at intended storage temperatures should be established. The influence of freeze-thaw cycles (a minimum of three cycles at two concentrations in triplicate) should be studied. The stability of the analyte in matrix at ambient temperature should be evaluated over a time period equal to the typical sample preparation, sample handling, and analytical run times.
- Reinjection reproducibility should be evaluated to determine if an analytical run could be reanalyzed in the case of instrument failure.
- The specificity of the assay methodology should be established using a minimum of six independent sources of the same matrix. For hyphenated mass spectrometry-based methods, however, testing six independent matrices for interference may not be important. In the case of LC-MS and LC-MS-MS-based procedures, matrix effects should be investigated to ensure that precision, selectivity, and sensitivity will not be compromised. Method selectivity should be evaluated during method development and throughout method validation and can continue throughout application of the method to actual study samples.

- Acceptance/rejection criteria for spiked, matrix-based calibration standards and validation QC samples should be based on the nominal (theoretical) concentration of analytes. Specific criteria can be set up in advance and achieved for accuracy and precision over the range of the standards, if so desired.

V. METHOD DEVELOPMENT: MICROBIOLOGICAL AND LIGAND-BINDING ASSAYS

Many of the bioanalytical validation parameters and principles discussed above are also applicable to microbiological and ligand-binding assays. However, these assays possess some unique characteristics that should be considered during method validation.

A. SELECTIVITY ISSUES

As with chromatographic methods, microbiological and ligand-binding assays should be shown to be selective for the analyte. The following recommendations for dealing with two selectivity issues should be considered:

1. Interference from Substances Physiochemically Similar to the Analyte

- Cross-reactivity of metabolites, concomitant medications, or endogenous compounds should be evaluated individually and in combination with the analyte of interest.
- When possible, the immunoassay should be compared with a validated reference method (such as LC-MS) using incurred samples and predetermined criteria for agreement of accuracy of immunoassay and reference method.
- The dilutional linearity to the reference standard should be assessed using study (incurred) samples.
- Selectivity may be improved for some analytes by incorporation of separation steps prior to immunoassay.

2. Matrix Effects Unrelated to the Analyte

- The standard curve in biological fluids should be compared with standard in buffer to detect matrix effects.
- Parallelism of diluted study samples should be evaluated with diluted standards to detect matrix effects.
- Nonspecific binding should be determined.

B. QUANTIFICATION ISSUES

Microbiological and immunoassay standard curves are inherently nonlinear and, in general, more concentration points may be recommended to define the fit over the standard curve range than for chemical assays. In addition to their nonlinear characteristics, the response-error relationship for immunoassay standard curves is a nonconstant function of the mean response (heteroscedasticity). For these reasons, a minimum of six non-zero calibrator concentrations, run in duplicate, is recommended. The concentration-response relationship is most often fitted to a 4- or 5-parameter logistic model, although others may be used with suitable validation. The use of anchoring points in the asymptotic high- and low-concentration ends of the standard curve may improve the overall curve fit. Generally, these anchoring points will be at concentrations that are below the established LLOQ and above the established ULOQ.

Whenever possible, calibrators should be prepared in the same matrix as the study samples or in an alternate matrix of equivalent performance. Both ULOQ and LLOQ should be defined by acceptable accuracy, precision, or confidence interval criteria based on the study requirements.

For all assays the key factor is the accuracy of the reported results. This accuracy can be improved by the use of replicate samples. In the case where replicate samples should be measured during the validation to improve accuracy, the same procedure should be followed as for unknown samples.

The following recommendations apply to quantification issues:

- If separation is used prior to assay for study samples but not for standards, it is important to establish recovery and use it in determining results. Possible approaches to assess efficiency and reproducibility of recovery are (1) the use of radiolabeled tracer analyte (quantity too small to affect the assay), (2) the advance establishment of reproducible recovery, (3) the use of an internal standard that is not recognized by the antibody but can be measured by another technique.
- Key reagents, such as antibody, tracer, reference standard, and matrix should be characterized appropriately and stored under defined conditions.
- Assessments of analyte stability should be conducted in true study matrix (e.g., should not use a matrix stripped to remove endogenous interferences).
- Acceptance criteria: At least 67% (4 out of 6) of QC samples should be within 15% of their respective nominal value, 33% of the QC samples (not all replicates at the same concentration) may be outside 15% of nominal value. In certain situations, wider acceptance criteria may be justified.
- Assay reoptimization or validation may be important when there are changes in key reagents, as follows:

Labeled analyte (tracer)
- Binding should be reoptimized.
- Performance should be verified with standard curve and QCs.

Antibody
- Key cross-reactivities should be checked.
- Tracer experiments above should be repeated.

Matrix
- Tracer experiments above should be repeated.

Method development experiments should include a minimum of six runs conducted over several days, with at least four concentrations (LLOQ, low, medium, and high) analyzed in duplicate in each run.

VI. APPLICATION OF VALIDATED METHOD TO ROUTINE DRUG ANALYSIS

Assays of all samples of an analyte in a biological matrix should be completed within the time period for which stability data are available. In general, biological samples can be analyzed with a single determination without duplicate or replicate analysis if the assay method has acceptable variability as defined by validation data. This is true for procedures where precision and accuracy variabilities routinely fall within acceptable tolerance limits. For a difficult procedure with a labile analyte where high precision and accuracy specifications may be difficult to achieve, duplicate or even triplicate analyses can be performed for a better estimate of analyte.

A calibration curve should be generated for each analyte to assay samples in each analytical run and should be used to calculate the concentration of the analyte in the unknown samples in the run. The spiked samples can contain more than one analyte. An analytical run can consist of QC samples, calibration standards, and either (1) all the processed samples to be analyzed as one batch or (2) a batch composed of processed unknown samples of one or more volunteers in a study. The calibration (standard) curve should cover the expected unknown sample concentration range in addition to a calibrator sample at LLOQ. Estimation of concentration in unknown samples by extrapolation of standard curves below LLOQ or above the highest standard is not recommended. Instead, the standard curve should be redefined, or samples with higher concentration should be diluted and reassayed. It is preferable to analyze all study samples from a subject in a single run.

Once the analytical method has been validated for routine use, its accuracy and precision should be monitored regularly to ensure that the method continues to perform satisfactorily. To achieve this objective, a number of QC samples prepared separately should be analyzed with processed test samples at intervals based on the total number of samples. The QC samples in duplicate at three concentrations (one near the LLOQ (i.e., #3 × LLOQ), one in midrange, and one close to the high end of the range) should be incorporated in each assay run. The number of QC samples (in multiples of three) will depend on the total number of samples in the run. The results of the QC samples provide the basis of accepting or rejecting the run. At least four of every six QC samples should be within 15% of their respective nominal value. Two of the six QC samples may be outside the 15% of their respective nominal value, but not both at the same concentration.

The following recommendations should be noted in applying a bioanalytical method to routine drug analysis:

- A matrix-based standard curve should consist of a minimum of six standard points, excluding blanks (either single or replicate), covering the entire range.
- Response Function: Typically, the same curve fitting, weighting, and goodness of fit determined during prestudy validation should be used for the standard curve within the study. Response function is determined by appropriate statistical tests based on the actual standard points during each run

in the validation. Changes in the response function relationship between prestudy validation and routine run validation indicate potential problems.

- The QC samples should be used to accept or reject the run. These QC samples are matrix spiked with analyte.
- System suitability: Based on the analyte and technique, a specific SOP (or sample) should be identified to ensure optimum operation of the system used.
- Any required sample dilutions should use like matrix (e.g., human to human) obviating the need to incorporate actual within-study dilution matrix QC samples.
- Repeat Analysis: It is important to establish an SOP or guideline for repeat analysis and acceptance criteria. This SOP or guideline should explain the reasons for repeating sample analysis. Reasons for repeat analyses could include repeat analysis of clinical or preclinical samples for regulatory purposes, inconsistent replicate analysis, samples outside of the assay range, sample processing errors, equipment failure, poor chromatography, and inconsistent pharmacokinetic data. Reassays should be done in triplicate if sample volume allows. The rationale for the repeat analysis and the reporting of the repeat analysis should be clearly documented.
- Sample Data Reintegration: An SOP or guideline for sample data reintegration should be established. This SOP or guideline should explain the reasons for reintegration and how the reintegration is to be performed. The rationale for the reintegration should be clearly described and documented. Original and reintegration data should be reported.

ACCEPTANCE CRITERIA FOR THE RUN

The following acceptance criteria should be considered for accepting the analytical run:

- Standards and QC samples can be prepared from the same spiking stock solution, provided the solution stability and accuracy have been verified. A single source of matrix may also be used, provided selectivity has been verified.
- Standard curve samples, blanks, QCs, and study samples can be arranged as considered appropriate within the run.
- Placement of standards and QC samples within a run should be designed to detect assay drift over the run.
- Matrix-based standard calibration samples: 75%, or a minimum of six standards, when back-calculated (including ULOQ) should fall within ±15%, except for LLOQ, when it should be ±20% of the nominal value. Values falling outside these limits can be discarded, provided they do not change the established model.
- Acceptance criteria for accuracy and precision as outlined in section IV.F, "Specific Recommendation for Method Validation," should be provided for both the intra-day and intra-run experiment.

- Quality Control Samples: Quality control samples replicated (at least once) at a minimum of three concentrations (one within 3× of the LLOQ (low QC), one in the midrange (middle QC), and one approaching the high end of the range (high QC)) should be incorporated into each run. The results of the QC samples provide the basis of accepting or rejecting the run. At least 67% (four out of six) of the QC samples should be within 15% of their respective nominal (theoretical) values; 33% of the QC samples (not all replicates at the same concentration) can be outside the ±15% of the nominal value. A confidence interval approach yielding comparable accuracy and precision is an appropriate alternative.

The minimum number of samples (in multiples of three) should be at least 5% of the number of unknown samples or six total QCs, whichever is greater.

- Samples involving multiple analytes should not be rejected based on the data from one analyte failing the acceptance criteria.
- The data from rejected runs need not be documented, but the fact that a run was rejected and the reason for failure should be recorded.

VII. DOCUMENTATION

The validity of an analytical method should be established and verified by laboratory studies, and documentation of successful completion of such studies should be provided in the assay validation report. General and specific SOPs and good record keeping are an essential part of a validated analytical method. The data generated for bioanalytical method establishment and the QCs should be documented and available for data audit and inspection. Documentation for submission to the agency should include (1) summary information, (2) method development and establishment, (3) bioanalytical reports of the application of any methods to routine sample analysis, and (4) other information applicable to method development and establishment and/or to routine sample analysis.

A. SUMMARY INFORMATION

Summary information should include:

- Summary table of validation reports, including analytical method validation, partial revalidation, and cross-validation reports. The table should be in chronological sequence, and include assay method identification code, type of assay, and the reason for the new method or additional validation (e.g., to lower the limit of quantitation).
- Summary table with a list, by protocol, of assay methods used. The protocol number, protocol title, assay type, assay method identification code, and bioanalytical report code should be provided.

- A summary table allowing cross-referencing of multiple identification codes should be provided (e.g., when an assay has different codes for the assay method, validation reports, and bioanalytical reports, especially when the sponsor and a contract laboratory assign different codes).

B. Documentation for Method Establishment

Documentation for method development and establishment should include:

- An operational description of the analytical method
- Evidence of purity and identity of drug standards, metabolite standards, and internal standards used in validation experiments
- A description of stability studies and supporting data
- A description of experiments conducted to determine accuracy, precision, recovery, selectivity, limit of quantification, calibration curve (equations and weighting functions used, if any), and relevant data obtained from these studies
- Documentation of intra- and inter-assay precision and accuracy
- In NDA submissions, information about cross-validation study data, if applicable
- Legible annotated chromatograms or mass spectrograms, if applicable
- Any deviations from SOPs, protocols, or GLPs (if applicable), and justifica tions for deviations

C. Application to Routine Drug Analysis

Documentation of the application of validated bioanalytical methods to routine drug analysis should include:

- Evidence of purity and identity of drug standards, metabolite standards, and internal standards used during routine analyses
- Summary tables containing information on sample processing and storage. Tables should include sample identification, collection dates, storage prior to shipment, information on shipment batch, and storage prior to analysis. Information should include dates, times, sample condition, and any deviation from protocols.
- Summary tables of analytical runs of clinical or preclinical samples. Information should include assay run identification, date and time of analysis, assay method, analysts, start and stop times, duration, significant equipment and material changes, and any potential issues or deviation from the established method.
- Equations used for back-calculation of results
- Tables of calibration curve data used in analyzing samples and calibration curve summary data

- Summary information on intra- and inter-assay values of QC samples and data on intra- and inter-assay accuracy and precision from calibration curves and QC samples used for accepting the analytical run. QC graphs and trend analyses in addition to raw data and summary statistics are encouraged.
- Data tables from analytical runs of clinical or preclinical samples. Tables should include assay run identification, sample identification, raw data and back-calculated results, integration codes, and/or other reporting codes.
- Complete serial chromatograms from 5–20% of subjects, with standards and QC samples from those analytical runs. For pivotal bioequivalence studies for marketing, chromatograms from 20% of serially selected subjects should be included. In other studies, chromatograms from 5% of randomly selected subjects in each study should be included. Subjects whose chromatograms are to be submitted should be defined prior to the analysis of any clinical samples.
- Reasons for missing samples
- Documentation for repeat analyses. Documentation should include the initial and repeat analysis results, the reported result, assay run identification, the reason for the repeat analysis, the requestor of the repeat analysis, and the manager authorizing reanalysis. Repeat analysis of a clinical or preclinical sample should be performed only under a predefined SOP.
- Documentation for reintegrated data. Documentation should include the initial and repeat integration results, the method used for reintegration, the reported result, assay run identification, the reason for the reintegration, the requestor of the reintegration, and the manager authorizing reintegration. Reintegration of a clinical or preclinical sample should be performed only under a predefined SOP.
- Deviations from the analysis protocol or SOP, with reasons and justifications for the deviations.

D. OTHER INFORMATION

Other information applicable to both method development and establishment and/or to routine sample analysis could include:

- Lists of abbreviations and any additional codes used, including sample condition codes, integration codes, and reporting codes
- Reference lists and legible copies of any references
- SOPs or protocols covering the following areas:
  - Calibration standard acceptance or rejection criteria
  - Calibration curve acceptance or rejection criteria
  - Quality control sample and assay run acceptance or rejection criteria

- Acceptance criteria for reported values when all unknown samples are assayed in duplicate
- Sample code designations, including clinical or preclinical sample codes and bioassay sample code
- Assignment of clinical or preclinical samples to assay batches
- Sample collection, processing, and storage
- Repeat analyses of samples
- Reintegration of samples

Validation:

Full validation: Establishment of all validation parameters to apply to sample analysis for the bioanalytical method for each analyte.

Partial validation: Modification of validated bioanalytical methods that do not necessarily call for full revalidation.

Cross-validation: Comparison validation parameters of two bioanalytical methods.

Appendix E: Toxicity and IARC Classification of Targeted Constitutes

Smoke Constitute	Molecular Formula	Molecular Weight (g/mol)	Melting Point (°C)/Boiling Point (°C)	CAS Number	Toxicity	IARC Categories
NNN	$C_9H_{11}N_3O$	177.20	47/154	16543-55-8	Ld50: 1000 mg/kg (oral, rat) NNN causes cancer in experimental animals.	Group 2B
NNK	$C_{10}H_{13}N_3O_2$	207.23	72/unknown	10028-15-6	Studies showed that with a dose of NNK of 100 mg/kg, several point mutations were formed in the RAR-β gene and that induced tumorigenesis in the lungs.	Group 2B
Acetaldehyde	C_2H_4O	44.05	−123/20	75-07-0	LD50:1930 mg/kg (rat, oral) Acetaldehyde is toxic when applied externally for prolonged periods, an irritant, and a probable carcinogen.	Group 2B
Acrolein	C_3H_4O	56.06	−88/53	107-02-8	LD50:46 mg/kg (rat, oral) Acrolein is toxic and is a strong irritant for the skin, eyes, and nasal passages. The main metabolic pathway for acrolein is the alkylation of glutathione. The WHO suggests a *tolerable oral acrolein intake of 7.5 μg/day/kg of body weight.*	Group 2A
Formaldehyde	CH_2O	30.03	−92/−19	50-00-0	Ld50: 800 mg/kg (oral, rat) Ingestion of 30 mL (1 oz.) of a solution containing 37% formaldehyde has been determined to cause death in an adult human. Water solution of formaldehyde is very corrosive, and its ingestion can cause severe injury to the upper gastrointestinal tract.	Group 2A
Benzene	C_6H_6	78.11	6/80	71-43-2	LD50:930 mg/kg (rat, oral) Substantial quantities of epidemiologic, clinical, and laboratory data link benzene to aplastic anemia, acute leukemia, and bone marrow abnormalities. The specific hematologic malignancies that benzene is associated with include acute myeloid leukemia (AML), aplastic anemia, myelodysplastic syndrome (MDS), acute lymphoblastic leukemia (ALL), and chronic myeloid leukemia (CML).	Group 1

(Continued)

Smoke Constitute	Molecular Formula	Molecular Weight (g/mol)	Melting Point (°C)/Boiling Point (°C)	CAS Number	Toxicity	IARC Categories
1,3-Butadiene	C_4H_6	54.09	−108/−4	106-99-0	LD50:548 mg/kg (rat, oral) Acute exposure results in irritation of the mucous membranes. Higher levels can result in neurological effects such as blurred vision, fatigue, headache, and vertigo. Exposure to the skin can lead to frostbite. Long-term exposure has been associated with cardiovascular disease. There is a consistent association with leukemia and weaker association with other cancers.	Group 2A
Acrylonitrile	C_3H_3N	53.06	−84/77	107-13-1	LD50:80 mg/kg (rat, oral) Symptoms observed after the administration of a toxic dose of AN include cholinomimetic signs, tremors, convulsions, hypothermia, asphyxial signs, and death. Necropsy has demonstrated the presence of edema, hyperemia, and damage to the brain, liver, kidney, and lung, as well as hemorrhagic necrosis of the stomach and adrenal.	Group 2B
Carbon monoxide	CO	28.01	−205/−191	630-08-0	It combines with hemoglobin to produce carboxyhemoglobin, which could result in seizure, coma, and fatality. Exposures to carbon monoxide may cause significant damage to the heart and central nervous system, especially to the globus pallidus, [28] often with long-term chronic pathological conditions.	Group 3
4-Aminobiphenyl	$C_{12}H_{11}N$	169.22	53/299	92-67-1	LD50:2340 mg/kg (rats, oral)	Group 1
Cadmium	Cd	48	321/767	7440-43-9	LD50:10.3 mg/kg (rats, oral) Intake of cadmium through diet associates to higher risk of endometrial, breast, and prostate cancer. Cadmium exposure is also a risk factor associated with early atherosclerosis and hypertension, which can both lead to cardiovascular disease.	Group 1

(Continued)

Smoke Constitute	Molecular Formula	Molecular Weight (g/mol)	Melting Point (°C)/Boiling Point (°C)	CAS Number	Toxicity	IARC Categories
Catechol	$C_6H_6O_2$	110.1	105/246	120-80-9	LD50:300 mg/kg (rats, oral) Catechol produces a striking synergistic genotoxic response with hydroquinone response in cultured human lymphocytes. Catechol inhibits lymphocyte mitogenesis and agglutination, processes dependent on microtubule and sulfhydryl integrity.	Group 2B
Crotonaldehyde	C_4H_6O	70.09	−76/104	123-73-9	LD50:206 mg/kg (rats, oral) Crotonaldehyde is an irritant. It is listed as an *extremely hazardous substance* as defined by the U.S. Emergency Planning and Community Right-to-Know Act.	Group 3
Hydrogen cyanide	CHN	27.03	−13/26	3017-23-0	LC50:357 mg/m³ (mouse, 5 min) A hydrogen cyanide concentration of 300 mg/m³ in air will kill a human within 10–60 min. A hydrogen cyanide concentration of 3500 PPM (about 3200 mg/m³) will kill a human in about 1 min.	Group 2A
Hydroquinone	$C_6H_6O_2$	110.11	172/287	123-31-9	LD50:190 mg/kg (rats, oral) Numerous studies have revealed that hydroquinone can cause exogenous ochronosis, a disfiguring disease in which blue–black pigments are deposited onto the skin.	Group 2B

(Continued)

Smoke Constitute	Molecular Formula	Molecular Weight (g/mol)	Melting Point (°C)/Boiling Point (°C)	CAS Number	Toxicity	IARC Categories
2-Aminonaphthalene	$C_{10}H_9N$	143.19	112/306	91-59-8	LD50:779 mg/kg (rats, oral) Investigation of the in vitro acetylation of arylamines by human liver cytosol has indicated that the carcinogenic arylamines such as 2-aminonaphthalene display a strong affinity for this polymorphically distributed enzyme system, and moreover, liver cytosol from rapid acetylator phenotypes effected an 8–12 times greater acetylation rate in vitro than that observed with liver cytosol from slow acetylator phenotypes.	Group 1
Nitrogen oxide	NO	30.01	−164/−152	10102-43-9	LC50:1068 mg/m^3 (rats, oral) Nitrogen oxide is a gas available in concentrations of only 100 ppm and 800 ppm. Overdosage with inhaled nitric oxide will be seen by elevations in methemoglobin and pulmonary toxicities associated with inspired NO_2. Elevated NO_2 may cause acute lung injury.	Group 3

Index